Causation, Coherence, and Concepts

BOSTON STUDIES IN THE PHILOSOPHY OF SCIENCE

VOLUME 256

For other titles published in this series, go to
www.springer.com/series/5710

Wolfgang Spohn

Causation, Coherence, and Concepts

A Collection of Essays

Springer

Wolfgang Spohn
Universität Konstanz
Germany

ISBN 978-94-007-8705-6 ISBN 978-1-4020-5474-7 (eBook)

Softcover re-print of the Hardcover 1st edition 2009

Printed on acid-free paper

9 8 7 6 5 4 3 2 1

springer.com

Meiner Mutter,
der versunkenen Selbstverständlichkeit,
und meinem Vater,
dem schließlich Gelassenen.

Preface

In this collection I present 16 of my, I feel, more substantial papers on theoretical philosophy, 12 as originally published, one co-authored with Ulrike Haas-Spohn (Chapter14), one (Chapter 15) that was a brief conference commentary, but is in fact a suitable appendix to Chapter 14, one as a translation of a German paper (Chapter 12), and one newly written for this volume (Chapter 16), which, however, is only my recent attempt to properly and completely express an argument I had given in two earlier papers. I gratefully acknowledge permission of reprint from the relevant publishers at the beginning of each paper.

In disciplinary terms the papers cover epistemology, general philosophy of science, philosophy of language, and philosophy of mind. The section titles *Belief, Causation, Laws, Coherence*, and *Concepts* and the paper titles give a more adequate impression of the topics dealt with. The papers are tightly connected. I feel they might be even read as unfolding a program, though this program was never fully clear in my mind and still isn't. In the *Introduction* I attempt to describe what this program might be, thus drawing a reconstructed red thread, or rather two red threads, through all the papers. This will serve, at the same time, as an overview over the papers collected.

When rereading all these papers for the purpose of this edition, I thought I can still stand to each of their claims and arguments, even of the older ones. This is not true of all of my papers. This was one criterion of exclusion. In one case, though, I regret this. I considered to include also my "Stochastic Independence, Causal Independence, and Shieldability" from the *Journal of Philosophical Logic* 9 (1980), 73–99, since it is the first specific articulation of the foundations of the theory of Bayesian nets and their causal interpretation (that is in fact contained as a section of my German dissertation in 1976). However, this paper is my most awkward and overformalized piece, and it contains, I think, false claims about the transitivity of causation that I have corrected only in my paper reprinted here as Chapter 2. Instead, I included Chapter 4 that indicates the content of that earlier paper and comments on its relation to the leaders of the meanwhile established theory of Bayesian nets and their causal interpretation.

That I can stand to all the papers collected does not mean, though, that they would satisfy me. Ever so often I was tempted to put them into clearer or simpler or new ways I had found in the meantime, to elaborate on thoughts I had only hinted at, and so forth. Obviously, this would have been an unending task, and so I did not even start. The only amendments I allowed myself consist in cross-references within brackets and in a few additional footnotes marked by a dagger † and explaining where later on I have elaborated on a sketchy idea, which significance some remarks have in relation to discussions emerging only afterwards, or how I have changed my terminology.

I fear my papers are not easy to read, since many of them make free use of formal methods. These methods are entirely natural for me, but I know, of course, that this is not a shared attitude. Are they a precondition of good philosophy? Emphatically no. The best and most important philosophers did not use them or could even not know what they are. Different fields are amenable to these methods to varying degrees. My predilection, of course, is for those fields that are so amenable, and my ambition is to extend those fields.

Where formal methods are applicable, they are certainly most useful. They open up a second layer of argument. There is then not only the level of informal argument and clarity, there is also the level of rigorous definition and proof *and*, this is crucial, the continuous translation between the two levels, establishing thorough checks and balances. A one-layered roof is fragile, but a two-layered roof with numerous crossbeams in between is incomparably more stable. In the end, I do not know of any better way, if feasible, to improve security in the deeply insecure fields of philosophy. (In Spohn 2005c I got the opportunity to expand a bit on the character of formal philosophy.)

This was the pathetic argument. There is, though, a more individual reason. There are four kinds of papers in relation to formal matters. Papers of the first category move exclusively on the formal level and are only interested in formal results; they tend to be unphilosophical unless firmly grounded in papers from the other categories. The second category consists of the formally explicit philosophical papers as I mostly conceive of mine. The third kind of philosophical papers are informal, but clearly indicate that the author has the formal version in the drawer. The final kind consists of the informal papers for which no formal version exists. The ideal papers, I find, are those of the third category, readable for everyone, but rich in program and perspective. There are masters of this category I greatly admire.

I feel, however, that the third category is unstable. In principle, there is a simple test for distinguishing between the third and the fourth category: simply try to produce the formal version by yourself! This is either easy or impossible. In fact, though, there is a thin line between the two categories. As an author you can only be sure to write within the third category, when you actually have a formal version; the mere hope or guess it could be produced is treacherous, and the thin line is easily crossed. If you actually have the formal version, it needs checking, by readers; thus you have landed the second category. At least, this is how I perceive the matter. I always wanted to be sure to never cross the thin line, and thus could not help going on writing formally explicit papers.

There are not so many occasions to express one's gratitude. Therefore I allow myself some length.

I have an abstract sense of gratefulness towards the times I am living in: peace and prosperity and the opportunity to study and study, at most hampered by one's own imperfections and never directly affected by the slow disasters and the sudden catastrophes ineradicable from history. In particular, I feel most privileged in having started studying philosophy in the philosophically most exciting decade of the last century. One might say that the logical revolution and thus analytic philosophy started with Frege (1879) (although it is probably more appropriate to see Frege rather as the culmination of a rich development in the 19th century – cf. Peckhaus 1997). Look at our trees and bushes, though. Many of them form their first tiny buds already in late autumn that start growing only after wintry latency and explode to blossom in spring. So it was with analytic philosophy and its execution of the logical revolution. Winter lasted till World War II. Logic was pushed forward rather in mathematics, with some radiation to the philosophy of mathematics, and those who saw its great potential in philosophy in general were few and confronted a hostile environment. Spring started only after World War II, when the intellectuals of the Vienna and Berlin circle, assisted by Quine, began their success story in United States and when Russell, Moore, and Wittgenstein and their associates began dominating British philosophy. Still, it took more than 20 years till this turned into a mass movement, relatively speaking. This required the post-war prosperity with its fast increasing numbers of students who reached their intellectual maturity only in the late 1960s. In any case, in my perception the tree of analytic philosophy was in full blossom only around 1969 – when I started studying philosophy in Munich. I think I sensed my luck every day, but it took some years to fully realize it, and only much later I started seeing the (partial) history of philosophy of the 20th century in this way.

Therefore, my greatest philosophical indebtedness is to Wolfgang Stegmüller. Around 1966, as a school boy, I read Stegmüller's *Hauptströmungen der Gegenwartsphilosophie* (1960), read several chapters on Heidegger, Hartmann, Häberlin, etc., neither liked nor understood them, and was then totally captivated by its ch. IX on Rudolf Carnap and the Vienna Circle. Ever since I wanted to study this kind of philosophy at Stegmüller's institute, and I did. In 1969, his institute reached its peak as well, with around eight positions for associated and assistant professors that he could maintain till his death in 1991. It must have been the largest single philosophy institute in Germany, and a unique one. There were hardly places in Germany for doing analytic philosophy at all, none nearly as large, and none so devotedly logically and systematically oriented. In fact, there is presently no longer any such place in the whole of Germany; it's a shame. At this institute, I got in touch with so many fields of analytic philosophy, either immediately or with a delay of a few years; it was a most exciting time. The importance of Wolfgang Stegmüller for 40 years of post-war philosophy in Germany and of his role in (re-) importing analytic philosophy into Germany cannot be overestimated. I am sad to feel that this importance is hardly recognized any more, as he had foreseen in his agonies. In any case, I am grateful that I could stay at this institute till 1986, and I still stand to these intellectual origins with some proud; I have remained a disciple of Wolfgang Stegmüller and of the great minds he mediated.

I am deeply grateful for the philosophical teachers I had there and for the philosophical friends and comrades I found there: Max Drömmer, Franz von Kutschera,

Eike von Savigny, Wilhelm Essler, Ulrich Blau, Walter Hoering, Peter Hinst, Ulrich Berk, Reinhard Kleinknecht, Godehard Link, Andreas Kamlah, Andreas Kemmerling, Georg Meggle, Michael Heidelberger, Matthias Varga von Kibéd, Wolfgang Balzer, Reinhard Werth, Felix Mühlhölzer, Ulrich Gähde, Carlos Ulises Moulines, Julian and Martine Nida-Rümelin, Arthur Merin, Christian Piller, Wilhelm Vossenkuhl, Wolfgang Benkewitz, Hans Koch, Anna Kusser, Hans Rott, and Ulrike Haas, my later wife. With many, philosophical exchange continues at least intermittently.

I found much more fruitful opportunities for discussion in the international scene than in Germany. I am particularly indebted in various ways to Richard Jeffrey, Nancy Cartwright, Patrick Suppes, Karel Lambert, Brian Skyrms, Carl Hempel, Daniel Hunter, Judea Pearl, John Perry, Isaac Levi, Clark Glymour, Kevin Kelly, Scott Sturgeon, David Papineau, Matthias Hild, Joseph Halpern and others.

There is no doubt that the English speaking market, in particular the American market dominates analytic philosophy. The fact that the United States have raised to the most powerful nation after World War II and that English has become the lingua franca in most academic disciplines has not failed to heavily affect philosophy as well: Therefore the continental analytic philosophers are prone to attend to the American (and British and Australian) market rather than to each other, not to their favor. This is fortunately changing in the last years. I gratefully acknowledge that I have profited a lot in various ways from my continental partners and friends, Maria Carla Galavotti, Domenico Costantini, Peter Gärdenfors, Wlodek Rabinowicz, Friedrich Stadler, Nenad Miscevic, Miklos Redei, Jacques Dubucs, and others.

My time in Regensburg 1986–91 was a time of latency (and a lot of work with editing the journal *Erkenntnis*). The same is true of my time in Bielefeld 1991–96. Strange; perhaps I was there for too brief a time; certainly I was also much devoted to my small kids. New valuable philosophical connections (besides the ones that were revived) developed there with Albert Newen, Bernd Buldt, Matthias Risse, and Peter Lanz, whose death in 1997 was a great loss.

In Konstanz, where I am since 1996, things changed considerably. My personal resources improved, I more strongly engaged in the organization of philosophical research in a fruitful competition with my colleagues, I attracted more Ph.D. students, and the University of Konstanz attracted more external philosophers. These have been eleven fertile years so far. On the other hand, I am increasingly shocked about the recklessness with which politics and administration rope in their scholars and scientists into their industry; this is, to say the least, an inconsiderate management of the most precious intellectual resources that desperately need care and shelter. This applies to Germany in general (and perhaps more widely), but Konstanz is no exception (how could it?). The problem has massively aggravated in the last seven years, and no change is in sight.

I deeply appreciate all the working relations and philosophical exchanges I had and continue to have there with my colleagues Jürgen Mittelstraß, Hubert Schleichert, Gottfried Seebaß, Peter Stemmer, and Gereon Wolters, and with my partners, collaborators, and Ph.D. students at the department and in several research groups, namely Peter Schroeder-Heister, Hans Kamp, Ede Zimmermann, André

Fuhrmann, Volker Halbach, Holger Sturm, Erik Olsson, Manfred Kupffer, Ludwig Fahrbach, Max Urchs, Michael Esfeld, Luc Bovens, Christoph Fehige, Jacob Rosenthal, Gordian Haas, Franz Huber, and Wolfgang Freitag.

So many names! This is why I have listed them indiscriminately and only once, though several would have deserved more prominence. Everyone mentioned, though, (and many not mentioned) helped me improving the content of my papers in some way or other. The relation between the amount of help and the size of its effect is entirely in my own responsibility. Further special acknowledgments are contained in my papers.

Almost last and certainly not least, I am most grateful to Liisa Kurz, my secretary in Bielefeld, and to Ruth Katzmarek, my secretary in all my time in Konstanz, for typing, correcting, and endless further work, to Alexandra Zinke for preparing the bibliography and the indices of this collection, and to Ulrich Riebe for proof-reading.

There is another great debt I would like to acknowledge. There is not only the personal influence, but also that from reading; it is perhaps especially important for philosophers who cultivate the inner dialogue with past centuries and millennia. It would be tedious to mention all the books and papers from which I have learned most. One author, however, whom I have briefly met only once 30 years ago, stands out: David Lewis. An anecdote perhaps best characterizes my relation to him. In Spring 1973 I submitted my master thesis that contained, among other things, an axiomatization of conditional deontic logic and a proof of its soundness (published in my first publication 1975) of which I was quite proud – until I discovered Lewis' *Counterfactuals* from the same year that proved the same in a much more general and elegant way. This was symptomatic. My interests largely overlap with his, independently, I feel, and not due to his influence (with the big exception of ontology, where he has fixed his views very early – one of his great strengths, but perhaps also a weakness – whereas I am still struggling). Therefore, his writings have been a tremendous continuous challenge for me. This challenge drives my papers much more than I make explicit. In the Chapters 3 and 8, though, I expressly take up the challenge concerning his central views on Humean supervenience and causation. My debt to him, in any case, is inestimable.

My wife Ulli lived with me through all the joy and misery of the papers in this volume, and through all the greater common joy in the past 27 years. She has been a continuous partner in life and in philosophy, much more than is expressed in our single joint paper reprinted here as Chapter 14. I guess I have never been fully aware of how much she carried me and still does. I can't make good for this with words.

I dedicate this volume to my parents: to my mother Dr. Ortrud Spohn, née Knopp (1911–1976), who could hardly see the beginnings. I lay the collection to her feet. And to my father Dr. Karl Spohn (1914–2003), who saw most of it. I am glad I promised him this dedication in his last year. Philosophy is the passion of my life; this was somehow determined when I was sixteen. It is unfathomable how much this determination owes to the guidance of my parents.

Konstanz, March 2007 Wolfgang Spohn

Contents

Introduction

The papers presented here do not form a systematic unity. Nor do they deal just with their individual separable topics. They cohere tightly, by sharing introductions, taking up issues left open in another paper, being combinable to one natural bigger paper. In this introduction I want to briefly explain what the connections are and thus to give a kind of preview to the collection. The connections are not retrospectively read into the papers. At no place, though, do I summarize them in such a stream-lined version. This is why I felt the introduction is required.

There are in fact two red threads through the collection, of somewhat uneven generality. The one is epistemology or rather, since this is ambiguous, the theory of (graded) belief and not that of knowledge. The other, despite its technical name more general one is two-dimensional semantics. The threads are in fact intertwined, in an intricate way that I hope to make clear at the end. So, let me start with the simpler red thread.

When it comes to the theory of belief – that is the more basic part of epistemology despite deep philosophical programs claiming primacy for the theory of knowledge – probability theory or Bayesianism is just perfect and proved it for 350 years – except that it is incomplete. It is intuitively incomplete since it does not talk about belief at all, but only about degrees of belief; and it is internally incomplete since it leaves probabilities conditional on null events undefined. (The incompleteness is more fully explained in Chapter 1.) What we need, hence, is a theory of belief (or plain belief or acceptance, that's all the same) not only in its static form of doxastic logic as perfectly developed by Hintikka (1962) or in its incomplete dynamic form as presented by belief revision theory (cf., e.g., Gärdenfors 1988), but endowed with a complete dynamics, as we find it in Bayesianism.

This aim is achieved in *Chapter 1* "Ordinal Conditional Functions: A Dynamic Theory of Epistemic States" (1988), which is in fact contained as sect. 5.3 of my Habilitationsschrift (1983a). I still believe that it achieves this aim in an optimal way. What I called ordinal (and natural) conditional functions there and in some later papers in order to have an unmistakably clumsy name, are nowadays called ranking functions, a much more elegant and still unmistakable name. (For a more recent survey see my forthcoming b.)

The point of having a complete dynamics of doxastic states, that I found so obvious that I explained it only in expository papers such as my (2000a) and (2005b), is that only a complete dynamics is equivalent to an account of induction and inductive inference (cf. also ch. 9, sect. 2). This equivalence is massively exploited in many of my papers. This is why Chapter 1 is basic for this collection; indeed, five papers (too many) require a brief introduction into ranking theory (which, of course, I did not eliminate). Don't be confused by the slight formal variations and the slightly changing terminology!

For David Hume inductive inference and causal inference were one and the same. Therefore, causation is a prime field of application for ranking theory. In fact, the genetic order was reverse. In my dissertation (1976) I worked on probabilistic causation (in the context of decision theory). I was attracted to this topic because probabilistic theories of causation and explanation were at that time in a much better and more sophisticated state than deterministic theories. The basic reason was that probability theory provided clear and adequate notions of relevance and conditional relevance and thus means for dealing with all the riddles of explanation and causation centering around these notions, riddles on which Hempel's deductive-nomological account of explanation had foundered: the problem of irrelevant law specialization, the distinction between causes and symptoms (or epiphenomena), Reichenbach's screening-off relation, etc. (cf. Salmon 1989). The only deterministic account that could hope to compete with this sophistication was Lewis' (1973b) counterfactual analysis of causation. The potential of this account was obvious from Lewis' paper, but it was little elaborated at that time (and still struggles with more sophisticated problem cases; see Collins et al. 2004). More importantly, it seemed dubious why we should try to elucidate such an intuitively clear notion as causation by something so unspecific and hardly comprehensible as a similarity relation between possible worlds.

Anyway, thus attracted I came up with a probabilistic analysis of causal dependence between variables now known as the causal interpretation of Bayesian nets (cf. my 1976/78, sect. 3.3 and my 1980). However, the probabilistic turn seemed perverse, in a way. After all, our primary notion of causation is deterministic, even though the history of physics has forcefully undermined this primacy. If it seems unavoidable, then, that the theory of causation bifurcates into a deterministic and a probabilistic branch, the branches should at least remain closely related, displaying what is substantially one notion of causation. This is why and how I came to think of ranking theory. Due to its pervasive formal analogy to probability it allows to construct the deterministic and the probabilistic theory of causation in perfect parallel. This is a central message of the papers on causation contained in this collection (and already of my Habilitationsschrift 1983a).

In *Chapter 2* "Direct and Indirect Causes" (1990a) I present the account of probabilistic causation to which I still adhere. As it should be, it takes causation between facts as the basic notion to be analyzed (or causation between events; I have discussed this issue in my 1983a, ch. 4, but not in my papers). Therefore, the account goes beyond my former attempts that, like almost all of the statistical and social science literature, analyze only causal dependence between variables. Of course, the explication of causation between facts entails the intended analysis of causal dependence between variables. Moreover, it presents the result of my long struggle with the riddles of indirect causation: that an indirect cause may be negatively relevant to its indirect effect, that a fact may be ambiguous by being an indirect cause as well as an indirect counter-cause of another fact, that a fact may be a relay or switch insofar as it is an indirect cause of another fact, while its negation would also have been an indirect cause of that fact, and so on. I give a theoretical argument that it is best to conceive causation as transitive, as Lewis (1973b) had assumed – a conclusion that had met a lot of skepticism within the literature on probabilistic causation (that had been mine, too) and that, ironically, is again critically considered in the recent literature on deterministic causation (cf., e.g., Hall 2000; Hitchcock 2001).

At the end of the introduction of Chapter 2 I remarked that all the considerations of that paper apply to deterministic causation as well, via ranking theory. Since this remark went unnoticed, I made it explicit, most recently in *Chapter 3* "Causation: An Alternative" (2006), that also explains how some paradigmatic problem cases the counterfactual analysis of causation is still fighting with, namely symmetric overdetermination and preemption by trumping, can be more naturally treated within my ranking-theoretic analysis.

Chapter 4 "Bayesian Nets Are All There Is to Causal Dependence" (2001a) is mainly an afterthought to Chapter 2, but alludes to Chapter 3 as well. It took me quite some time to realize – because the result was so perplexing – that those fully developing the theory of Bayesian nets and their causal interpretation, namely Pearl (1988, 2000) and Glymour et al. (1987) and Spirtes et al. (1993), had quite a different over-all picture of causation than I had. How could this be on the basis of almost identical theories? In this paper I try to clarify the issue and to argue for my view. An essential point is that the authors mentioned have a simpler theory of deterministic causation in the background, whereas I think via Chapter 3 that the dialectic situation repeats itself at the deterministic level. In this paper I also affirm my commitment to the impossibility of Salmon's (1980) interactive forks that lies at the basis of the causal interpretation of Bayesian nets. I admit I am still disturbed by Nancy Cartwright's insistence on the existence of such forks (cf., e.g., Cartwright 2001, 2003) – an argument that in my view could be resolved only within a continuous version of Bayesian nets, obviously an ambitious subject at which Martel (2003) is the only attempt I have seen.

Another difference I have not only with the above-mentioned authors lies in my basically subject-relative understanding of causation. Probabilistic theories of causation were still ambiguous between credence and chance, between a subjectivistic and an objectivistic interpretation – if one only knew what chances are. By contrast, ranking functions are explained only as representing doxastic states. Thus, when I explain causation relative to a ranking function, I explain it relative to some subject's doxastic state. Hume did so as well when he claimed that "the idea of necessity", i.e., causal necessity, is "deriv'd from some internal impression, or impression of reflexion" (1739, Book I, Part IV, sect. XIV, p. 165). Of course, he sensed the absurdity of this claim only a few paragraphs later and ended up in ambiguity. I was depressed by the absurdity, too (this was a major reason why I did not publish my Habilitationsschrift in which I did not yet know how to get rid of it). The reason to entertain it nevertheless was that ranking theory provides notions of relevance and conditional relevance as adequate as the probabilistic ones, so that many problems faced by objectivistic theories simply dissolved. However, I must admit, and want to be able to assert, that the causal relations are what they are, independently of any observer. *Chapter 5* "Causal Laws Are Objectification of Inductive Schemes" (1993a) tries to do justice to the objectivistic intuition. It does so by specifying the conditions under which ranking functions can be objectified, i.e., be said to be uniquely determined by, and thus to correspond to, objective truths or facts. To the extent this objectification works causation, too, can be conceived as an objective relation. This paper is my only attempt elaborating this idea, and it is perhaps that paper of the collection most wrapped up in itself. Still, it is a cornerstone of my account of causation.

All this is my way to establish causation as a "covertly epistemological notion", as I express it in the opening sentence of Chapter 1. Another such covertly epistemological notion is the notion of a law. This is suggested by the metaphorical account of laws as inference tickets or by taking inductive support, explanatory power, or counterfactual strength as the marks of lawlikeness. Within my framework it is most natural to take the first mark, the role of laws in confirmation, as a starting point of analysis. And so I do in *Chapter 6* "Laws, Ceteris Paribus Conditions, and the Dynamics of Belief" (2002). The thesis I argue for is surprisingly simple. Just as a statistical law is, in the simplest case, a set of independent and identically distributed random variables or a Bernoulli measure over the space generated by these variables, so a deterministic law is a ranking function according to which the variables considered (the individual applications of the law) are independent and identically distributed. This thesis fits surprisingly well. According to it, a possible law is a particularly persistent doxastic attitude, which, this is important, is objectifiable in the sense of Chapter 5.

How laws in this sense can be confirmed by single instances – that was supposed to be their characteristic feature – is not obvious since confirmation applies to hypotheses or propositions in the first place and not to doxastic attitudes or ranking functions. The story is only indicated in Chapter 6, but fully elaborated in *Chapter 7* "Enumerative Induction and Lawlikeness" (2005a). It is just de Finetti's story about statistical laws. De Finetti showed – although he would not have it expressed in this way – that any symmetric probability measure for an infinitely repeated chance set-up corresponds to a unique mixture of the possible statistical laws for that chance set-up and that increasing evidence makes our opinion (almost surely) converge to the true statistical law. Likewise, apart from some niceties, any symmetric ranking function for an infinite set of cases to which one of a set of alternative laws in the sense explained might apply corresponds to a unique mixture of these possible laws, and again increasing evidence makes our opinion converge to the true law or possibly disconfirms all possible laws. At least, this is proved in Chapter 7 for the simplest possible case, but I know that it also holds for more complex cases. Van Fraassen (1989) wanted to abandon laws in favor of symmetry. However, if Chapter 7 is correct, the two notions remain wedded.

Chapter 6 moreover addresses the issue of ceteris paribus laws or laws subject to a ceteris paribus condition (since it was written for a collection of papers on ceteris paribus laws). This is a bewildering topic; the only options seem to be to deny the phenomenon or to say something wrong or something non-committal about it. In my view, this trilemma is an effect of inadequate means of analysis. The logic of ceteris paribus conditions is the logic of defeasible reasoning, which in turn is a contested subject matter, but well accounted for by ranking theory. Given the ranking-theoretic account of strict laws, a uniform treatment thus seems feasible. Or so I argue in Chapter 6.

This program of uncovering covertly epistemological notions amounts to a rejection of Humean supervenience and to the development of a counter-program for which Paul Grice and Simon Blackburn have coined the term "Humean projection". The issue is how to understand modality: not metaphysical necessity, which is a different matter, not intentionality and intensionality, which belong to the philosophy of mind and language, but, to choose a neutral term, empirical or natural modality like nomic and causal necessity, chance or objective probability, or, in other words, full and partial determination, and counterfactuals (although I stay away from a linguistic analysis of this most intricate idiom). David Lewis contended that all empirical or natural modal truths supervene on the totality of (local or individual) non-modal facts, where supervenience is a kind of metaphysical modality. This is his doctrine of Humean supervenience, which he wisely restricts to a contingent supervenience (although the nature of his restrictions is not particularly clear – see sect. 8.5). What I have suggested above is that it is more helpful to understand these natural modalities as "covertly epistemological", as objectifications or projections of our doxastic attitudes.

I explicitly settle my argument with David Lewis with respect to objective probabilities in *Chapter 8* "Chance and Necessity: From Humean Supervenience to Humean Projection" (to appear). Chance is the "big bad bug" Lewis (1986a, p. xiv) feared; he thought to get rid of it in his (1994b) by replacing his old Principal Principle by a slightly, but importantly modified one. Chapter 8 is a critical discussion of the ensuing literature, arguing that the new Principal Principle does not make sense in the desired way and that his claims that are intended as purely ontological still hide epistemological ingredients. The big bad bug stays with him. Alternatively, I attempt to spell out in detail what a projectivistic understanding of objective probability might be; this is basically de Finetti's story brought to the height of current philosophical sophistication. The crucial point, though, is that this attempt would be insulated as such; it acquires its full force only in the context of my other papers on laws and causation.

Is this a program carried by an ultimately idealistic spirit and offering only fake objectivity? No, I do not think that any such allegation would be appropriate. It is rather an attempt to disentangle the ontological-epistemological entanglement of which the epistemological turn of the Enlightenment has so forcefully made us aware, an attempt to pay epistemology its due and at the same time to grant the realist properly understood mind-independent objectivity not only of particular non-modal facts, but also of those natural modalities.

Here we are on the verge of connecting the one red thread explained so far with the other red thread of this collection. However, the epistemological thread is not yet fully laid out. No epistemological story can be complete without attending to the difficult notion of apriority or epistemic necessity. I think the determinately dynamic view on epistemology that I have taken in Chapter 1 and that has proved so fruitful in dealing with inductive and causal inference and natural modalities also gives us a vantage point for dealing with apriority.

My primary notion of apriority is more general than the usual one. Any feature of a doxastic state is a priori if and only if each doxastic state must have it. Thus, a proposition is a priori if and only if it must be believed in each doxastic state. This is the traditional notion of apriority, but there are more doxastic features than the belief in propositions.

Now it should have been clear that when I, as a philosopher, talk about doxastic states, their static, and their dynamics, I am talking only about rational doxastic states, their rational static, and their rational dynamics. What the laws of theoretical and practical rationality are – only the former, not the latter are discussed in this collection – is not pre-decided. It is rather the result of an on-going normative discussion that is intensely led in philosophy. For instance, my definition of ranking functions in Chapter 1 is based on such normative principles for belief and belief change. Or when Rudolf Carnap proposed his versions of inductive logic, he was arguing for principles of rationality going beyond the basic probability axioms. The point is whatever the laws of rationality we settle on, if they hold for all doxastic states, they describe a priori features of them.

The dynamic perspective makes clear that there in fact are two notions of apriority. I have already introduced the first one. Since it applies to all possible doxastic states, it applies to all changes of doxastic states as well. Hence, I also call it unrevisable apriority. As said, this is the traditional notion expressing epistemic necessity. Alternatively, we may define a feature of a doxastic state to be a priori if and only if each initial doxastic state must have it. Since such a feature may be lost through learning, I call this defeasible apriority. The classical example is the principle of insufficient reason that requires, for instance, to start with an equal distribution over the possible results of a throw of an unknown die, but, of course, allows discovering its possible asymmetries. Similar things are described in the literature as prima facie rules or rules of presumption or as weak apriority, and I sense an increasing awareness of the importance of this notion. Even the traditional explanation of the a priori as that which is known before or independent of all experience displays this ambiguity, even though it has focused then on the unrevisable reading.

The critical point of defeasible apriority is, of course, the notion of an initial doxastic state. Where does a rational dynamics begin? The idea, unsurprisingly, is to relativize doxastic states to the conceptual spaces on which they operate and to define a doxastic state as initial relative to given conceptual means if and only if the doxastic state contains nothing beyond that what is required for possessing those conceptual means; each concept thus is associated with its a priori content. This entails, for instance, the defeasible apriority of Euclidean geometry for physical space, as long as we had no other way of conceptualizing space. This relativization is also needed for unrevisable apriority; that bachelors are unmarried, to take a worn-out example, is unrevisably believed not in all doxastic states whatsoever, but only in all doxastic states possessing the concepts of a bachelor and marriage. I am well aware that this tentative definition of "initial" may be interpreted on varying philosophical backgrounds and sounds problematic on most. Chapter 15 better explains some of my background. The matter definitely needs further attention, but despite its difficulties I am convinced that defeasible as well as unrevisable apriority are philosophically most significant notions.

In fact, what I do in Chapters 9–12 is to start a larger, though unfinished investigation into both kinds of apriority. In my view this can be carried through much more fruitfully in terms of ranking-theoretic epistemology than in terms of Bayesianism, simply because the former, in contrast to the latter, contains the notion of belief and thus more squarely connects with traditional epistemology. The aim is to make more substantial claims about what is a priori (in either of the two senses) that go beyond analytical, mathematical, or Cartesian truths ("I exist now") that still dominate the present discussion.

More specifically, I inquire into the a priori structure of reasons. Being a reason is explicated as speaking for or being positively relevant in either the probabilistic or the ranking-theoretic sense. And since we rationally learn through reasons, the structure of reasons must be such as to assure our ability to learn. *Chapter 9* "A Reason for Explanation: Explanations Provide Stable Reasons" (1991) is so far my deepest inquiry into that structure. After explicating causal explanation in ranking-theoretic terms on the basis of my account of causation, it starts with such innocent principles like "for every assumption or proposition there is a reason" or "for every true assumption or proposition there is a true reason". However, due to the precise formal sense of these principles one can study their relation among each other and to other principles. For instance, some entail at least a weak principle of causality like "each fact has a cause or an effect". This is much more than nothing. Another consequence is expressed in the title of the paper.

Chapter 10 "Two Coherence Principles" (1999c) strengthens the coherence principles of Chapter 9 in another direction. After explaining that some other arguments do not succeed, I show there, in an almost formalizable way, that simple learnability principles and a basic theory of perception entail a general coherence principle saying, as it were, that the world cannot be separated into two epistemologically independent parts. A bit pathetically, I call this the unity of science.

Chapters 11 and 12 are concerned with a priori reasons in the defeasible sense. Besides more fully explaining my use of the notions of apriority, *Chapter 12* "A Priori Reasons: A Fresh Look at Disposition Predicates" (1997c) reconsiders disposition predicates and their associated reduction sentences and argues that the latter are more adequately understood as a defeasibly a priori reason relation between a disposition and its manifestation given the test situation. In particular, the defeasible apriority is able to adequately account for the ceteris paribus constraint on the reduction sentence – a point that closely links Chapter 12 to Chapter 6.

Chapter 11 "How to Understand the Foundations of Empirical Belief in a Coherentist Way" (1997/98) is an application of Chapter 12. Every part of reality has the disposition to appear to us in a certain way. Thus, the observations of Chapter 12 generalize to what I call the Schein-Sein (appearance-being) principle saying that for observable propositions *p* there is a defeasibly a priori reason relation between *p* and the proposition that it appears to a given subject as if *p*. This is a familiar idea, but I argue that this is its most adequate expression. Of course, the principle has consequences for the issue of foundationalism vs. coherentism. What it says in effect is that, for observable propositions *p, p* and "it appears to me as if *p*" are equally foundational, and thus the foundation of empirical belief is rather given by a coherentist link that is not strictly foundational due to its defeasibility.

The last two paragraphs suggest that I should have reversed the order of Chapters 11 and 12. However, Chapter 12 is the first that (next to) explicitly moves within the framework of two-dimensional semantics, as do the other papers in the final section of this collection, whereas the earlier papers in this collection refer to it at best implicitly or not at all. My thinking about this framework began only around 1988, and only dimly and slowly. Let me sketch how I presently see the significance of this framework.

That philosophy of language and thus the notion of meaning moved into the center of (theoretical) philosophy was, no doubt, the most important achievement of the first six decades of 20th century philosophy. But it was burdened with an original sin. Meaning has an ontological aspect, since it comprises reference; with our words we describe, and refer to, what is. And meaning has an epistemological aspect, since it is more or less synonymous with cognitive significance; with our words we express our beliefs about what is. (Moreover, we *do* a lot of things with words; but this is not in my present focus.) These two aspects were hopelessly confused, however, in the first 80 years of philosophy of language (say, since Frege's *Sinn und Bedeutung* 1892). The confusion shows up in the continuous double purpose intensions and propositions had to, but could not serve, in the continuous indecision between verifiability (or assertibility) and truth conditions, and at many other places.

The radical change came with Kripke and Putnam (and those preparing the ground like Dagfinn Føllesdal and Ruth Barcan Marcus), ironically not because they really cleared up the deep confusion – Searle (1958) was not wrong about names, Kripke (1972) only talked at cross-purposes with him; and the same holds, say, for Putnam (1965) and Feyerabend (1962) with respect to theoretical terms –, but because they most forcefully pushed the ontological reading of "intension" and related terms. After that one could simply no longer stick to the confusion.

The hallmark of the change is Kripke's reform of modalities; this was, by the way, my reason in the preface for dating the full blossom of analytic philosophy around 1970. There is (metaphysical) necessity (and possibility), there is apriority or epistemic necessity (and possibility), and the two are independent; analyticity is down-graded to a derivative notion and defined as a priori metaphysical necessity. However, necessity and apriority were not yet on a par; modal logic and intensional semantics were then reserved for the ontological aspect, and there was at first no corresponding theorizing for apriority.

This changed only with Kaplan (1977) and Stalnaker (1978), the birth of two-dimensional semantics in my view. The grand picture that thus emerged is this: There is the set of epistemic possibilities, there is the set of ontic possibilities, and there is a correspondence mapping epistemic onto ontic possibilities. Since it will acquire some importance, let's call it the *EO-map*. In the simplest case both kinds of possibilities are just possible worlds, and the EO-map is identity. Together, these two sets span a two-dimensional space of possibilities.

Now, every word or phrase receives, in a recursive way, a two-dimensional meaning that assigns a type-adequate extension to each point of this two-dimensional space. This sounds abstract and formalistic, but it is most substantial. The two-dimensional meaning first provides an ontic intension for each epistemic possibility or situation, and it provides an epistemic intension or a cognitive significance. This epistemic intension that assigns a type-adequate extension to each epistemic possibility derives from the two-dimensional meaning by diagonalization; that is, the extension of a phrase in an epistemic possibility is just its two-dimensional meaning evaluated in that epistemic possibility and at its EO-map. We might also read this conversely at least for some words or phrases: we may start with the phrase's epistemic intension, then project its ontic intension in a given epistemic possibility from its extension in that possibility, and thus arrive at its two-dimensional meaning. (This is Kaplan's theory of direct reference; see also the modal extension principle of Peacocke 1997.) In any case, the EO-map and diagonalization are indispensable features of the framework.

Kripke's pair of modalities is well accounted for within this scheme. A sentence expresses an (unrevisably) a priori truth if its epistemic intension is true in each epistemic possibility. There is no way for such a sentence to turn out false. And in a given epistemic situation a sentence expresses a metaphysical necessity if its ontic intension in this situation is true at each ontic possibility. In that situation such a sentence could not be false.

This picture offers a grand promise. There are ontology and epistemology, the two basic disciplines of theoretical philosophy. They span the space of meaning, the third core topic. Thus, two-dimensional semantics promises to clearly separate ontological and epistemological aspects of meaning and at the same to articulate their relation, in terms of the EO-map and diagnalization. There is hardly anything deeper to accomplish in theoretical philosophy. Chalmers (2006) speaks no less emphatically of the golden triangle of meaning, reason, and modality.

I am convinced that this formal frame is basically correct and by itself already a great advance. There is always the danger to distort phenomena in order to squeeze them into a given frame. However, as with ranking functions, my continuous experience is reverse, namely that the two-dimensional frame enormously helps to get clear about the phenomena.

One must grant, though, that the interpretations of the framework are multifarious and vacillating. For Kaplan (1977), epistemic possibilities were just contexts, and thus he offered a semantics of indexicality or context-dependence. At the same time, he heavily restricted the relevance of the framework by denying it to account for the cognitive significance of proper names. Stalnaker (1978) was the first to consider the first (or vertical) dimension in a properly epistemological way and to emphasize the importance of diagonalization, thus accounting for cognitive significance. However, in his (1989) and (1990) he turned out denying that the diagonals would deliver any such thing as cognitive significance in the sense of narrow contents, which others thought the framework must do. Evans (1979) and Davies and Humberstone (1981) were further early contributions enriching the spectrum of interpretations. Haas-Spohn (1995) generalized Kaplan's theory so as to comprise Putnam's (1975) hidden indexicality and explained how Kaplan's linguistic two-dimensional meanings (his characters) and Stalnaker's more subjective two-dimensional meanings (his propositional concepts) can be understood within one frame. Moreover, she took great care to specify the lexical meaning rules of two-dimensional semantics. At the cost of the latter Chalmers (1996) and Jackson (1998) turned away from the contextualist understanding of the epistemic possibilities and deepened the epistemological side of two-dimensional semantics. And so forth. In a way, the usefulness for linguistic purposes and the adequacy for epistemological needs remains the essential tension for the whole approach. Chalmers (2006) lists a dozen possible interpretations of two-dimensional semantics, and even the ones to be taken seriously are disturbingly many. No wonder that what I here called ontic and epistemic intension has received many different names by different authors (that I do not list here).

There is also a worry about the EO-map. For Stalnaker it was just identity. For Kaplan, it was truncation; that is, for him epistemic possibilities or contexts were lists of indices, and ontic possibilities or circumstances of evaluation were simply shorter lists. As far as I see, Chalmers was the first to suspect that the EO-map may not be trivial at all (see Chalmers 2006, but this paper as well as his insight are much older). He prefers to describe epistemic possibilities as so-called scenarios, which basically are descriptions, and then he needs and specifies a substantial EO-map from scenarios to possible worlds (as ontic possibilities). I shall return to this issue.

In view of all this one may give up on the framework, totally confused. Before one yields to this inclination, however, one must realize how much of current theoretical philosophy is at least implicitly couched in two-dimensional terms. For instance, I am aware of only very few places where David Lewis uses the term "two-dimensional". Yet his philosophizing is imbued by the framework. The referential/attributive distinction can only be understood within that frame. Rigidification and derigidification have become common terms of art that acquire precise sense only in the two-dimensional frame. I understand the distinction between the role and the realizer property denoted by a suitable predicate also as a two-dimensional one. The notion of response-dependency that has gained some currency seems to me to refer to that framework as well. And so forth. I find the philosophical evidence overwhelming that the task is to make as good sense of the framework as possible and not to get rid of it.

Chapters 12–16, in any case, work at better making sense of it. My interest – or, as far as Chapter 14 is concerned, our interest – was both, the general philosophy and the specific two-dimensional meaning of interesting word classes:

Chapter 12 that I had already mentioned elaborates on the two-dimensional meaning of disposition predicates, a most pervasive class. At that time I had found only one paper that had more or less explicitly addressed the two-dimensional meaning of disposition predicates, namely Prior et al. (1982), and I disagreed with it. (There were more, though. Already Mellor (1974), for instance, is aware of the two dimensions, and in a way one may say that even Armstrong (1968, sect. 6.VI) struggles with them.) Contrary to Prior et al. (1982), I follow the traditional view that the ontic intension of a disposition predicate is the (categorical) base of the disposition, whereas the epistemic intension, as already mentioned, is essentially characterized by the defeasibly a priori reason relation expressed by the pertinent reduction sentence.

Chapter 13 "The Character of Color Terms: A Materialist View" (1997) addresses the two-dimensional meanings, characters in Kaplan's terminology, of color terms, i.e., more specifically, of both phrases, "*x* is red" and "*x* appears red to *y*". In the upshot, I argue that both phrases are not essentially different from "*x* is water". They are both hidden indexicals; and Chisholm's (1957) three senses of "looks" or "appears", the phenomenal, the comparative, and the epistemic sense, do not point to ambiguity, but may each be appropriate depending on the context, the epistemic possibility one is in. Before and after, I have seen quite a number of papers dealing with this issue, but I find mine still fully adequate.

Chapter 14 "Concepts Are Beliefs About Essences" (2001) is of a more general nature and wears its core thesis in its title. The basic problem it addresses is this: As Haas-Spohn (1995) has made clear, the two-dimensional scheme duplicates itself on a communal and on a subjective level. On the communal level it captures Kaplanian characters, linguistic meanings associated with words and phrases of a given language or rather language stage. However, only on the subjective level it is able to capture what we have in our minds, the concepts and narrow contents. But what are they? One danger is to reduce concepts to mere words or morphosyntactic forms that are then loaded with information, even with information about what they might mean in one's linguistic community; but this information is not part of the concept. Such emptiness seems unacceptable. This is a danger Haas-Spohn (1995) and others have succumbed to. The other danger often considered to be unavoidable is to take a concept as the totality of its connections to other concepts; but so much holism is intolerable. In this paper we propose and defend a reasonable middle course, which, we feel, has still a lot of latent potential for the architecture of two-dimensional semantics.

Chapter 15 "Changing Concepts" (2004) is a brief, but important supplement to Chapter 14. What is discussed in Chapter 14 are not really concepts, but rather concept stages, and we would like to be able to say that concept stages are stages of one concept. Chapter 15 discusses what might hold together the stages to form one concept. The main answer refers back to the inquiries about which unrevisably or defeasibly a priori truths are associated with concepts.

Chapter 16 "The Intentional Versus the Propositional Structure of Contents", finally, is a paper prepared for this volume, but goes back to my (1997a) and (1998). It is another paper about the basic structure of two-dimensional semantics. The most common view about epistemic and ontic possibilities is that the former are centered possible worlds and the latter simply possible worlds. Perhaps a view about what a possible world is added; and perhaps there is a sense, as Chalmers

(2006) has emphasized, that "possible world" means two different things in the two connections. In Chapter 16 I argue that all this will not do; rather, both kinds of possibilities must be complemented by a sequence of objects (or a variable assignment). In linguistic semantics, there is a long-standing awareness of this requirement, though for semantic reasons. In philosophy of language, the point, though first noticed by Perry (1979), had, as far as I can see, no deeper repercussions – certainly a mistake. In this paper I try to give a strictly epistemological and thus a new argument for this requirement. And I emphasize that it is not a mere formality, but a reform at the foundations of two-dimensional semantics. The possibilities now are models or relations (the objects in the sequence are related as they are in the world at issue), and this has proliferating consequences. For instance, in my view it undermines the primacy of sentence meanings so dear to many philosophers of language.

So much for the survey over the contents of the papers collected here. In particular those about two-dimensional semantics remain a patchwork, I feel. They promise a coherent picture of two-dimensional semantics, but they do not realize it. One reason is that the basic architecture of two-dimensional semantics is still not clear; and without it the rest is bound to hang in the air. So, let me use the final pages of this introduction at least for a sketch of what I presently take this basic architecture to be.

This will also allow me to say how the two red threads I have outlined are intertwined. In principle the connection is simple; if two-dimensional semantics is to combine epistemology and ontology, then all the detailed epistemological considerations feed in into two-dimensional semantics. The connection is deeper, though, as we shall see below.

The basis of two-dimensional semantics is the nature of epistemic and ontic possibilities and of the correspondence between them, i.e., the EO-map. I stated that most were content with assuming these possibilities to be (centered) worlds in some sense and the EO-map to be trivial in some way. The question of precisely understanding possible worlds could then be left to the metaphysicians. Above I stated this in order to put forward the point of Chapter 16 that objects must play a more explicit role in these possibilities. However, even apart from this point the prevailing attitude will not do. Chalmers (2006) has seen that the EO-map is not trivial at all and that one must proceed much more thoughtfully at that point. However, he then heads into a different direction. For him, ontic possibilities are possible worlds that he always understands in the sense of Lewis (1986b), whereas epistemic possibilities or scenarios preferably are maximal hypotheses, linguistic constructions in an idealized language (cf. Chalmers 2006, pp. 83ff.).

I shall not start an argument with Chalmers' ideas here, but my idea is quite different. I think that (complete) epistemic possibilities are Lewisian possible worlds, maximal objects endowed with some space-time analogous extension relative to which maximality makes sense at all. This extension could be anything but Euclidean. Hence it need not conform to the Kantian a priori forms of intuition, but the epistemological role is similar. What might count as space-time analogous is not clear, however; Lewis (1986b, pp. 71ff.) who started such speculation remains inconclusive, too. Probably, connected topological spaces are already too general a structure.

Such a Lewisian world is fully determinate; everything there is to it is essentially so; each difference makes for another possible world. However, such a maximal object is a maximal black box for us; it is entirely unknown, indeed unconceived. It is the point of departure of our epistemic endeavor. We confront such a possibility, whatever it is, and we try to make sense of it. Or rather, we are contained in such a possibility, and therefore these possibilities must be centered Lewisian worlds. Or still better, in view of Chapter 16 we have to add a sequence of objects that might so far be any parts of the space-time analogous extension of the Lewisian world. However, this only expresses the a priori conception that whatever the world might turn out to be it is a world of objects.

We might also call such an epistemic possibility a Kantian noumenal world, if we avoid the association of there being an inaccessible or unknowable reality. It is rather the as yet unaccessed and unknown working material of our cognitive efforts.

Now, confronted with such a Lewisian world we develop concepts and form beliefs. Concept formation has presuppositions. Worlds that would not stimulate our senses do not yield even to purely perceptual concepts. Deferential concepts require the embedding into a linguistic community to defer to. And so forth. The concepts we actually have would not fit most of the worlds, and the beliefs we have exclude still much more. But we might have other beliefs and even other concepts, depending on the epistemic possibility we encounter. I assume that many possibilities would be completely dark and barren – unless we exclude them on a priori grounds and take the sensibility and the conceptualizability of an epistemic possibility not as a harmony actually pre-established by evolution, but as an a priori truth.

In any case, the concepts we develop facing such a possibility have some a priori content, as I tried to explain in Chapters 12 and 15. As indicated earlier in this introduction, this apriority is relative to such concepts being formed at all. And the beliefs we form are a priori constrained by normative principles of rationality, as I tried to explain in Chapters 9–11. Building on such beginnings we try to do ever better. The evolution of our beliefs is a central topic in this collection, and how that account may be continued to cover also the evolution of concepts was at least envisaged in Chapters 14 and 15.

What, then, is the goal of this process? A goal that we shall never reach and that is never reachable by all human standards? Of course, we always move in the middle of this process, far from the beginnings and much farther from its end. It is obvious that I am not drawing a picture of the actual ontogenesis or phylogenesis of our cognitive life as individuals or as a species. The purpose of my far-fetched speculations is rather at all to gain a frame for describing the process we are always amidst, a frame I take two-dimensional semantics to be providing.

So, to repeat, what is the ideal end of the process of concept and belief formation? In the end we have fully investigated the Lewisian world and have completed our judgment about it; all the evidence, even if only counterfactually available, is acquired, and all even only counterfactual ways to improve our judgment according to our rules of rationality are exhausted. Then we have reached a state of omniscience, no proposition remains undecided, we know the nature of every object and of every property and relation, and we know all properties of and all relations among all objects existing in this world.

What we have thus determined is, I contend, an ontic possibility, a totality of coexisting states of affairs, a possible world as Wittgenstein has conceived in his *Tractatus* (1922) and as Armstrong has repeatedly explained it (e.g., in his 1997). I tend to think of such a Wittgensteinian world in an essentialist way. Each object is individuated by its possibly or usually relational essence, properties and relations are individuated by (metaphysically) necessary equivalence, states of affairs are built from objects, properties, and relations, and a Wittgensteinian world is a (in a sense to be specified) maximal collection of states of affairs in which the objects have properties and relations within their ranges of contingency. Indeed, such a Wittgensteinian world is the essence of the corresponding Lewisian world; all states of affairs obtaining in a Lewisian world do so necessarily. (This is not to say, of course, that these states of affairs are necessary themselves.) Certainly, the essentialist picture agrees with Chapter 14 that suggested that we have fully conceived a (small or big) object or a property when and only when we have finally discovered its essence.

Thus, the EO-map is far from trivial; it embodies nothing less than the full transformation of a Lewisian into a Wittgensteinian world by a complete process of concept and belief formation. The papers in this collection may therefore as well be understood as working at the details of the EO-map. In any case, I think that such a grasp of the EO-map lies at the basis of a proper understanding of two-dimensional semantics.

In a way, we might understand an ontic possibility also as a phenomenal world in the Kantian sense, when fully conceptualized and judged. I am certainly not entitled to engage here in Kant exegesis. Also, we should not enter Kant's elaborated, but foreign theory of concept formation or any of his idealistic verbiage. I think, however, that when Kant is pondering about noumena and phenomena he is partly struggling with similar issues as we find at the foundations of two-dimensional semantics.

The two kinds of possibilities are moreover associated with two notions of truth. Ontic possibilities or Wittgensteinian worlds are governed by the correspondence notion of truth. The ontic intension of a sentence or the wide content of a belief, relative to a given epistemic situation, is a (complex) state of affairs that may or may not correspond to, i.e., be contained in a given totality of coexisting states of affairs. By contrast, epistemic possibilities or Lewisian worlds are governed by a pragmatist, or coherentist, or evaluationist notion of truth. This is something much more elusive, alluding to our principles of epistemic rationality, or, if you like, to our weighing of epistemic values, etc., and to the results they yield in the Peircean limit of inquiry. The various adjectives point to various doctrines trying to grasp this elusive matter, but their intent is, I think, the same. Thus, even the long-standing debate about different notions of truth seems to be resolved within the two-dimensional meaning of "true".

What, finally, about the empirical or natural modalities in ontic possibilities or Wittgensteinian worlds? Are they mere collections of individual states of affairs (and combinations thereof)? Or do they contain causation, determination, laws, and chances? Emphatically yes, according to Armstrong (1983, 1997), and I concur. How they are so contained should, however, be understood in the two-dimensional way outlined. We approach a Lewisian world with our inductive powers, conjecture causal relations and projectible regularities, apply statistical methodology, etc., and the more complete our inquiry, the better we can settle on objectifiable deterministic and stochastic laws, on objective causal relations. This is how the earlier papers in this collection belaboring "covertly epistemological" notions and their program countering Humean supervenience fit into the broader two-dimensional picture; the Humean projection referred to above is part of the EO-map. This is the deep way how the two red threads of this collection intertwine.

Note here the strict analogy between de Finetti's account of subjective probabilities as mixtures of objective probabilities and the account of doxastic intensions as diagonals of two-dimensional meanings. We might well call de Finetti (1937) the inventor of diagonalization and not Stalnaker (1978) or Kaplan (1977). This is not a mere play of words, as it would have been, if I had alluded, say, to Cantor's diagonalization. It is precisely diagonalization in the two-dimensional sense of establishing a fundamental relation between epistemology and ontology that is prepared by de Finetti.

Alas, all this is figurative and speculative. I would not have dared writing any such pages in a paper. Devoting four pages to speculation before entering 340 pages of pedantic argument and theory construction seems excusable, though. If I had waited with this collection till I am fully clear about these speculations and can substantiate every piece of it, who knows whether it would ever have appeared?

Part I
Belief

Chapter 1
Ordinal Conditional Functions: A Dynamic Theory of Epistemic States [†1,*]

1.1 Introduction

Many of the philosophically most interesting notions are overtly or covertly epistemological. Overtly epistemological notions are, of course, the concept of belief itself, the concept of subjective probability, and, presumably the most important, the concept of a reason in the sense of a theoretical reason for believing something. Covertly epistemological notions are much more difficult to understand; maybe, they are not epistemological at all. However, a very promising strategy for understanding them is to try to conceive of them as covertly epistemological. One such notion is the concept of objective probability[1]; the concept of explanation is another. A third, very important one is the notion of causation, which has been epistemologically problematic ever since Hume. Finally, there is the notion of truth. Many philosophers believe that there is much to be said for a coherence theory of truth or internal realism; they hold some version of the claim that something for which it is impossible to get a true reason cannot be true, and that truth is therefore covertly epistemological.

Now, if one wants to approach these concepts in a more formal way in order to understand them more clearly and more precisely, the first step will be to try to get

---

[†1] This paper was originally published in: William Harper and Brian Skyrms (eds.), *Causation in Decision, Belief Change, and Statistics*, Kluwer, Dordrecht, 1988, pp. 105–134. This book is the second volume of the proceedings of the NSF conference *Probability and Causation* held at the University of California at Irvine on July 15–19, 1985. The paper is essentially a translation and elaboration of Spohn (1983, sect. 5.3).

[*] Many have helped. I am very much indebted: first to Godehard Link for spending a Christmas vacation making various helpful remarks and suggestions; to Wolfgang Stegmüller and Max Drömmer for fruitful discussion; to Kurt Weichselberger for decisive help in Section 1.7; to Brian Skyrms and Bill Harper for giving me the opportunity to present this paper at their conference; to Peter Gärdenfors for further fruitful discussion; to Isaac Levi for some enlightening controversies and for drawing my attention to the work of Shackle; not last and not least to Joe Lambert for carefully going through the whole stuff with me twice more and almost forgetting dinner about it; to Jeremy Adler for being so kind to descend from investigating Goethe's German to correcting and improving my English; and finally to the Wissenschaftskolleg zu Berlin for giving me the leisure for preparing the final version of this paper.

[1] Which is clearly conceived as covertly epistemological by Lewis (1980a), for instance.

a formal grip on epistemology. Here, I am concerned only with this first step.[2] Considering the impressive amount of work in formal epistemology, two general points arise.

The first is very familiar, though it still strikes me as somehow odd; it consists in the fact that formal epistemology, i.e. the formal representation of epistemic states, may be divided into a probabilistic and a deterministic branch (and some things which don't quite fit into the scheme). In a deterministic epistemology, as I call it, one talks about a proposition *being simply believed true or false or neither* by some epistemic subject. The formal machinery established for this works with belief sets, truth in all doxastic alternatives, or similar things well known from epistemic logic.[3] In a probabilistic epistemology, belief is more finely graded and comes in numerical degrees. The formal machinery appropriate to it is, of course, probability theory,

This dichotomy is naturally prepared for on the intuitive level. All the intuitive notions we have for subjective and objective probability fall on the probabilistic side. Plain belief, of course, belongs to the deterministic side. And so does truth; the simplest reason for this is, I think, that an arbitrary, perhaps uncountable conjunction of truths is still a truth – this being a formal property of truth which cannot be modelled probabilistically. However, the dichotomy is not complete on the intuitive level. The concept of a reason is certainly neutral between the two forms of epistemology. The same holds for the concept of explanation, as we have learned from Hempel, and for the concept of causation, as has been stressed by many who take probabilistic causation seriously.

Of course, one would like to get rid of this dichotomy, i.e. to reduce one side of it to the other; and this can only mean reducing deterministic to probabilistic epistemology. However, this is not so easy, as is highlighted by the famous lottery paradox. Indeed, the different behaviour of conjunction in deterministic and probabilistic formalisms seems to entirely exclude such a reduction. Then, we should do the second best, i.e. we should develop both forms of epistemology as far as possible and then look what we can say about their relations.

Now, however, we have to consider the second point, namely that deterministic epistemology is in a much poorer shape than probabilistic epistemology. One important aspect is that probabilistic epistemology is well entrenched in a behavioral theory, i.e. decision theory; and this is hardly counterbalanced by the fact that

---

[2] In my (1983b), I have proposed a way of progressing from the notion of reason to the notion of cause. I have refrained there from introducing the formal machinery developed in this paper; footnote 18 of that paper marks the point where what I there called selection functions and shall call here simple conditional functions should be replaced by the ordinal conditional functions to be defined – for reasons more fully explained here in Section 1.3.

[3] I am not happy with the term "deterministic epistemology", but I could not find a better one. It derives from the natural and familiar distinction between deterministic and probabilistic causation which, in my opinion, is closely related to the different forms of epistemology.

a deterministic epistemology can be more easily used in a theory of language.[4] What is more important, however, is that the inner functioning of deterministic epistemology is so much poorer. Usual probabilistic conditionalization and the generalized conditionalization of Jeffrey (1965, ch. 11), give a plausible account of rational epistemic changes. Probability theory also provides a good model for the impact of evidence and counter-evidence on our beliefs, for the weighing of reasons and counter-reasons; it provides, in other words, a good explication for relevance, potential or conditional relevance, and irrelevance in the epistemic sense. As far as I can see, deterministic epistemology can, in the present state, not produce equivalent achievements.

That is precisely what this paper is about; I shall try to raise deterministic epistemology to the level of probabilistic theorizing. More specifically, I shall try to give a more satisfying account of rational changes, i.e. of the dynamics of deterministic epistemic states. It is to be expected, and will become evident, that this brings advance also on the other scores mentioned. Moreover, it will turn out that the problems I am concerned with are in fact present and unsolved at the probabilistic side as well; thus the paper will also add something to probabilistic epistemology.

This being my focus, I greatly simplify my business by proceeding from the obsolete view that belief is a strictly propositional attitude, i.e. that the objects of belief are complete propositions as expressed by eternal sentences. I thereby neglect other serious problems with epistemic states such as the de-re/de-dicto distinction, the fact that belief is most likely neither propositional nor sentential, but something mid-way, and the observation that belief seems to be as heavily indexical as language itself. But there is no agreed formal epistemology for handling these problems, and our dynamic problem is certainly intricate enough; hence, I comply with that old view and its associated method of possible world talk.

Having thus laid out the general setting, I shall proceed in the following way. First of all, I'd like to keep separate the story I have to tell and the comments relating it to existing ideas and conceptions. My reason for this is not the novelty of the story (only one feature is really new, as far as I know); rather, I wish to do so because: I think that the story is simple and self-contained; I do not want anything read into it which is not explicitly written into it; and the danger of misreading is the greater, the sooner one mixes up this story with similar, but not completely congruent stories. Thus, I defer all comparative remarks to the final Section 1.8. The story I want to tell starts in Section 1.2 with a presentation of what I take to be the essentials of the received deterministic conception of epistemic states. In Section 1.3, I shall state a crucial problem and argue that it cannot be adequately treated within that received conception. In Section 1.4, I shall introduce my proposal for a solution of this problem i.e. the concept of an ordinal conditional function, and in Sections 1.5 and 1.6 the theory of ordinal conditional functions is

[4] Think, e.g., of the disquotation principle saying that if *X* sincerely and seriously utters "*p*", then *X believes* that *p*. This is an important, though not generally true linguistic fact; and it is hard to see what a probabilistic version of it could look like.

developed up to a point where it may not be too much to say that this theory offers a genuine qualitative counterpart to probability theory.[5] Finally, Section 1.7 explains why the whole story also has a considerable bearing for probabilistic epistemology.

1.2 Simple Conditional Functions

Having made things simple by assuming belief to be propositional, we shall work with the common, technically convenient framework of possible worlds. Thus, throughout this paper, W is to denote a non-empty set of possible worlds (or a sample space, in probabilistic terms).[6] A *proposition* then is just any subset of W.

The most straightforward deterministic representation of an epistemic state is, of course, as a set of propositions, namely those propositions believed true in that state. Will any set of propositions do? No. Usually, it is required, as conditions of rationality, that such a set of propositions be consistent and deductively closed. One might object that this requires an unattainable logical perfection rather than a form of rationality. Indeed; but the logical perfection is already assumed by taking belief to be propositional. For, taking belief to be propositional means that, for any two sentences having the same content, i.e. expressing the same proposition, an epistemic subject should recognize them to have the same content. Thus, it means that epistemic subjects have perfect semantic knowledge which embraces perfect logical knowledge. And given that, the conditions of rationality seem perfectly acceptable; any indication that a subject violates these conditions is also evidence that his semantic knowledge is not perfect.[7]

Formally, these conditions amount to this: If $\mathcal{B}$ is a set of propositions, then $\mathcal{B}$ is consistent iff $\bigcap \mathcal{B} \neq \emptyset$, and $\mathcal{B}$ is deductively closed iff we have $A \in \mathcal{B}$ whenever there is a $\mathcal{B}' \subseteq \mathcal{B}$ with $\bigcap \mathcal{B}' \subseteq A$.[8] From this, it follows immediately that, for consistent and deductively closed $\mathcal{B}$, $A \in \mathcal{B}$ iff $\bigcap \mathcal{B} \subseteq A$. Thus, we can represent an epistemic state simply by a single non-empty proposition C, and the set of

[5] This is not to be confused with what is ordinarily called qualitative probability which is a relational, comparative concept.

[6] Where I don't at all oppose construing a possible world as a maximal consistent set of sentences of a given language, as a valuation of that language, or the like.

[7] This consideration suggests that the idealization of belief as propositional should be overcome not by seeking for a stricter objective individuation of the objects of belief, but by getting a grip on the subjective imperfections of semantic knowledge.

[8] One might argue about whether $\mathcal{B}'$ should here be assumed to be countable or finite or neither. With my definition of deductive closure I have assumed what has been called the generalized consequence principle; cf. e.g. Pollock (1976, pp. 19f.) or Gärdenfors (1981, p. 308). I do so, because I find this principle convincing given the idealization of perfect semantic knowledge and because it makes things technically much simpler. However, as far as I see, everything I say in this paper could be adapted to a weaker assumption without essential complications.

propositions believed true in that state is $\{A \mid C \subseteq A\}$. We shall call this proposition C the *net content* of that epistemic state.

If we represent epistemic states simply by their net contents, what can we say about their temporal change? To begin with, it is clear that epistemic changes may have many causes: experiences, forgetfulness, wishful thinking, drugs, etc. And it is also clear that from our armchair position we can at best hope to say something about *rational* epistemic changes on the ground of experience, information and the like. So, suppose that the epistemic state of the subject X at time t has the net content C and that the proposition A represents all the information X gets and accepts between t and t'. What then is the net content C' of X's epistemic state at t', provided X is not subject to arational influences? We have to distinguish two cases here.

First, consider the case where $C \cap A \neq \emptyset$, i.e. where the new information is compatible with the old beliefs of X. In this case, it is reasonable to assume that $C' \subseteq C \cap A$, since the new information, because of its compatibility with C, does not force X to give up any of his old beliefs. And it is also reasonable to assume that $C \cap A \subseteq C'$; otherwise, X would at t' believe some proposition not implied by his old beliefs and the new information, and there is no good reason for doing so. Thus, rational belief change is in this case characterized by $C' = C \cap A$.

The other case to consider is that $C \cap A = \emptyset$, i.e. that the new information contradicts the old beliefs. This is a very common case; we often learn that we were wrong. And usually, it is an undramatic case; the rearrangement of beliefs usually takes place without much difficulty. However, all attempts to spell out objective principles for the rearrangement of beliefs in this case have failed. The only thing that can at present be confidently said about this case is that X arrives at *some* new epistemic state which includes the belief in A (since A was supposed to be accepted information), i.e. that $\emptyset \neq C' \subseteq A$.

We are thus left with an incomplete account of rational belief change. How can we improve upon the situation? Well, I shall not try to say anything more substantial about the last critical case – as so many have tried to do by invoking such things as lawlike sentences, modal categories, similarity, epistemic importance, informational value, etc., which may appear to be antecedently understandable. Rather, the only thing I shall try to do is to turn what appears to be a partially undetermined process on the surface level of the net contents of epistemic states into a completely determined process on some suitable deeper level. Thus, all the notions introduced in the course of my story are only meant to provide a theoretical substructure to this surface level which derives its meaning exclusively from what it says about the surface level (which I indeed assume to be antecedently understandable). In a sense, we shall only go beneath and not beyond what we have already said. I stress this point, because it seems to involve changing the usual tactics towards our question.

So, what can be done along these lines? Since the above observations about epistemic changes hold for any possible information, we can, as a first reasonable step, define a function which collects all the possible changes of the net contents of epistemic states brought about by all possible pieces of information. Such functions are defined in:

Definition 1: The function g is a *simple conditional function* (*SCF*) iff g is a function from the set of all non-empty subsets of W into the set of all subsets of W such that the following conditions hold for all non-empty $A, B \subseteq W$:

(a) $\emptyset \neq g(A) \subseteq A$,
(b) if $g(A) \cap B \neq \emptyset$, then $g(A \cap B) = g(A) \cap B$.

The interpretation of SCFs is clear: If we use an SCF g for describing X at t, it says that, if A is the information X accepts by $t' > t$, $g(A)$ is the net content of X's epistemic state at t; or briefly: X believes at t B conditional on A iff $g(A) \subseteq B$. This includes that the net content of X's epistemic state at t itself is given by $g(W)$, since the tautological information W leaves X's epistemic state unchanged; hence, X believes B at t iff $g(W) \subseteq B$. An SCF thus provides a *response scheme* to all possible pieces of information.

It is also clear that an SCF should have the properties fixed in Definition 1: The exclusion of the empty set from the domain of an SCF reflects the fact that a contradiction is not an acceptable information. Clause (a) says that, whatever information is accepted, the beliefs remain consistent and include the information. And clause (b) is a natural generalization of what we have said about the case where the new information is compatible with the old beliefs: Our above consideration concluded that, in the present terms, $g(B) = g(W) \cap B$, if $g(W) \cap B \neq \emptyset$; and if we take not, as we did, $g(W)$, but rather the state informed by A, i.e. $g(A)$, as the starting point of that consideration, we just get clause (b).[9]

An SCF is, we understand, a response scheme to all possible pieces of information. Now, a natural further step, which has not been made so far, is to assume that the response scheme which holds for a subject X at some time t is already embodied in the epistemic state of X at t. This means, however, that we give up representing epistemic states simply by their net contents. Rather, we now conceive them as more complicated things representable by SCFs. This is an advance; we can now state a rule for the dynamics of belief which is completely determinate: If the SCF g represents the epistemic state of X at t and if A is the information X accepts between t and t', then X believes B at t' iff $g(A) \subseteq B$ (provided X is not subject to arational influences).

Is this the end of the story? No, for a very simple reason which will be introduced in the next section. Before that, let me introduce an intuitively and technically very

[9] Some, e.g. Lewis (1973a, p. 58), prefer to replace the condition that $g(A) \neq \emptyset$ (which is tantamount to universality) by the condition that $A \subseteq B$ and $g(A) \neq \emptyset$ imply $g(B) \neq \emptyset$. However, that's much of a muchness. The only difference is this: With the alternative definition one can prove that there is a $D \subseteq W$ such that $g(A) = \emptyset$ iff $A \subseteq D$. Now alter g by putting $g(A) = A$ for $A \subseteq D$ and leave it otherwise unchanged (thus, there is no change, if $D = \emptyset$); then g complies with our definition. In both cases, accepting a piece of information $A \subseteq D$ leads to complete epistemic collapse. In our case, only the information is then believed; all induction ceases. In the alternative case, everything is then believed (if this makes sense); induction goes crazy. I find our description of that desperate situation a bit more pleasant; besides, SCFs in our sense are more easily generalized to the ordinal conditional functions introduced in Section 1.4.

useful concept which is equivalent to that of an SCF. Here as well as in all later sections, $\alpha, \beta, \gamma, \ldots, \zeta$ will always be used to denote ordinal numbers.

Definition 2: The sequence $(E_\alpha)_{\alpha<\zeta}$ is a *well-ordered partition*, a *WOP* (of *W*) iff we have for all $\alpha, \beta < \zeta$: $E_\alpha \neq \emptyset$, $E_\alpha \cap E_\beta = \emptyset$ for $\alpha \neq \beta$, and $\bigcup_{\alpha<\zeta} E_\alpha = W$.

Definition 3: If $(E_\alpha)_{\alpha<\zeta}$ is a WOP and *g* an SCF, we say that $(E_\alpha)_{\alpha<\zeta}$ *represents g* iff for each non-empty $A \subseteq W$ $g(A) = E_\beta \cap A$, where $\beta = \min \{\alpha \mid E_\alpha \cap A \neq \emptyset\}$.

Theorem 1: *Each SCF is represented by exactly one WOP, and each WOP represents exactly one SCF.*

Proof: Let *g* be an SCF. Define by transfinite recursion: $E_\beta = g(W \setminus \bigcup_{\alpha<\beta} E_\alpha$. Let ζ be the smallest α for which $E_\alpha = \emptyset$. It is obvious that $(E_\alpha)_{\alpha<\zeta}$ is a WOP. Does it represent *g*? Yes, as may be seen thus: Let *A* be a non-empty subset of *W* and $\beta = \min \{\alpha \mid E_\alpha \cap A \neq \emptyset)$. Then we have with the help of clause (b) of Definition 1: $g(A) = g((W \setminus \bigcup_{\alpha<\beta} E_\alpha) \cap A) = g(W \setminus \bigcup_{\alpha<\beta} E_\alpha) \cap A = E_\beta \cap A$.

Conversely, let $(E_\alpha)_{\alpha<\zeta}$ be a WOP. Let the function *g* be defined for all non-empty $A \subseteq W$ as in Definition 3. It is obvious that *g* then satisfies clause (a) of Definition 1. Now suppose that $g(A) \cap B \neq \emptyset$. This means that $E_\beta \cap A \cap B = \emptyset$, where $\beta = \min \{\alpha \mid E_\alpha \cap A \neq \emptyset)$. Hence, we also have $\beta = \min \{\alpha \mid E_\alpha \cap A \cap B \neq \emptyset)$. This implies that $g(A \cap B) = g(A) \cap B$. Thus, *g* also satisfies clause (b) of Definition 1, i.e. is an SCF.

Finally, the uniqueness claims of Theorem 1 again are rather obvious. Q. E. D.

A WOP $(E_\alpha)_{\alpha<\zeta}$ is easily interpretable as an *ordering of disbelief* in possible worlds; E_0 contains the possible worlds not disbelieved at all, E_1 contains the least disbelieved worlds, E_2 the second least disbelieved, and so on.[10] The rule for changing beliefs then takes a very simple form: If you now have the ordering $(E_\alpha)_{\alpha<\zeta}$ of disbelief, then you now believe that the true world is among the not disbelieved worlds, i.e. in E_0; thus, E_0 is the net content of your present state. And if you get information *A*, then you believe that the true world is among the least disbelieved within that information, i.e. in your new net content $E_\beta \cap A$, where $\beta = \min \{\alpha \mid E_\alpha \cap A \neq \emptyset)$. What Theorem 1 shows is that response schemes (SCFs) are equivalent to such orderings of disbelief; so we may, and shall indeed, carry through the following considerations in terms of WOPs.

1.3 A Problem with Simple Conditional Functions

So far, we have arrived at conceiving epistemic states as SCFs or WOPs. But there is a problem; the rule for epistemic change we have stated is simply insufficient. In this rule, the old epistemic state was represented by an SCF, but the ensuing epistemic state was still represented in the former way by its net content. This will not

[10] Continuously using these negative terms is a somewhat clumsy and contorted mode of expression. But Isaac Levi has convinced me that this is precisely the intuitively appropriate terminology.

do, of course. Having decided to represent epistemic states by SCFs, we must represent *all* epistemic states we are talking of in this way; that is, we must also represent the ensuing state by some SCF, and we must say which SCF that is. The problem becomes pressing, if we consider several successive epistemic changes. The above rule explains the first of these changes; but after that we are back on the surface level of net contents, where we cannot apply the above rule to account for the further changes.

The problem is obvious and grave; but it has received surprisingly little attention. In fact, the only place I found where the problem is explicitly recognized in this way is in Harper (1976, pp. 95ff.), where he tries to solve its probabilistic counterpart with respect to Popper measures.[11] What can we do about it? Well, let's at least try to solve it within our representation of epistemic states. If this should fail, as it will, we shall at least see more clearly what is missing.

It will be intuitively more transparent in this attempt to work with orderings of disbelief, i.e. WOPs. Thus, let the old epistemic state be represented by the WOP

$$E_0, E_1, E, \ldots, E_\zeta$$

(which we suppose only for illustrative reasons to have a last term), and let A be the information to be accepted and $\beta = \min\{\alpha \mid E_\alpha \cap A \neq \varnothing)$. Some new epistemic state ensues which should also be represented by a WOP. Can we determine this new WOP in a reasonable way?

A first proposal might be this: It seems plausible to assume that, after information A is accepted, all the possible worlds in A are less disbelieved than the worlds in $\bar{A}$ (where $\bar{A}$ is the relative complement $W \setminus A$ of A). Further, it seems reasonable to assume that, by getting information only about A, the ordering of disbelief of the worlds within A remains unchanged, and likewise for the worlds in $\bar{A}$. Both assumptions already determine uniquely the new ordering of disbelief; it is given by the sequence

$$E_\beta \cap A, \ldots, E_\zeta \cap A,\ E_0, \ldots, E_{\beta\text{-}1}, E_\beta \cap \bar{A}, \ldots, E_\zeta \cap \bar{A}.$$

where – this is important – all empty terms must still be deleted; otherwise, we wouldn't have a WOP. Wasn't that a quick solution? Well, it isn't a good one. Let me point out three shortcomings.

First, according to this proposal, epistemic changes are not reversible; there is no operation of the specified kind which reinstalls the old ordering of disbelief. In fact, there is in general no way at all, even if we know β, to infer from the new WOP what the old one was. The technical reason for this is just the deletion of empty terms, since after they have been deleted, we no longer know where they have been deleted. However, it is certainly desirable to be able to account for the reversibility of epistemic changes.

---

[11] I shall say in Section 1.7 how Popper measures relate to the present subject.

Secondly, according to this proposal, epistemic changes are not commutative. If A and B are two logically independent propositions, it is easily checked that getting informed first about A and then about B leads to one WOP, getting informed first about B and then about A leads to another WOP, and getting informed at once about $A \cap B$ leads to still another WOP. This is definitely an inadequacy. To be sure, one wouldn't always want epistemic changes to commute. The two pieces of information may somehow conflict, in which case the order in which they are received may matter. But the normal case is certainly that information just accumulates, and in this case the order of information should be irrelevant. However, according to our proposal it is irrelevant only in trivial cases.

Thirdly, the assumption that, after getting informed about A, all worlds in $\overline{A}$ are more disbelieved than all worlds in A seems too strong. Certainly, the first member, i.e. the net content of the new WOP, must be a subset of A; thus, at least some worlds in A must get less disbelieved than the worlds in $\overline{A}$. But it is utterly questionable whether even the most disbelieved world in A should get less disbelieved than even the least disbelieved world in $\overline{A}$; this could be effected at best by the most certain information.

This last consideration suggests a second proposal. Perhaps one should put only the least disbelieved and not all worlds in A at the top of the new WOP which then looks thus:

$$E_\beta \cap A, E_0, \ldots, E_{\beta\text{-}1}, E_\beta \setminus A, E_{\beta+1}, \ldots E_\zeta.$$

Here again, empty terms still have to be deleted ($E_\beta \setminus A$ may be empty). However, that's no good, either. This proposal does not fare better with respect to the reversibility and commutativity of epistemic changes, as may be easily verified. Moreover, we have now gone to the other extreme. The information A is now treated as only minimally reliable; it is given up as soon as only a single consequence of the things believed together with A, i.e. of $E_\beta \cap A$, turns out to be false.

One may try further: But I think that the case already looks hopeless. There is no good solution to our problem within the confines of SCFs or WOPs. Nevertheless, there are two important conclusions to be drawn from these efforts.

One conclusion is this: In the first proposal the information A was accepted maximally firmly; in the second it was accepted minimally firmly. We considered both extremes undesirable. But then no degree of firmness is the right one for all cases. Rather, the natural consequence is that, in order to specify the new epistemic state, we must say not only which information it is that changes the old state; we must also specify with which firmness this information is incorporated into the new state. This consequence is most important; it means that we have so far neglected a parameter which plays a crucial role in epistemic changes. No wonder that we tried in vain.

The other conclusion is this: We discovered that the reversing of epistemic changes was impossible because of the deletion of empty terms. This suggests that we should generalize the concept of a WOP to the effect that such a partition may contain empty terms. This is what we shall do. Technically, this is a small trick which will, however, make all the difference. Note that this has another important consequence. There may then be two such generalized partitions which order the possible worlds in exactly the same way and which thus differ only by having empty terms at different places. These two partitions should be viewed as two different epistemic states; and this implies that not only the ordering of worlds, but also their relative distances in these partitions are relevant. Mathematically, this means that we have to consider not only the order, but also the arithmetical properties of ordinals.

Now we are well prepared. We only have to adhere to these conclusions. The first conclusion will be developed in Section 1.5, the second right now.

1.4 Ordinal Conditional Functions

It is more convenient to formalize such generalized partitions as functions from possible worlds to ordinals. Moreover, we shall explicitly relativize these functions to a given field of propositions. So far, there was no need for this relativization; but now, when things get more technical, it will prove very useful. The same is done in probability theory, where it is important to compare or relate probability measures on different σ-fields. So, let us define:

Definition 4: Let $\mathcal{A}$ be a complete field of propositions over W (i.e. a non-empty set of subsets of W closed under complementation and arbitrary union and intersection). Then we call κ an *$\mathcal{A}$-measurable ordinal conditional function* (*$\mathcal{A}$-OCF*),[†2] if and only if κ is a function from W into the class of ordinals such that $\kappa^{-1}(0) \neq \emptyset$ and for all atoms[12] A of $\mathcal{A}$ and all $w, w' \in A$ $\kappa(w) = \kappa(w')$. Moreover, we define for any $A \in \mathcal{A} \setminus \{\emptyset\}$ $\kappa(A) = \min \{\kappa(w) \mid w \in A\}$.[13]

It is obvious that OCFs generalize WOPs and thus SCFs. The measurability condition is also obvious; it demands that an $\mathcal{A}$-OCF does not discriminate possible worlds which are not discriminated in $\mathcal{A}$.

Two simple observations will be permanently used:

Theorem 2: *Let κ be an $\mathcal{A}$-OCF. Then we have*

(a) *for each $A \in \mathcal{A} \setminus \{\emptyset, W\}$, $\kappa(A) = 0$ or $\kappa(\bar{A}) = 0$ or both,*

[†2] Following the advice of Goldszmidt and Pearl (1992), I call OCFs, or rather their restriction to the range of natural numbers, ranking functions since the late 1990s and nowadays even negative ranking function (since they represent *dis*belief).

[12] A is an atom of $\mathcal{A}$ iff $A \neq \emptyset$ and there is no $B \in \mathcal{A}$ with $\emptyset = B \subset A$. Complete fields of sets are always atomic.

[13] This latter function for propositions is the more important one. The corresponding notion at the level of WOPs is the function assigning to each proposition A the number $\{\alpha \mid E_\alpha \cap A \neq \emptyset\}$, which we have frequently used, though not explicitly introduced. Note, by the way, that it is our acceptance of the generalized consequence principle (cf. Note 8) which is in the end responsible for the possibility of reducing the propositional function to a function defined for possible worlds.

(b) *For all* $A, B \in \mathcal{A} \setminus \{\emptyset\}$, $\kappa(A \cup B) = \min\{\kappa(A), \kappa(B)\}$.

Intuitively, an OCF is not only an ordering, but a *grading of disbelief* in possible worlds. It is clear how such a grading of disbelief is to be understood as a deterministic epistemic state: In state κ, the true world is always believed to be in $\kappa^{-1}(0)$; thus $\kappa^{-1}(0)$ is the net content of the epistemic state κ, and hence the stipulation that $\kappa^{-1}(0) \neq \emptyset$. *A* is then *believed* in the state κ iff $\kappa^{-1}(0) \subseteq A$, i.e. $\kappa(\overline{A}) > 0$. (Beware: $\kappa(A) = 0$ only means that *A* is not believed to be false in state κ; and this leaves open the possibility that also $\kappa(\overline{A}) = 0$, i.e. that *A* is also not believed true in state κ.)

Relative to an OCF κ, we may also introduce degrees of firmness of belief (and thereby slightly reduce the contorted talk of disbelief). If we, for a moment, also allow for negative ordinals, we may say that *A is believed with firmness* α *relative to* κ iff either $\kappa(A) = 0$ and $\alpha = \kappa(\overline{A})$ or $\kappa(A) > 0$ and $\alpha = -\kappa(A)$. Thus, in state κ one believes or disbelieves *A* iff, respectively, one believes *A* with positive or negative firmness; firmness 0 means that one is neutral to *A*. And we might also say that *A is more plausible than B* iff *A* is believed with greater firmness than *B*, i.e. iff $\kappa(\overline{A}) > \kappa(\overline{B})$ or $\kappa(A) < \kappa(B)$.[14]

It is clear that the role of taking the minimum corresponds to the role addition has in probability theory; compare the definition of κ(*A*) with the probabilistic formula $P(A) = \sum_{w \in A} P(\{w\})$. The two sides of the correspondence differ, however, in a very characteristic way. To put it somewhat metaphorically.

In probability theory, epistemically interpreted, possible worlds have a probability *mass*. They compete for their share of the total mass available; and in epistemic changes these shares get redistributed. Thus, this competition may be conceived as a sort of territorial fight where the parties aim at getting as large as possible. A proposition may then be conceived as a team consisting of its members; and each such team is as weighty and fares as well in this competition as the sum of the masses of its members.

In the theory of OCFs, possible worlds have, by contrast, *grades* of disbelief. They compete for grades, 0 being the top grade above an unending sequence of lower grades; and in epistemic changes their grades will get rearranged. Thus, this competition may be conceived as a sort of race where the parties aim at reaching the top. A proposition may again be conceived as a team consisting of its members; but in this race, each such team is just as good as its best members.[15]

Exactly how do the grades get rearranged in epistemic changes? This is the subject of the next section.

---

[14] All this is a sort of exercise in intuitively interpreting OCFs. Degrees of firmness could also have been introduced relative to WOPs; but this would have been misleading, because, relative to WOPs, numbers have a purely ordinal meaning.

[15] Still, it would be inappropriate to say that only the best members count. Imagine a proposition having only one member with a good grade, the rest being very far behind. Then, if this good member fell back very badly, so would the whole proposition. If, however, the rest were not so bad, the top member could fall back without disastrous consequences for the team. In this sense the rest matters, too.

1.5 Conditionalization and Generalized Conditionalization

In what follows I shall make use of the somewhat uncommon *left-sided subtraction* of ordinals[16] which is defined in the following way: Let α and β be two ordinals with $\alpha \le \beta$; then $-\alpha + \beta$ is to be that uniquely determined ordinal ξ, for which $\alpha + \xi = \beta$.[17]

Moreover, we shall throughout use the following auxiliary concept:

Definition 5: Let κ be an $\mathcal{A}$-OCF and $A \in \mathcal{A} \setminus \{\emptyset\}$. Then the *A-part* of κ is to be that function $\kappa(\cdot \mid A)$[18] defined on A for which for all $w \in A$ $\kappa(W \mid A) = -\kappa(A) + \kappa(w)$. For $B \in \mathcal{A}$ with $A \cap B \neq \emptyset$ we also define $\kappa(B \mid A) = \min \{\kappa(w \mid A) \mid w \in A \cap B\}$ $= -\kappa(A) + \kappa(A \cap B)$.

Thus, if $\mathcal{A}' = \{A \cap B \mid B \in \mathcal{A}\}$, $\mathcal{A}'$ is a complete field of subsets of A, and $\kappa(\cdot \mid A)$ is an $\mathcal{A}'$-OCF. One might say that the A-part of κ is the restriction of κ to A shifted to 0, i.e. in such a way that the minimum taken is 0. It will soon become clear why I have here chosen the same notation as is used in probability theory.

With the aid of this concept we can define the notion central to the dynamics of epistemic states:

Definition 6: Let κ be an $\mathcal{A}$-OCF, $A \in \mathcal{A} \setminus \{\emptyset, W\}$, and α an ordinal. Then $\kappa_{A,\alpha}$ is to be that $\mathcal{A}$-OCF for which

$$\kappa_{A,\alpha}(W) = \begin{cases} \kappa(w \mid A), & \text{if } w \in A \\ \alpha + \kappa(w \mid \overline{A}), & \text{if } w \in \overline{A} \end{cases}.$$

We call $\kappa_{A,\alpha}$ the *A,α-conditionalization of* κ.

Thus, the A,α-conditionalization of κ is the union of the *A-part* of κ and of the $\overline{A}$ -part of κ shifted up by grades. Trivially, we have $\kappa_{A,\alpha}(\overline{A}) = 0$ and $\kappa_{A,\alpha}(\overline{A}) = \alpha$ hence, $\overline{A}$ is believed in $\kappa_{A,\alpha}$ with firmness α. By having introduced the parameter α, we have now taken account of the first conclusion of Section 1.3.

Definition 6 conforms to the intuitive requirement that getting informed only about $\overline{A}$ does not change the epistemic state restricted to A, or $\overline{A}$, i.e. the *grading* of disbelief within A, or $\overline{A}$. In other words, the $\overline{A}$,α-conditionalization of κ leaves the A-part as well as the $\overline{A}$ -part of κ unchanged; they are only shifted in relation to one another. Thereby, we have finally also made use of and given meaning to the relative distances of possible worlds in an OCF, as was implied by our second conclusion of Section 1.3.

---

[16] It would be a natural idea to restrict the range of OCFs to the set of natural numbers. In fact, much of the following could thereby be simplified since usual arithmetic is simpler than the arithmetic of ordinals. For the sake of formal generality I do not impose this restriction. But larger ranges may also be intuitively needed. For example, it is tempting to use OCFs with larger ranges to represent the stubbornness with which some beliefs are held in the face of seemingly arbitrarily augmentable counter-evidence.

[17] For details cf. Klaua (1969, p. 173).

[18] This is a short notation for the function assigning to each w in the domain the value $\kappa(W \mid A)$.

The failure of WOPs may now be seen to have a simple mathematical reason: it's just that the set of all WOPs is not closed under all the above shiftings; therefore, no reasonable conditionalization could be defined for them. With the OCFs, this problem disappears; the class of all OCFs *is* closed under all these shiftings.[19]

The A,α-conditionalization of κ should not always be interpreted as the change of κ which results from obtaining the information A with positive firmness. There are two exceptional cases. For the first case, suppose that $\kappa(\bar{A}) = \beta > 0$; thus, A is believed already in κ. Now, if $\alpha = \beta$, there is no change at all; if $\alpha > \beta$, then one has got additional reason for A whereby the belief in A is strengthened; and if $\alpha < \beta$, then one has got some reason against A whereby the belief in A is weakened, though not destroyed. The second case is the A,0-conditionalization of κ. This may best be described as the neutralization of A and $\bar{A}$, since in $\kappa_{A,\alpha}$ neither A nor $\bar{A}$ is believed. In both cases, it would be inappropriate to say that one was informed about A. But the epistemic changes described in them may certainly be found in reality and are thus properly covered by Definition 6.[20,21]

The problems we had with our proposals in Section 1.3 no longer trouble us. Of course, epistemic changes according to Definition 6 are reversible:

Theorem 3: *Let* κ *be an* $\mathcal{A}$-OCF *and* $A \in \mathcal{A} \setminus \{\emptyset, W\}$ *such that* $\kappa(A) = 0$ *and* $\kappa(\bar{A}) = \beta$. *Then we have* $(\kappa_{A,\alpha})_{A,\beta} = (\kappa_{\bar{A},\alpha})_{A,\beta} = \kappa$.

Moreover, accumulating information commutes. Here, as in the sequel, we shall say that two ordinals α and β *commute* iff $\alpha + \beta = \beta + \alpha$.

Theorem 4: *Let* κ *be an* $\mathcal{A}$-OCF *and* $A, B \in \mathcal{A} \setminus \{\emptyset, W\}$ *such that* $\kappa(A \cap B) = \kappa(A \cap \bar{B}) = \kappa(\bar{A} \cap B) = 0$, *and let* α *and* β *be two commuting ordinals. Then we have* $(\kappa_{A,\alpha})_{B,\beta} = (\kappa_{B,\alpha})_{A,\alpha}$.

Proof: Set $C_1 = A \cap B$, $C_2 = A \cap \bar{B}$, $C_3 = \bar{A} \cap B$, and $C_4 = \bar{A} \cap \bar{B}$, and for $n = 1,\ldots,4$, $\kappa(C_n) = a_n$, $\kappa_{A,\alpha}(C_n) = b_n$, $(\kappa_{A,\alpha})_{B,\beta}(C_n) = c_n$, $\kappa_{B,\beta}(C_n) = d_n$, and $(\kappa_{B,\beta})_{A,\alpha}(C_n) = e_n$. It suffices to show that $c_n = e_n$ for $n = 1, \ldots,4$: We have assumed that $a_1 = a_2 = a_3 = 0$. By Definition 6 we now get:

$b_1 = 0$, $b_2 = 0$, $b_3 = \alpha$, $b_4 = \alpha + a_4$, and

[19] This was pointed out to me by Godehard Link.

[20] It is easy to link Definition 6 with Gärdenfors (1984). Gärdenfors there discusses contractions and minimal changes of what he calls belief sets, where these belief sets are essentially equivalent to our net contents. Keeping in mind that $\kappa^{-1}(0)$ is the net content of state κ, we may define the minimal change of $\kappa^{-1}(0)$ needed to accept A as $\kappa^{-1}_{A,\alpha}(0)$ for some $\alpha > 0$ (this does not depend on which $\alpha > 0$ we choose). And we may define the contraction of $\kappa^{-1}(0)$ with respect to A as $\kappa^{-1}(0)$, if $\kappa(\bar{A}) = 0$, and as $\kappa^{-1}_{A,\alpha}(0)$, if $\kappa(\bar{A}) > 0$. It is then easy to prove that contractions and minimal changes so defined have all the properties (1)–(21) Gärdenfors (1984, pp. 140–142) wants them to have.

[21] A self-comment: In my (1983b), I explicated the notion that A is a reason for B relative to SCFs (which I there called selection functions). This has now turned out to be inadequate, but it is easily repaired: *A is a reason for B* in the state κ iff B is believed in κ with greater firmness given A than given $\bar{A}$, i.e. iff $\kappa(B \mid A) > \kappa(\bar{B} \mid \bar{A})$ or $\kappa(B \mid A) < \kappa(B \mid \bar{A})$. The rest of the paper is easily adapted to this new definition. (Instead of "A is a reason for B in a given epistemic state" one may also say that A *means* B in that state. This is, it seems to me, the most basic meaning of meaning on which other (linguistic) concepts of meaning may be built.)

$d_1 = 0, d_2 = 0, d_3 = \beta, d_4 = \beta + a_4$.

Again applying Definition 6, we get from this:

$c_1 = 0, c_2 = \beta, c_3 = \alpha, c_4 = \beta + \alpha + a_4$, and
$e_1 = 0, e_2 = 0, e_3 = \beta, e_4 = \alpha + \beta + a_4$.

Thus $c_n = e_n$ for $n = 1, 2, 3$, and also $c_4 = e_4$, since α and β commute. Q. E. D.

The conclusion of Theorem 4 holds also under more general conditions. These, however, are not so illuminating as to justify the clumsy calculations needed.

We may further generalize our topic. As is well known, Jeffrey (1965, ch. 11) made a substantial contribution to the dynamics of probabilistic epistemic states by discovering generalized conditionalization. There, a probability measure P is conditionalized not by some proposition A, but rather by a probability measure Q on some set of propositions. Q represents here some new state of information with respect to these propositions, and the generalized conditionalization of P by Q describes how the total epistemic state P changes because of this new state of information. Nobody seems to have even thought of doing the same for deterministically conceived epistemic states; but here, the parallel extends in quite a natural way:

Definition 7: Let $\mathcal{B}$ be a complete subfield of $\mathcal{A}$, κ an $\mathcal{A}$-OCF, and λ a $\mathcal{B}$-OCF. Then κ_λ is to be that $\mathcal{A}$-OCF for which for all atoms B of $\mathcal{B}$ and all $w \in B$ $\kappa_\lambda(w) = \lambda(B) + \kappa(w \mid B)$. We call κ_λ the *λ-conditionalization of* κ.

Definition 6 is only a special case of Definition 7:

Theorem 5: *Let* κ *be an* $\mathcal{A}$-OCF, $A \in \mathcal{A} \setminus \{\varnothing, W\}$, *and* λ *that* $\{\varnothing, A, \bar{A}, W\}$*-measurable* OCF *for which*

$$\lambda(w) = \begin{cases} 0 \text{ for } w \in A, \\ \alpha \text{ for } w \in A. \end{cases}$$

Then, $\kappa_\lambda = \kappa_{A,\alpha}$.

Of course, generalized conditionalization is reversible, too:

Theorem 6: *Let* $\mathcal{B}$, κ, *and* λ *be as in Definition* 7; *and let* κ' *be the* $\mathcal{B}$*-measurable coarsening of* κ *defined by* $\kappa'(w) = \kappa(B)$ *for all atoms* B *of* $\mathcal{B}$ *and all* $w \in B$. *Then* $(\kappa_\lambda)_{\kappa'} = \kappa$.

With the aid of Definition 7 we can state our most general rule for rational epistemic change: Let X's epistemic state at time t with respect to the field $\mathcal{A}$ of propositions be represented by the $\mathcal{A}$-OCF κ. Suppose further that the experiences between t and t' directly affect only X's attitude towards propositions in the field $\mathcal{B}$ and cause him to adopt the $\mathcal{B}$-OCF λ as epistemic state with respect to $\mathcal{B}$. Then κ_λ represents X's epistemic state at t' with respect to $\mathcal{A}$ (provided X is not subject to arational influences).

This formulation of the rule brings out a fact which seems by now to be well accepted in epistemology in general. It was realized in probabilistic epistemic modelling with Jeffrey's generalized conditionalization (this was its revolutionary point), but it does not seem to have been clearly recognized in deterministic epistemic modelling: I mean the fact that what is described by rules of epistemic change are never rational inner reactions to outward circumstances or happenings, but always rational adjustments of the overall epistemic state to inner epistemic changes in particular quarters; how these initial epistemic changes come about is in any case a matter to which a rationality assessment cannot be reasonably applied and which therefore falls outside the scope of investigations like this one. This fact is formally mirrored, here as in Jeffrey, by the fact that epistemic states, probability measures or OCFs, are conditionalized by things of their own kind; talking of conditionalization by propositions (or events), albeit technically correct, has been intuitively very misleading.

1.6 Independence and Conditional Independence

Related to conditionalization, there is another important topic in probability theory in particular, but also in epistemology in general: namely dependence and independence. I know of no reasonable definition of independence for deterministic representations of epistemic states. Logical independence will not do, of course, since almost everything is logically independent of almost everything. The best we can do within the domain of SCFs is to say that A is epistemically independent of B relative to the SCF g, if and only if $g(B) \subseteq A$ iff $g(\overline{B}) \subseteq A$ and $g(B) \subseteq \overline{A}$ iff $g(\overline{B}) \subseteq \overline{A}$, i.e. iff acceptance of B, or of $\overline{B}$, does not matter to whether A or $\overline{A}$ or neither is believed. However, this implies, for example, that each A believed true in state g is independent of each B believed neither true nor false in g; and this is certainly much too much independence.

Not surprisingly, there is no problem with independence with respect to OCFs:

Definition 8: Let κ be an $\mathcal{A}$-OCF and $\mathcal{B}$ and $\mathcal{C}$ two complete subfields of $\mathcal{A}$. Then *$\mathcal{C}$ is independent of $\mathcal{B}$ with respect to* κ iff for all atoms B of $\mathcal{B}$ and all atoms C of $\mathcal{C}$ $B \cap C \neq \emptyset$ and $\kappa(B \cap C) = \kappa(B) + \kappa(C)$. *$\mathcal{B}$ and $\mathcal{C}$* are *independent* (with respect to κ) iff $\mathcal{C}$ is independent of $\mathcal{B}$ and $\mathcal{B}$ is independent of $\mathcal{C}$. Moreover, if $A, B \in \mathcal{A} \setminus \{\emptyset, W\}$, A is *independent of* B (with respect to κ) iff $\{\emptyset, A, \overline{A}, W\}$ is independent of $\{\emptyset, B, \overline{B}, W\}$, and *$A$ and B* are *independent* (with respect to κ) iff A is independent of B and B is independent of A.

Definition 8 copies probabilistic independence concepts as far as possible. Note that independence with respect to OCFs need not be symmetric, simply because addition of ordinals is not commutative; therefore the distinction between "A is independent of B" and "A and B are independent".

Independence so defined has the properties we would expect.

Theorem 7: *If $\mathcal{C}$ is independent of $\mathcal{B}$ with respect to κ, then for all $B \in \mathcal{B} \setminus \{\emptyset\}$ and all $C \in \mathcal{C} \setminus \{\emptyset\}$*: $\kappa(B \cap C) = \kappa(B) + \kappa(C)$.

Proof: Let $B' \in \mathcal{B} \setminus \{\emptyset\}$ and $C' \in \mathcal{C} \setminus \{\emptyset\}$. Let further $\mathcal{B}'$ be the set of atoms of $\mathcal{B}$ which are subsets of B' and $\mathcal{C}'$ the set of atoms of $\mathcal{C}$ which are subsets of C'. Hence,

$B' = \bigcup \mathcal{B}'$ and $C' = \bigcup \mathcal{C}'$, and moreover, $\kappa(B') = \min\{\kappa(B) \mid B \in \mathcal{B}'\}$ and $\kappa(C') = \min\{\kappa(C) \mid C \in \mathcal{C}'\}$. Then we have

$$\begin{aligned}&\kappa(B' \cap C') = \min\{\kappa(B \cap C) \mid B \in \mathcal{B}', C \in \mathcal{C}'\} = \\ &= \min\{\kappa(B) + \kappa(C) \mid B \in \mathcal{B}', C \in \mathcal{C}'\} = \\ &= \min\{\kappa(B) \mid B \in \mathcal{B}'\} + \min\{\kappa(C) \mid C \in \mathcal{C}'\} = \\ &= \kappa(B') + \kappa(C'). \text{ Q. E. D.}\end{aligned}$$

The converse of Theorem 7 is obviously true. An immediate consequence of Theorem 7 and Definitions 5 and 7 is:

Theorem 8: *The following three assertions are equivalent*:

(a) *$\mathcal{C}$ is independent of $\mathcal{B}$ with respect to κ.*
(b) *For all $B \in \mathcal{B} \setminus \{\varnothing\}$ and $C \in \mathcal{C} \setminus \{\varnothing\}$ $\kappa(C \mid B) = \kappa(C)$ holds true.*
(c) *For each $\mathcal{B}$-OCF λ and each $C \in \mathcal{C} \setminus \{\varnothing\}$ $\kappa_\lambda(C) = \kappa(C)$ holds true.*

Theorem 8 particularly clearly shows the intuitive adequacy of Definition 8.

The parallel to probability theory may be extended further. In probability theory, one also defines independence for families of subfields. This can be done here as well.

Definition 9: Let $(\mathcal{B}_\alpha)_{\alpha<\beta}$ be a sequence of complete subfields of $\mathcal{A}$ and κ an $\mathcal{A}$-OCF. Then $(\mathcal{B}_\alpha)_{\alpha<\beta}$ is called *independent with respect to* κ iff for all atoms B_α of $\mathcal{B}_\alpha$ $(\alpha < \beta)$ $\bigcap_{\alpha<\beta} B_\alpha \neq \varnothing$ and $\kappa(\bigcap_{\alpha<\beta} B_\alpha) = \sum_{\alpha<\beta} \kappa(B_\alpha)$.

The connection to Definition 8 is stated in:

Theorem 9: *$(\mathcal{B}_\alpha)_{\alpha<\beta}$ is independent iff for all $\gamma < \beta$ the complete field generated by $\bigcup_{\gamma\leq\alpha<\beta} \mathcal{B}_\alpha$ is independent of the complete field generated by $\bigcup_{\alpha<\gamma} \mathcal{B}_\alpha$.*

Proof: Define $\mathcal{C}_\gamma$ and $\mathcal{D}_\gamma$ to be, respectively, the complete field generated by $\bigcup_{\alpha<\gamma} \mathcal{B}_\alpha$. and $\bigcup_{\gamma\leq\alpha<\beta} \mathcal{B}_\alpha$. Now suppose first that for all atoms B_α of $\mathcal{B}_\alpha (\alpha < \beta)$ $\kappa(\bigcap_{\alpha<\beta} B_\alpha) = \sum_{\alpha<\beta} \kappa(B_\alpha)$. This implies that for all atoms B_α of $\mathcal{B}_\alpha (\alpha < \gamma)$ $\kappa(\bigcap_{\alpha<\gamma} B_\alpha) = \sum_{\alpha<\gamma} \kappa(B_\alpha)$, and similarly, that for all atoms B_α of $\mathcal{B}_\alpha$ $(\gamma \leq \alpha < \beta)$ $\kappa(\bigcap_{\gamma\leq\alpha<\beta} B_\alpha) = \sum_{\gamma\leq\alpha<\beta} \kappa(B_\alpha)$. Thus, we have $\kappa(\bigcap_{\alpha<\beta} B_\alpha) = \kappa(\bigcap_{\alpha<\gamma} B_\alpha) + \kappa(\bigcap_{\gamma\leq\alpha<\beta} B_\alpha)$, and this means that $\mathcal{D}_\gamma$ is independent of $\mathcal{C}_\gamma$.

Conversely, suppose that for all $\gamma < \beta$ $\mathcal{D}_\gamma$ is independent of $\mathcal{C}_\gamma$. For $\gamma = 1$, this says that for all atoms B_α of $\mathcal{B}_\alpha$ $(\alpha < \beta)$ $\kappa(\bigcap_{\alpha<\beta} B_\alpha) = \kappa(B_0) + \kappa(\bigcap_{1\leq\alpha<\beta} B_\alpha)$. This implies in particular that for all atoms B_α of $\mathcal{B}_\alpha$ $(\alpha = 0, 1)$ $\kappa(B_0 \cap B_1) = \kappa(B_0 + B_1)$. For $\gamma = 2$, we therefore get that for all atoms B_α of $\mathcal{B}_\alpha (\alpha < \beta)$ $\kappa(\bigcap_{\alpha<\beta} B_\alpha) = \kappa(B_0 \cap B_1) + \kappa(\bigcap_{2\leq\alpha<\beta} B_\alpha) = \kappa(B_0) + \kappa(B_1) + \kappa(\bigcap_{2\leq\alpha<\beta} B_\alpha)$. Continuing this line of reasoning by transfinite induction till β then leads to the desired result. Q. E. D.

An immediate consequence of Theorem 9 is:

Theorem 10: *Let $(\mathcal{B}_\alpha)_{\alpha<\beta}$ be independent. Let $(\Gamma_\gamma)_{\gamma<\delta}$ be a partition of $\{\alpha \mid \alpha < \beta\}$ such that we have for all $\gamma, \gamma' < \delta$:* if *$\gamma < \gamma'$, then $\alpha < \alpha'$ for all $\alpha \in \Gamma_\gamma$ and $\alpha' \in \Gamma_{\gamma'}$.*

Let finally C_γ *be the complete field generated* $\bigcup_{\alpha\in\Gamma_\gamma} \mathcal{B}_\alpha$. *Then the sequence* $(C_\gamma)_{\gamma<\delta}$ *is also independent.*

In probability theory, the corresponding theorem is known as the theorem of the composition of independent fields.

As a last topic, let me take up conditional independence. It is well known that conditional independence is central for a probabilistic theory of causality. Thus, this topic will become important, when one turns to deterministic theories of causality. Here, however, I take it up only for demonstrating the parallel between probability measures and OCFs a bit further.

Definition 10: Let $\mathcal{B}$ and $\mathcal{C}$ be two complete subfields of $\mathcal{A}$, κ an $\mathcal{A}$-OCF, and $A \in \mathcal{A}\setminus\{\emptyset\}$. Then $\mathcal{C}$ is *independent* of $\mathcal{B}$ *conditional on A* (or *given A*) *with respect to* κ iff for all atoms B of $\mathcal{B}$ and all atoms C of $\mathcal{C}$ with $A \cap B \cap C \neq \emptyset$ $\kappa(B \cap C \mid A) = \kappa(B \mid A) + \kappa(C \mid A)$. If $\mathcal{D}$ is another complete subfield of $\mathcal{A}$, then $\mathcal{C}$ is *independent of* $\mathcal{B}$ *conditional on* $\mathcal{D}$ (or *given* $\mathcal{D}$) (with respect to κ) iff for each atom D of $\mathcal{D}$ $\mathcal{C}$ is independent of $\mathcal{B}$ given D. Further phrases may be defined in analogy to Definition 8.

The intuitive interpretation of Definition 10 should be clear and is supported by the fact that Theorems 7 and 8 hold correspondingly for conditional independence. The following theorems are more interesting; the expression "$\mathcal{B} + \mathcal{C}$" used in them is meant to denote the complete field generated by $\mathcal{B} \cup \mathcal{C}$.

Theorem 11: *Let* $\mathcal{B}$, $\mathcal{C}$, $\mathcal{D}$, *and* $\mathcal{E}$ *be four complete subfields of* $\mathcal{A}$. *Suppose that* $\mathcal{C}$ *is independent of* $\mathcal{B}$ *given* $\mathcal{D} + \mathcal{E}$ *and that* $\mathcal{D}$ *is independent of* $\mathcal{B}$ *given* $\mathcal{E}$. *Then* $\mathcal{C} + \mathcal{D}$ *is independent of* $\mathcal{B}$ *given* $\mathcal{E}$.

Proof: Let B, C, D, and E be variables for atoms of $\mathcal{B}$, $\mathcal{C}$, $\mathcal{D}$, *and* $\mathcal{E}$, respectively. The first assumption says that for all B, C, D, and E with $B \cap C \cap D \cap E \neq \emptyset$:

$$\begin{aligned}\kappa(B\cap C \mid D\cap E) &= \kappa(B \mid D\cap E) + \kappa(C \mid D\cap E), \text{ i.e. by Definition 5}\\ &= -\kappa(D \mid E) + \kappa(B\cap C\cap D \mid E)\\ &= -\kappa(D \mid E) + \kappa(B\cap D \mid E) + (-\kappa(D \mid E) + \kappa(C\cap D \mid E)), \text{ i.e.}\end{aligned}$$

$$\kappa(B\cap C\cap D|E) = \kappa(B\cap D|E) + (-\kappa(D|E) + \kappa\,(C\cap D|E)$$

The second assumption states that for all B, D, and E with $B \cap D \cap E = \emptyset$:

$$\kappa(B\cap D \mid E) = \kappa(B \mid E) + \kappa(D \mid E).$$

The last two equations together yield that for all B, C, D, and E with $B \cap C \cap D \cap E) \neq \emptyset$

$$\kappa(B\cap C\cap D \mid E) = \kappa(B \mid E) + \kappa(C\cap D \mid E)).$$

and that's what we had to prove. Q. E. D.

In the same way, the result symmetric to Theorem 11 may be proved:

Theorem 12: *If $\mathcal{B}$ is independent of $\mathcal{C}$ given $\mathcal{D} + \mathcal{E}$ and independent of $\mathcal{D}$ given $\mathcal{E}$, then $\mathcal{B}$ is independent of $\mathcal{C} + \mathcal{D}$ given $\mathcal{E}$.*

Moreover, we have

Theorem 13: *If $\mathcal{B}$ is independent of $\mathcal{C} + \mathcal{D}$ given $\mathcal{E}$ and independent of $\mathcal{C} + \mathcal{E}$ given $\mathcal{D}$, then $\mathcal{B}$ is also independent of $\mathcal{C} + \mathcal{D} + \mathcal{E}$ given $\mathcal{D} \cap \mathcal{E}$.*

Proof: Let B, C, D, and E be as in the proof of Theorem 11, and let F be a variable for the atoms of $\mathcal{D} \cap \mathcal{E}$ (which, to be sure, is also a complete field). The first premise says that for all B, C, D, E, and F with $B \cap C \cap D \cap E \neq \varnothing$ and D, $E \subseteq F$:

$\kappa(C \cap D \cap B \mid E \cap F) = \kappa(C \cap D \mid E \cap F) + \kappa(B \mid E \cap F)$, i.e. by Definition 5,
$= -\kappa(E \mid F) + \kappa(C \cap D \cap E \cap B \mid F)$
$= -\kappa(E \mid F) + \kappa(C \cap D \cap E \mid F) + (-\kappa(E \mid F) + \kappa(E \cap B \mid F)$, i.e.

(1) $\kappa(C \cap D \cap E \cap B \mid F) = \kappa(C \cap D \cap E \mid F) + (-\kappa(E \mid F) + \kappa(E \cap B \mid F))$.

Likewise, the second premise says that for all B, C, D, E, and F with $B \cap C \cap D \cap E \neq \varnothing$ and D, $E \subseteq F$:

$\kappa(C \cap E \cap B \mid D \cap F) = \kappa(C \cap E \mid D \cap F) + \kappa(B \mid D \cap F)$, i.e. as before

(2) $\kappa(C \cap D \cap E \cap B \mid F) = \kappa(C \cap D \cap E \mid F) + (-\kappa(D \mid F) + \kappa(D \cap B \mid F))$.

(1) and (2) imply that for all D, $E \subseteq F$

$$-\kappa(E \mid F) + \kappa(E \cap B \mid F) = -\kappa(D \mid F) + \kappa(D \cap B \mid F).$$

This in turn implies that for all atoms E, E' of $\mathcal{E}$ with E, $E' \subseteq F$

(3) $-\kappa(E \mid F) + \kappa(E \cap B \mid F) = -\kappa(E' \mid F) + \kappa(E' \cap B \mid F)$.

Now, there must be an atom E_0 of $\mathcal{E}$ with $E_0 \subseteq F$ such that $\kappa(E_0 \mid F) = 0$, since $0 = \kappa(F \mid F) = \min\{\kappa(E \mid F) \mid E \subseteq F\}$. Thus, we have, using (3), $\kappa(E_0 \cap B \mid F) = \min\{\kappa(E \cap B \mid F) \mid E \subseteq F\} = \kappa(B \mid F)$. Using (3) once more, this yields that for all $E \subseteq F$

$$-\kappa(E \mid F) + \kappa(E \cap B \mid F) = \kappa(B \mid F).$$

Substituting this result in (1), we finally get

$$\kappa(C \cap D \cap E \cap B \mid F) = \kappa(C \cap D \cap E \mid F) + \kappa(B \mid F)$$

for all B, C, D, E, and F with $B \cap C \cap D \cap E \neq \varnothing$ and D, $E \subseteq F$. Q. E. D.

The assertion symmetric to Theorem 13 does not necessarily hold.

These theorems are as analogous to probabilistic theorems as can be.[22,†3] Here I would like to end for the time being. I think there can be no doubt that OCFs are vastly superior to SCFs or WOPs.

1.7 Connections with Probability Theory

So far, all this has been a story wholly within deterministic epistemology. But there is in fact an exact probabilistic duplicate of our story progressing from net contents to OCFs. The probabilistic counterparts to our net contents are probability measures. With net contents we had the problem that we could say nothing about the new net content resulting from information incompatible with the old net content; this problem induced us to introduce the SCFs. The corresponding problem is that probabilities conditional on propositions having probability 0 are not defined in standard probability theory; we can say nothing about the new probability measure resulting from information having probability 0 in the old epistemic state. One solution to this problem, perhaps the most prominent one, consists in introducing Popper measures.[23] These are indeed the probabilistic counterparts to our SCFs; it is known that it is an SCF which, if adapted to the algebraic framework of probability theory (which operates with σ-fields instead of complete fields), represents the 0–1-structure of a Popper measure P, i.e. the relation $\{\langle A, B\rangle \mid P(B \mid A) = 1\}$.[24] This means, however, that Popper measures are as insufficient for a dynamic theory of epistemic states as SCFs are; this was very clearly pointed out by Harper (1976, pp. 95f). Hence, the probabilistic story calls for continuation, too. It is quite obvious what this should look like; just define probabilistic counterparts to OCFs which would be something like functions from propositions to ordered pairs consisting of an ordinal and a real between 0 and 1. I won't now pursue this in technical detail, since this fusion of probability theory and the theory of OCFs appears to me to be fairly straightforward. But the advantage of such probabilified OCFs over Popper measures is quite clear; it is the same as that of OCFs over SCFs.

---

[22] Cf. e.g. Spohn (1980, Theorem l(d) and (e)). Indeed, I wonder how far the mathematical analogy could be extended. What I have shown is that the probabilistic theory of dependence, independence, and conditionalization can be carried over to OCFs. The Definition 7 of generalized conditionalization suggests that the concept of a *mixture* may also be meaningfully carried over from probability measures to OCFs. This might be worth exploring. One essential point of dissimilarity is that, as far as I see, there is no meaning to a theory of integration within the theory of OCFs.

[†3] In fact, Theorems 11–13 were intended to show that conditional independence with respect to OCFs satisfies the graphoid axioms as they were later on called by Pearl (1988, p. 88). The idea of a mixture mentioned in the previous footnote is finally pursued in Chapter 7.

[23] Cf. e.g. van Fraassen (1976).

[24] Cf. Harper (1976, pp. 87ff.), or my (1986). The dimensionally well-ordered families of probability measures introduced in the latter paper are the counterparts to our WOPs; and these families represent Popper measures just as WOPs represent SCFs according to Theorem 1.

One point, however, is still open; we do not yet have an *explanation* of why OCFs behave so much like probability measures (which is, of course, not given by the proposed fusion of probability theory and the theory of OCFs). But this explanation may be made, I think, along the following lines within the framework of nonstandard probability theory[25]: Let P be a nonstandard probability measure for which there is an infinitesimal i such that for each A $P(A)$ is of the same order as i^n for some (nonstandard) natural number n (i.e. $P(A)/i^n$ is finite, but not infinitesimal). Now define $\kappa\{B \mid A) = n$ iff $P(B \mid A)$ is of the same order as i^n. Then κ is like an OCF within this framework. Indeed, we have thereby defined a homomorphism from a class of nonstandard probability measures onto the class of (nonstandard) OCFs which maps, first, addition of probabilities into taking the minimum of OCF-values and which maps, secondly, multiplication and division of probabilities into addition and subtraction of OCF-values. More specifically, whenever A and B are independent according to P, they are so according to κ; and for the κ so defined, we have $\kappa\{B \mid A) = \kappa\{B \cap A) - \kappa(A)$. This would explain why OCFs obey the same laws as probability measures concerning independence and conditionalization.[26]

1.8 Discussion

I see three points where the foregoing story should be related to the actual state of discussion.

The first point is that all concepts introduced in Section 1.2 are, of course, absolutely standard. SCFs are better known under the label of (class) selection functions, which play a central role in conditional logic; in this context, a number of slightly different concepts of a selection function have been proposed, and it is well known that nearly every semantics for conditional logic is based on some such concept.[27] Moreover, if $(E_\alpha)_{\alpha<\zeta}$ is a WOP, then the sequence $(\bigcup_{\alpha<\beta} E_\alpha)_{\beta\leq\zeta}$ is a (universal) system of similarity spheres (at one possible world) in the sense of David Lewis. (In general, a system of spheres need not to be well-ordered, of course.) Thus, Theorem 1 can be already found in Lewis (1973a, pp. 58f.) and in other places.

Why, then, did I define the SCFs in the way I did? Well, I have stated my reasons for doing so fully in Section 1.2. These reasons are debatable, but I have the impression that the slight differences between the various concepts of a selection function

[25] The idea is essentially due to Kurt Weichselberger. I have merged his idea with an idea I found in Skyrms (1983, p. 158).

[26] One may perhaps conclude that I should have carried through the whole business of OCFs within a nonstandard framework form the start. However, I am happier with the standard version presented, and I did not want to burden my theory with nonstandard number models.

[27] Cf. e.g. Nute (1980, chs. 1 and 3).

are motivated rather by differing opinions about conditional logic than about the dynamics of belief; and I was exclusively concerned with the latter which must not be mixed up with the former. (That was one reason why I deferred this comment.) To be sure, I completely side with Ernest W. Adams, Brian Ellis, Peter Gärdenfors, and others in maintaining that the various uses of the conditional can only be correctly and uniformly understood by relating them to a dynamic theory of epistemic states. But this relation is, I think, not yet sufficiently understood.

To be a bit more specific: If one accepts something like the straight thesis that the sentence "if *A*, then *B*" is accepted in (or true relative to) some epistemic state if and only if *B* is accepted in the revision of that state by *A*,[28] then one is bound to strain one or other side of this biconditional. A clear case, in my view, is provided by the very common causal conditionals. For, as I in effect argue in my (1983b), if the concept of revising epistemic states is only to tell how beliefs change, then, according to this thesis, the conditional "if *A*, then *B*" only states something about the evidential relations between *A* and *B*, i.e. about *A*,*s* being a reason for *B*, and thus does not yet express a causal relation between *A* and *B*. But let's not go further into this; my remark should only show why I want to confine myself to the dynamics of epistemic states and to leave aside the complicated relations to conditional logic.

The second point is this: If the central problem stated in Section 1.3 has been known at least since Harper (1976), what has been done to solve it? Surprisingly, not very much; and one reason for this is, it seems to me, that the issue has been obscured by what I have just complained about, i.e. by not clearly separating the dynamics of belief and conditional logic. In fact, I have found only three ideas which are addressed to this issue or can be so understood; and I shall deal with them, for the sake of brevity, only at a strategic level.

The first idea is that our problem of accounting for iterated belief changes appears to be analogous to the problem of providing a semantics for a language with iterated conditionals. The standard solution to the latter problem is to associate an SCF, a selection function, a system of similarity spheres, or whatever with *every* possible world (so that each conditional sentence has again a set of possible worlds as its truth condition). There is no need now to assess the semantic problem and its solution, though I always had the impression that in iterated intensional constructions the syntactic horse bolts with the semantic rider. The main point is that I don't see how the seeming analogy could be brought to bear; for, how should such a function from possible worlds to SCFs or whatever be interpreted as an epistemic state?

A second related idea is this: Enrich the language in which propositions are expressed by a conditional and thus by conditional sentences and propositions, and then exploit this new structural richness of the epistemic objects for a solution of our problem. This is, very roughly, the strategy applied by Harper (1976, pp. 95ff.). Ellis (1979, pp. 53ff.) and Gärdenfors (1979 and 1981, sects. II and III), seem to endorse it as well. However this strategy is brought to work in detail, it seems to be wrong from the start for two reasons: Our problem with SCFs shows,

[28] Gärdenfors (1981, p. 207), e.g., explicitly accepts this thesis.

one should think, that it is the characterization of epistemic states as SCFs and not the structure of the epistemic objects which is too poor; one would expect that a dynamic theory of epistemic states does not force us to make special assumptions about the underlying structure of the epistemic objects. So, this strategy seems to focus on the wrong point (whereas our OCFs conform to this expectation). Moreover, there is the problem of how the conditional is interpreted within this strategy. In order to keep within the spirit of their approach, Harper and the others want to interpret it in terms of the dynamics of belief so far elaborated. This, however, amounts in fact to assuming second or higher order epistemic states which are partially about propositions describing properties of lower order epistemic states. Interesting as this may be, this move is uncalled for; one would expect that the problem with SCFs can be solved strictly at the level of first-order epistemic states. Thus, this second idea seems to be an unconvincing mixture of the dynamics of belief and conditional logic.

The third and last approach is found in Gärdenfors (1984). The machinery developed there consists of belief sets, which are essentially equivalent to our net contents, and a relation of epistemic importance between sentences or propositions. With this machinery, Gärdenfors is able to describe successive changes of belief sets and thus gives a solution to our problem with SCFs – *provided* that the relation of epistemic importance is kept fixed. But why should it be so? An ordering of disbelief in our sense does essentially the same job as a belief set plus a relation of epistemic importance (though our respective interpretations of the two things do not match precisely); thus, that relation should be viewed as a part of an epistemic state which may change, too. Gärdenfors' approach therefore seems to me to provide only a restricted solution to our problem.

All this considered, there is enough reason to look for a solution to our problem elsewhere, as I have done in Sections 1.3–1.5.

The final point in need of a comment is that our OCFs look rather familiar; our degrees of disbelief seem more or less identical with the degrees of potential surprise in Shackle (1961/69). Indeed, the similarity is amazing, and the more so as Shackle developed his ideas long ago (before there was any conditional logic) and in quite a different scientific department. Since in particular his intuitive explanation of his functions of potential surprise perfectly fit my OCFs, it may be worthwhile to identify the points of difference, although this comparison is bound to be forced and somewhat unfair just because of the very different setting of his work.

According to Shackle (1961/69, p. 80),[29] a *function y of potential surprise* (an *FPS*) may be defined to be a function from a given field of propositions into the closed interval [0, 1] such that for all propositions A and B

(1) $y(\emptyset) = 1$
(2) either $y(A) = 0$ or $y(\bar{A}) = 0$ or both
(3) $y(A \cup B) = \min\{y(A), y(B)\}$

[29] Cf. also Levi (1980, p. 7).

(1) is the arbitrarily chosen maximal degree of potential surprise which is taken at least by ø, and (2) and (3) are identical with my Theorem 2 (Section 1.4). Thus, there seem to be hardly any differences between FPSs and OCFs; but there are four.

One point is that OCFs satisfy the generalization of (3) to arbitrary unions; but, as expressed in Notes 8 and 13, I do not attach much importance to this. Since in this generalization min is not weakened to inf, it forces the range of OCFs to be well-ordered; and then ordinals are the natural values for OCFs. Thus, the difference with respect to (3) also accounts for the differing ranges of FPSs and OCFs; but we shall see that there is more to the difference in the ranges.

Another difference is about the maximal degree of potential surprise. I also could have introduced a number larger than any ordinal as the OCF-value for ø; but this did not look nice, and so I preferred to make qualifications to the effect that ø is not imported into the domain of an OCF. The important point here is that I therefore do not allow any other proposition to take the maximal value. The reason is that, once a proposition were disbelieved to the maximal degree, it would always be disbelieved to the maximal degree, at least according to my rules of belief change; rational belief change could then no longer be treated within my framework. This was something I wanted to avoid. Shackle, by contrast, makes free use of the maximal degree of potential surprise. And Levi (1980, p. 7) explicitly assigns it to each proposition that is incompatible with what he calls a corpus of knowledge, and he therefore has trouble, e.g. in (1983), with specifying rules for changing such corpora of knowledge.

The essential point is that Shackle has no precise and workable account of conditional degrees of potential surprise, of changes of FPSs, etc. This becomes apparent in his handling of conjunctions. In his (1961/69, pp. 80ff., 199ff.), he sticks to the postulate that

(4) $y(A \cap B) = \max\{y(A), y(B \mid A)\}$

(where I have adapted the notation and where $y(B \mid A)$ is in fact undefined). In contrast to this, our Definition 5, which is fundamental for our Sections 1.5 and 1.6, is equivalent to

(5) $\kappa(A \cap B) = \kappa(A) + \kappa(B \mid A)$.

Shackle has obviously considered accepting something like (5) instead of (4); but he says little about why he finally rejected it. In his (1961/69, p. 205), he says only that (4) would be simpler and less unrealistic than something like (5).

A final significant difference may be inferred from (1)–(5). Shackle (1961/69, chs. XV–XVII) clearly intends his FPSs to be measurable on a ratio scale. But it is hard to see precisely how this scale is established and where it is really used; it seems that we may conceive FPSs as purely ordinal concepts.[30] In any case, FPSs as displayed by (1)–(4) are purely ordinal, as may be seen from the exclusive use of mathematical operations like max and min. But if this is so, FPSs correspond to our WOPs (or the functions definable by WOPs according to Note 13). This would mean that the decisive step towards OCFs is perhaps intended, but not really taken by FPSs.

[30] On p. 188, Shackle (1961/69) says that the assumption of the cardinality of his tool is "by no means indispensible to its main purpose".

Part II
Causation

Chapter 2
Direct and Indirect Causes[†1,*]

2.1 Introduction

Everybody agrees that the distinction between direct and indirect causation is important. And it seems easy to draw, *if* an analysis of causation in general is available. The causal influence of one event on another is direct, if it is not mediated by other events in between; otherwise it is indirect. The trouble is with the proviso. Indeed, I contend that the order of analysis must be reversed because the distinction is required for a successful analysis of causation. Such an analysis perhaps proceeds best in two steps: the first analyses direct causation, and the second extends the analysis to indirect causation and thus to causation in general. Such a strategy is at least plausible. For direct causation is a very special case and so may be supposed to be more easily explicable. Then, one might say that the relation "*A* is a cause of *B*" is just the transitive closure of the relation "*A* is a direct cause of *B*", thus completing the full analysis of causation.[1] The complete story is not so simple; but the idea will turn out to be right. In (1983b), I dealt mainly with the first step – direct causation. Here, I deal mainly with indirect causation.

Section 2.2 introduces the conceptual machinery required throughout the paper. Section 2.3 recapitulates my (1983b) explication of direct causes. Section 2.4 considers the circumstances of direct causal relations in greater detail. Section 2.5

[†1] This paper was originally published in: *Topoi* 9 (1990) 125–145. It is an elaboration and partial revision of the accounts of probabilistic causation I had given in Spohn (1980 and 1983a, ch. 6).

* I am indebted to the University of California at Irvine for giving me the opportunity to present much of the material during a visiting professorship during the Winter term 1988, to Maria Carla Galavotti for giving me another opportunity at Bagni di Lucca in October 1988, to Nancy Cartwright and Brian Skyrms for discussion and encouragement, and to Karel Lambert and Hans Rott for very carefully checking the manuscript and considerably improving style and content.

[1] This is precisely how Lewis (1973b) proceeds.

presents the main difficulties with indirect causes. Section 2.6, finally, proposes a strategy for dealing with these difficulties and shows that it will work.

The whole enterprise is subject to two major constraints. First, I shall discuss only causation of single events. The hope, of course, is that it will emerge from knowledge of causation in the single case what causal laws are. This procedure seems to me to be more likely to succeed than the reverse strategy endorsed by several philosophers,[2] though I shall not argue the point here.

Secondly, I am concerned here only with probabilistic causation because this is the context in which all the problems dealt with here have been raised and have been discussed most extensively. It should be mentioned, however, that each consideration, definition, and theorem of the present paper can routinely be extended to deterministic causation with the help of the theory of ordinal conditional functions (OCFs) I proposed in (1988); only some marginal adjustments may be needed. This follows because deterministic conditional independence defined for OCFs obeys essentially the same laws as probabilistic conditional independence.[3] It is thus possible to unify the deterministic and the probabilistic approach.[4,†2]

2.2 The Conceptual and Formal Framework

Each discussion of probabilistic causation proceeds from an explicitly given probability space: let I be a non-empty set of variables or factors. Each variable $i \in I$ is associated with a set Ω_i of at least two possible values i may take. The cross product Ω of all the Ω_i, is the set of all functions ω defined on I such that, for each $i \in I$, $\omega_i \in \Omega_i$; intuitively, each ω represents a possible course of events – a possible world in philosophers' talk, or a possible path in the mathematician's terminology. I, each Ω_i, and hence Ω are assumed to be finite. This severe restriction has several advantages. One of these is that there is no need to worry about measurability because each subset of Ω may be assumed to represent a state of affairs or an event in the mathematicians' sense, but not the philosophers'. Moreover, we assume a probability measure P assigning a probability to each state of affairs, i.e. to each subset of Ω. This completes the description of the underlying probability space.

This explicitness has an important philosophical consequence: namely, that everything said about causation is relative to the descriptive frame given by the set I of variables. Many discussions of examples suffer, I think, from an inadequate recognition of this relativization. It is essential because the causal relations may indeed vary

[2] For instance by Cartwright (1979), Giere (1980), and all those who take probability in causal contexts as a statistical property of event types or classes or the like. However, Cartwright herself attacks the reverse strategy in her (1988). See also Davis (1988) for a discussion of this point.

[3] Compare Theorem 2 below with sect. 6 of Spohn (1988) [here: sect. 1.6].

[4] Cf. Spohn (1988, sect. 7) [here: sect. 1.7].

[†2] These remarks were apparently not noticed. Thus, I finally made them explicit in Chapter 3.

with the frame. Consider, e.g., a series of throws of a die by a machine: relative to a coarse probabilistic description which contains only variables representing the throws, no throw will be causally relevant to the next one. Relative to a finer description, however, which, for each time, allows for variables representing the mechanical state of the whole system (but which may still be probabilistic, say, because of neglect of air resistance), each throw will be causally relevant to the subsequent ones. One is, perhaps, inclined to think of causation as an absolute notion. However, from the current starting point the only way to get rid of the relativization is via the most fine-grained descriptive frame embracing *all* variables whatsoever. I am not sure whether such a move makes sense; it is at least philosophically problematic. Here, I will be content with the relativized notion of causation.

The relativization of causes is even more apparent in the distinction between direct and indirect causation. A state of affairs which is a direct cause relative to a coarse descriptive frame not mentioning the mediating links may well turn out to be an indirect cause relative to a more complete descriptive frame.

If time is continuous and if variables are associated with points and not with intervals of time, then, presumably, direct causes either do not exist or are simultaneous with their direct effects. In either case, the strategy of explicating causation via direct causation would not work because, in either case, causation would certainly not be the transitive closure of direct causation. So the strategy of analysis here demands a descriptive frame with discrete time. The idea is that the results obtained for discrete time may be generalized to continuous time in a fashion similar to the way in which the theory of stochastic processes has been extended, and the hope is that this will raise only well-known mathematical, but not new conceptual or philosophical problems. I shall not attempt here, however, any such generalization.

I assume a weak order[5] $\leq$ on the set I of variables which represents the order of the times at which the variables are realized; $<$ is to denote the corresponding irreflexive order relation. Since I is finite, time is bound to be discrete. By assuming the order to be weak, simultaneous variables are in general allowed; the few exceptions will be explicitly noted. However, I shall not consider simultaneous causation; I am not sure whether this would be desirable.[6] And I plainly exclude backwards causation; it will be clear that this is vital to the theory to be proposed here.

An analysis of causation faces a number of well-known and unsolved problems relating to variables which have more than two possible values.[7] One may evade these problems by considering only binary variables. But there is a hitch to this restriction. The causal theorems to be proved essentially derive from the laws of conditional probabilistic independence, and there is one such law peculiar to binary

[5] This means that $\leq$ is transitive and complete.

[6] In (1980) I allowed for simultaneous causation in a way which preserved continuity with the restricted case. I am not sure whether the same procedure would work here.

[7] What is discussed with respect to more-than-two-valued variables is usually only causal relevance simpliciter and not positive or negative causal relevance. An exception is Suppes (1970, pp. 60ff.), but it has not been further discussed, as far as I know.

variables (see Theorem 2(f) below) which may have unforeseen and undesired consequences.[8] Therefore, variables will be assumed to be binary only when required, and the problems with variables with more then two values will be neglected.

Finally, I shall assume that the probability measure P is strictly positive, i.e. that $P(\{\omega\}) > 0$ for all $\omega \in \Omega$; hence, the conditional probability $P(B \mid A)$ is defined for each $A \neq \emptyset$. Since Ω is finite, this assumption is unproblematic. The reason for it is that all probabilistic theories of causation run into serious trouble with the limiting probabilities 0 and 1.[9]

How are probabilities to be understood in the present context? Any way you like. For instance, if one takes probability objectively, preferably in a propensity interpretation, then the definitions below attempt to explicate causation as it objectively is. If, however, probabilities are understood epistemically as those of a certain subject at a certain time, then these definitions account for the causal conception of that subject at that time.

For philosophical reasons, I prefer the second understanding of probability. There are two main reasons. First, objective probability is the much more problematic notion, and it seems to be heavily intertwined with causality.[10] The most promising attempt to understand it is, I think, via subjective probability.[11] This suggests to me that the appropriate order is to start with subjective probability, to explicate causation within the subjectivistic framework, and then to try to objectivize both.

Secondly, I have general reservations about too realistic an understanding of causation. There is a need for explaining the most pervasive and prominent epistemological role which the notion of causation plays. If one takes causation simply as a constituent of the real world, then the only explanation one can give seems to be this: "Causation is, of course, a fundamental and pervasive trait of reality; thus it is small wonder that the notion of causation plays a fundamental and pervasive role in our picture of reality". However, the same argument would hold, say, for quarks or electromagnetic forces. Thus, this kind of explanation assimilates the epistemological role of the notion of causation to that of our notions of other important things like quarks or electromagnetic forces. This seems to me to be a distortion; according to the views of Hume, Kant, and other philosophers,[12] the notion of causation has not only an important, but a peculiar epistemological role which cannot be sufficiently explained from a realistic point of view. However, this essay is deliberately neutral

[8] Within the theory of OCFs there is no such peculiar law and thus no technical difference between binary and other variables.

[9] This is clearly displayed by Otte (1981) who criticized Suppes (1970) essentially on this account. I have argued in (1980, pp. 92f.), that the trouble-maker is essentially the fact that in standard probability theory there are no conditional probabilities for conditions having probability 0. The problem evaporates in the unification mentioned at the end of the introduction.

[10] Cf., e.g., Salmon (1988a) who argues that propensities are best understood as probabilistic causes and that other objective probabilities are derived from propensities.

[11] Here, I refer to Lewis (1980a) and Skyrms (1984, ch. 3); see also Spohn (1987).

[12] The most eloquent at present is Putnam who repeatedly argues against a naturalistic conception of causation, e.g. in (1983b).

with respect to these deep and crucial philosophical issues. Its focus is on the *logic* of causation, and it is intended to inform the philosophy of causation.

The following notation will be used throughout: variables, i.e. elements of I, will be denoted by i, j, k, and l, subsets of I by J, K, L, M, and N (with or without subscripts). (i, j) refers to the open interval between i and j, i.e. to $\{k \in I \mid i < k < j\}$, and $[i, j]$ to the closed interval $\{k \in I \mid i \le k \le j\}$; $\{< j\}$ denotes the past of j, i.e. $\{k \in I \mid k < j\}$, and $\{< j - K\}$ the past of j except K, i.e. $\{< j\} - K$.[13] Instead of $\{< j - \{i_1, \ldots, i_n\}\}$ we simply write $\{< j - i_1, \ldots, i_n\}$.

Possible paths, i.e. elements of Ω, will be denoted by υ and ω, states of affairs, i.e. subsets of Ω, by A, B, C, D, and E. We often have to refer to partial paths or, rather, to the set of their completions, which are states of affairs: for each $\omega \in \Omega$ and $J \subseteq I$ we define ${}^{\omega}J = \{\upsilon \in \Omega \mid \upsilon(i) = \omega(i) \text{ for all } i \in J\}$,[14] and I write ${}^{\omega}i$ instead of ${}^{\omega}\{i\}$. In general, states of affairs which are concerned only with variables in some set J are called J-measurable states or simply J-states; mathematicians also call them J-cylinders. The formal definition is that A is a J-state iff, for all υ and ω agreeing on J, $\upsilon \in A$ iff $\omega \in A$. Thus, A is a J-state iff $A = \bigcup\{{}^{\omega}J \mid \omega \in A\}$; and in particular each ${}^{\omega}J$ is a J-state.

The laws of conditional probabilistic independence lie at the bottom of the whole inquiry and therefore need at least to be stated.

Definition 1: The states of affairs A and B are *independent conditional on* C, i.e. $A \perp B / C$, iff $P(A \cap B \mid C) = P(A \mid C)\, P(B \mid C)$. And the sets K and L of variables are *independent conditional on* the set M of variables, i.e. $K \perp L / M$, iff, for all K-states D, L-states E, and $\omega \in \Omega$, $D \perp E / {}^{\omega}M$. I shall often mix the two notations, i.e., more precisely: $K, A \perp L, B/M, C$ is to mean that, for all K-states D, L-states E, and $\omega \in \Omega$, $A \cap D \perp B \cap E / C \cap {}^{\omega}M$.

The independence of states of affairs obeys:

Theorem 1:

(a) If $A \perp B/C$, then $B \perp A/C$,
(b) if $P(C) \neq 0$ and $C \subseteq A$, then $A \perp B/C$,
(c) if A and A' are disjoint and $A \perp B/C$, then $A \cup A' \perp B/C$ iff $A' \perp B/C$,
(d) if $A \perp C/D$, then $A \perp B \cap C/D$ iff $A \perp B/C \cap D$.

The independence of sets of variables obeys:

Theorem 2:

(a) If $K \perp L/M$, then $L \perp K/M$,
(b) if $K \subseteq M$, then $K \perp L/M$,
(c) if $K' \subseteq K \cup M$, $L' \subseteq L \cup M$, $M \subseteq M' \subseteq K \cup L \cup M$, and $K \perp L/M$, then $K' \perp L'/M'$,

[13] The hyphen denotes set theoretic difference.

[14] I choose this notation because the restricted domain needs to be more salient than the path itself.

(d) if $J \perp K/L \cup M$ and $J \perp L/M$, then $J \perp K \cup L/M$,
(e) if K and L are disjoint, $J \perp K/L \cup M$, and $J \perp L/K \cup M$, then $J \perp K \cup L/M$ – provided P is strictly positive,
(f) if i is a binary variable, $K \perp L / M$, and $K \perp L / M \cup \{i\}$, then $K \cup \{i\} \perp L / M$ or $K \perp L \cup \{i\}/M$.

For proofs see, e.g., Dawid (1979) or Spohn (1980). In particular Theorem 2(e) will be important; this is a further reason for assuming a strictly positive probability measure.[15] This list of properties of conditional independence is not complete,[16] but Geiger and Pearl (1988) present a number of interesting partial completeness results.

Concerning causal notation, three things must be observed. First, the causal relata are always states of affairs which are states of a single variable and thus are, so to speak, logically simple; I do not see the need to consider logically complex states of affairs as causes or effects.[17] Second, whether A is a cause of B depends, of course, on the given world or path; there may well be two worlds such that A causes B only in one world, but not in the other. This path-relativity will be made explicit in the notation. Third, only facts can be causes or effects; A can cause B in ω only if A and B obtain in ω, i.e. if $\omega \in A \cap B$.

$A \xrightarrow[\omega]{+} B$ is to mean that A is a direct cause of B in ω; and $A \xrightarrow[\omega]{+\ldots+} B$ is to mean that A is a (direct or indirect) cause of B in ω. This notation, and all the notation to follow, always carries the presupposition that $\omega \in A \cap B$ and that there are variables i and j such that A is an i-state, B is a j-state, and $i < j$. A similar notation for counter-causation, causal relevance and irrelevance, etc. will be introduced later on.

2.3 Direct Causes

A is a cause of B iff A precedes B and raises the epistemic or metaphysical rank of B under the obtaining circumstances. This is the basic conception of causation which has found the widest agreement. In the deterministic case, it covers regularity theories and counterfactual approaches as well as analyses in terms of necessary and/or sufficient conditions which all differ on the relevant meaning of "raises the

[15] Theorem 2(f) does not hold for ordinal conditional functions. If their range is restricted to natural numbers, they satisfy the laws (a)–(e) without further qualification. Cf. Spohn (1988, sect. 6) [here: sect. 1.6].

[16] Studeny (1989) and Geiger and Pearl (1988, sect. 6), mention further properties, and there are still more.

[17] This is so because causes in the intuitive sense are partial causes as opposed to total causes and because I want to account directly for causes without considering total causes. Particularly in the context of deductive-nomological explanation philosophers have been attracted by the idea that, conversely, the notion of a total cause is the central one which has to be explicated first. This strategy, I think, has been rejected for good reasons.

epistemic or metaphysical rank". My proposal is still different, namely to explicate this phrase in terms of OCFs. In the probabilistic case, however, there is only one interpretation of this phrase, that is, that *A* raises the probability of *B*.

The phrase "the obtaining circumstances" is also unclear; it is, in a sense, the subject of the whole paper. For direct causation, however, there is a particularly simple definition. This indeed is the main reason for splitting the account of causation into two steps. As I have argued in (1980, pp. 79ff.) (1983b, pp. 384ff.), and (1983c, pp. 80ff.), each fact preceding the direct effect *B* and differing from the direct cause *A* is to count among the obtaining circumstances of the direct causal relation between *A* and *B*; whenever judgment about that relation is based on less, it may be just the neglected facts which would change the judgment. This means that the obtaining circumstances consist of the whole past of *B* with the exception of *A*. I shall turn this into a formal definition and briefly compare it with other proposals.

Definition 2: Let *A* be an *i*-state, *B* a *j*-state, $i < j$, and $\omega \in A \cap B$. Then, *A is a direct cause of B in* ω, i.e. $A \xrightarrow[\omega]{+} B$ iff $P(B \mid A \cap {}^{\omega}\{<j-i\}) > P(B \mid \bar{A} \cap {}^{\omega}\{<j-i\})$.[18] *A is a direct counter-cause of B in* ω, i.e. $A \xrightarrow[\omega]{-} B$ iff $P(B \mid A \cap {}^{\omega}\{<j-i\}) < P(B \mid \bar{A} \cap {}^{\omega}\{<j-i\})$. *A is directly causally relevant to B in* ω, i.e. $A \xrightarrow[\omega]{\pm} B$ iff $A \xrightarrow[\omega]{+} B$ or $A \xrightarrow[\omega]{-} B$. Finally, *A is directly causally irrelevant to B in* ω, i.e. $A \xrightarrow[\omega]{0} B$, iff not $A \xrightarrow[\omega]{\pm} B$.

In a way, Definition 2 proposes a radical solution to Simpson's troublesome paradox. If one conditionalizes on the whole past of the effect, there is no further subdivision of that past which could change the conditional probabilities. Of course, this is true only relative to a fixed descriptive frame; but this only emphasizes the importance of relativization.

Suppes (1970, pp. 41f.), moves from his own definition of *prima facie* causes toward Definition 2 by acknowledging the legitimacy and usefulness of relativizing his definitions to some background information. However, he does not expand on this suggestion. One may think that one need not mention the background as long as it is constant. But according to Definition 2, different direct causal relations refer to different backgrounds, to different obtaining circumstances. So it is mandatory to make the reference explicit.

When discussing Simpson's paradox, Suppes (1984) doubts that the problem posed by it is solvable absolutely. He says, for example, that "there is no end to the analysis of data in a practical sense" (p. 56). I agree. But surely there is a natural end to the analysis of data within a fixed descriptive frame, a point with which Suppes, in turn, seems to agree (p. 57). This is captured by Definition 2.

Good's theory (1961–63) differs from Definition 2 in several ways, but the crucial point is that in defining the tendency of *A* to cause *B* Good considers other conditional probabilities. He conditionalizes on the whole past of the cause and on

[18] $\bar{A}$ denotes the complement of *A* relative to *Q*.

all true laws of nature,[19] whereas I conditionalize on the whole past of the direct effect. I have not found a clear argument for the appeal to the true laws.[20] Indeed, as far as I am concerned, it spoils much of the philosophical interest of the whole enterprise; for, the hope is that a better grasp of laws of nature will emerge from the analysis of singular causation.

The main question is whether to conditionalize on the past of the cause or on the past of the effect. Definition 2 would obviously be inadequate as a general account of causation, and one might therefore favor Good's account. My theoretical reasons for not doing so will emerge later on. The basic objection, however, is provided by a simple example:

Take a two-person game; each of the players makes a choice, and the outcome is determined accordingly. So the outcome is caused by both of the choices, but these choices, let us suppose, are causally independent of each other. Indeed, we may assume that their temporal order is totally irrelevant to the whole set-up.[21] What is the causal efficacy of the choice of the first player to the outcome? On Good's account, it varies with the temporal order: if the second player makes his choice first, that choice must be conditionalized on, otherwise not. This seems unacceptable because the causal set-up is not really changed by changing the temporal order. On the other hand, if the choices are taken as direct causes of the outcome, Definition 2 judges the causal efficacy of one choice by considering the probabilities conditional on the other *irrespective* of their temporal order.

One might point out that there is no difference between Good's conditionalization proposal and mine if direct causes immediately precede their direct effects, i.e. if there are no temporally intermediate variables in the given descriptive frame. And one might think that it would indeed be reasonable to assume that direct causes immediately precede their direct effects.[22] At the present stage, however, this assumption is quite unreasonable. It is a strong assumption which implies that, in case I is linearly ordered by $\leq$, the given probability space is so well-behaved as to form a Markov chain. But attention should not be confined just to Markov processes; there are many examples of causal processes which can at present not be modelled as Markovian. Indeed, from a theoretical point of view, it would be disastrous to start by assuming well-behaved causal processes. What is needed is a *general* account of causation in terms of which the virtues of the various forms of

[19] Cf. Good (1961, pp. 308f.), and (1988, p. 27).

[20] In Good (1988), he only argues on p. 27 that his way of conditionalization yields the desired result that a falling barometric reading has no tendency at all to cause a storm. But this result may already be obtained by conditionalization with respect to the past of the spurious cause; no reference to laws of nature is required for this example.

[21] Thus, even if one player chooses first, the other does not know. Though many examples have the same structure, the irrelevance of the temporal order seemed to me to be particularly perspicuous in this game-theoretic case.

[22] Good (1961) indeed makes a similar assumption on p. 45 when he requires neighbors in causal chains to be contiguous in space and time.

well-behaved causal processes can be characterized. Hence, the strong assumption should be investigated at a later stage.

Cartwright (1979) is interested in causal laws rather than in singular causation. Nevertheless, it is instructive to compare her views with Definition 2. She argues that all the variables influencing *B* but not influenced by *A* constitute the obtaining circumstances of the causal relation between *A* and *B* and that conditionalization with respect to them tells us whether *A* is a cause of *B*. Though I conditionalize on much more, the conflict is less important than it seems. Cartwright rightly insists that one must not conditionalize with respect to variables mediating between cause and effect; indeed, if their values are given, the cause cannot be expected to be positively relevant to the effect. But in the special case of direct causation there are no mediating variables; and the difference is then reduced to the fact that I conditionalize also with respect to all variables which precede, but do not influence the effect, whereas she does not.

I think that the more extensive conditionalization proposal is harmless, but she does not. She says on p. 432 that "partitioning on an irrelevancy can make a genuine cause look irrelevant, or make an irrelevant factor look like a cause" and goes on to illustrate this alleged possibility. Eells and Sober (1983), who also conditionalize on irrelevant factors, argue on p. 42 that this illustration does not support Cartwright's restricted form of conditionalization; and I agree with them.

Maintaining Definition 2 has an important consequence. The upshot of Cartwright's paper is that there is no non-circular characterization of causation in probabilistic terms. But if I am right, this is certainly not true of direct causation. Hence, there is hope that Cartwright's skeptical view is not true of causation in general.

2.4 The Circumstances of Direct Causes

The foregoing defense notwithstanding, it must be admitted that Definition 2 does not embody the *only* possible explication of obtaining circumstances. There are five further explications; and it is important to clarify them and to see the extent to which they are equivalent.

Definition 2 was based on the observation that each fact preceding the direct effect *B* and differing from the direct cause *A* is relevant as a circumstance. Here, "relevant" was used in the widest possible sense, namely as "possibly relevant solely on the basis of temporal relations", which is fixed in:

Definition 3a: Let ω, *A*, *B*, *i*, and *j* be as in Definition 2. *Then the temporally possibly relevant circumstances of* (the direct causal relation between) *A and B in* ω are defined as $C^{++}_{\omega}(A,B) = {}^{\omega}\{<j - i\}$.

This widest sense of "relevant" yields, as is to be expected, the narrowest circumstances. But there is a stricter sense of "possibly relevant". Whether a variable is relevant to the relation between *A* and *B* may also depend on the probabilities involved. To specify this idea, we will need:

Definition 4: $R_\omega(B)$ is to denote the set of all variables directly causally relevant to B in ω, i.e. $R_\omega(B) = \{k \in \{<j\} \mid \text{not } B \perp k / {}^{\omega}\{<j-k\}\}$. And $R(B)$ is to denote the set of all variables directly causally relevant to B in some world, i.e. $R(B) = \bigcup_{\omega \in \Omega} R_\omega(B) = \{k \in \{<j\} \mid \text{not } B \perp k / \{<j-k\}\}$.

These sets will play an important role. A first crucial observation is:

Theorem 3: $R(B)$ is the smallest subset R of $\{<j\}$ such that $B \perp \{<j-R\}/R$.

By Definition 4, we have $k \in \{<j-R(B)\}$ iff $B \perp k/\{<j-k\}$; and from this Theorem 3 follows with the help of Theorem 2(e). Thus, $R(B)$ is the minimal set of variables preceding B which screens off all the other preceding variables from B; i.e. given their values, B is probabilistically independent of all the rest of the possible past of B. This yields another sense of "relevant", namely, "possibly relevant on the basis of temporal relations and probabilities alone":

Definition 3b: *The probabilistically possibly relevant circumstances of* (the direct causal relation between) *A and B in* ω are defined as $C^{+}_{\omega}(A,B) = {}^{\omega}(R(B) - \{i\})$.

What one usually has in mind, however, is not possible, but actual relevance; intuitively, it should suffice to consider only the actually relevant circumstances. Here is a first attempt of explication: Definition 2 can be interpreted metalinguistically as giving the truth conditions of the sentence "*A is a direct cause of B*", i.e. as specifying when this sentence is true in a world ω. Viewed in this way, it seems plausible to say that the actually relevant circumstances of *A*'s being a direct cause of *B* just consist in the fact that *A* is a direct cause of *B*, i.e. in the set of all the worlds which relate *A* and *B* in this way; likewise for "direct counter-cause" and "direct causal irrelevance". To render this idea precise we need the signum function for reals defined as $sgn(0) = 0$ and $sgn(x) = x/|x|$ for $x \neq 0$.

Definition 3c: *The actually relevant circumstances of* (the direct causal relation between) *A and B in* ω *in the widest sense*[23] are defined as $C''_{\omega}(A,B) = \{\upsilon \mid sgn[P(B \mid A \cap {}^{\upsilon}\{<j-i\}) - P(B \mid \bar{A} \cap {}^{\upsilon}\{<j-i\})] = sgn[P(B \mid A \cap {}^{\omega}\{<j-i\}) - P(B \mid \bar{A} \cap {}^{\omega}\{<j-i\})]\}$.

The deterministic analogue of this definition is not uninteresting, but the probabilistic concept is quite useless because it is not generally true that $sgn[P(B \mid A \cap C''_{\omega}(A,B)) - P(B \mid \bar{A} \cap C''_{\omega}(A,B))] = sgn[P(B \mid A \cap {}^{\omega}\{<j-i\}) - P(B \mid \bar{A} \cap {}^{\omega}\{<j-i\})]$; that is, if one conditionalizes on the circumstances in this widest sense, one may even get different causal conclusions. So the widest sense is too wide.

Here is a modification: The inadequate proposal holds that the actually relevant circumstances of *A*'s being a direct cause of *B* just consist in the fact that *A* is a direct cause of *B*. Now it seems that they rather consist in the fact that *A* is a direct cause of *B* in the way it actually is – where this additional clause refers to the specific numerical change of the probability of *B* which is actually due to *A*. The idea is captured in:

[23] It is now "circumstances", not "relevance" which is taken in its widest sense.

Definition 3d: *The actually relevant circumstances of* (the direct causal relation between) *A and B in* ω *in the wide sense* are defined as $C'_\omega(A,B) = \{\upsilon \mid$ for each $A' \in \{A, \bar{A}\}$ $P(B \mid A' \cap {}^{\upsilon}\{<j-i\}) = P(B \mid A' \cap {}^{\omega}\{<j-i\})\}$.

As can be easily shown, for each $\{<j-i\}$-measurable $D \subseteq C'_\omega(A,B)$, $P(B \mid A' \cap D) = P(B \mid A' \cap {}^{\omega}\{<j-i\})$ and hence $B \perp \{<j-i\} / A' \cap D$ for $A' \in \{A, \bar{A}\}$; in fact, $C'_\omega(A,B)$ is the largest $\{<j-i\}$-measurable set for which this is true. Thus, $C''_\omega(A,B)$ represents the widest circumstances such that conditionalization on them agrees with conditionalization on any more narrowly taken circumstances of necessity and not by accident because of lucky averaging.[24] This strongly indicates that we have hit upon a reasonable explication.

So let me study $C'_\omega(A,B)$ a bit more closely. One valuable piece of information concerns which cylinders are subsets of $C'_\omega(A,B)$. It is given by:

Theorem 4: Let ω, *A*, *B*, *i*, and *j* be as in Definition 2. For each $\upsilon \in C'_\omega(A,B)$ and $K \subseteq \{<j-i\}$ we then have ${}^{\upsilon}\{<j-K \cup \{i\}\} \subseteq C'_\omega(A,B)$ iff $B \perp K / A' \cap {}^{\upsilon}\{<j-K \cup \{i\}\}$ for each $A' \in \{A, \bar{A}\}$.

For proof it is sufficient to consider Definitions 1 and 3d.

The theorem points to a useful distinction in $C'_\omega(A,B)$. Each $\upsilon \in C'_\omega(A,B)$ differs from ω on some variables. The only interesting differences are in $\{<j-i\}$, because outside $\{<j-i\}$ the members of $C'_\omega(A,B)$ may vary arbitrarily, anyway. Thus, let $K = \{k \in \{<j-i\} \mid \upsilon(k) \neq \omega(k)\}$. Now the distinction is this: one case is that υ is in $C'_\omega(A,B)$ because all variations of ω on *K* are in $C'_\omega(A,B)$, i.e. because ${}^{\omega}\{<j-K \cup \{i\}\} \subseteq C'_\omega(A,B)$ or, equivalently, $B \perp K / A' \cap {}^{\omega}\{<j-K \cup \{i\}\}$ for $A' \in \{A, \bar{A}\}$. The other case is that these conditional independencies do not hold. In this case, υ is, in a sense, only accidentally in $C'_\omega(A,B)$, i.e. not because the variables in *K* do not matter to *B* given ${}^{\omega}\{<j-K \cup \{i\}\}$ and *A* or $\bar{A}$. Rather, the variables in *K* do matter; it is only that in some particular realizations of *K* the relevant conditional probabilities come out the same as for ω and that υ represents one such realization of *K*.

This suggests that the actually relevant circumstances of *A* and *B* in ω should be conceived a bit more narrowly, namely as comprising only all the arbitrary variations of ω in $C'_\omega(A,B)$ (*A*,*B*).

Definition 3e: *The actually relevant circumstances of* (the direct causal relation between) *A and B in* ω *in the narrow sense* are defined as $C_\omega(A,B) = \bigcup\{{}^{\omega}\{<j-K \cup \{i\}\} \mid K \subseteq \{<j-i\}$ and $B \perp K/A' \cap {}^{\omega}\{<j-K \cup \{i\}\}$ for each $A' \in \{A, \bar{A}\}\}$.

It will soon become clear why this is the preferred sense of the obtaining circumstances of a direct causal relation.

The five concepts of "obtaining circumstances" introduced so far are related in the following way:

[24] Equivalently we may say in Skyrms' terms (1980, part IA) that $C'_\omega(A,B)$ makes the probability of *B* given *A* or $\bar{A}$ maximally resilient over the rest of the past of *B*.

Theorem 5: $C^{++}_\omega(A,B) \subseteq C^{+}_\omega(A,B) \subseteq C_\omega(A,B) \subseteq C'_\omega(A,B) \subseteq C''_\omega(A,B)$; and if D and D' are any of these circumstances except $C''_\omega(A,B)$, then $P(B \mid A' \cap D) = P(B \mid A' \cap D')$ for each $A' \in \{A, \bar{A}\}$.

One may object that the most obvious suggestion has been ignored. Isn't it very natural to think that the actual circumstances of the direct causal relation between *A* and *B* are just all of the other actual direct causes and counter-causes of *B*? Indeed. This is precisely the proposal of Cartwright (1979) restricted to direct causes; Mellor (1988, p. 234), explicitly endorses it, too; and it seems to be a natural "actualization" of Definition 3b where the circumstances of *A* and *B* in ω that are possibly relevant in the probabilistic sense are defined as the conjunction of all the facts in ω which are possibly directly causally relevant to *B*. This suggestion is fixed in:

Definition 3f: *The ideal circumstances of* (the direct causal relation between) *A and B in* ω are defined as $C^*_\omega(A,B) = \bigcap \{D \text{ is a } k\text{-state for some } k \neq i \text{ and } D \xrightarrow[\omega]{\pm} B\} = {}^{\omega}(R_\omega(B) - \{i\})$.[25]

For the moment "ideal" means something bad. The basic trouble is that we cannot prove that $C^*_\omega(A,B) \subseteq C'_\omega(A,B)$. This means that the relevant probabilities conditional on the ideal circumstances may well differ from those conditional on the circumstances in the senses accepted so far. How can this happen? This is made clearer by a more positive result:

Theorem 6: $C_\omega(A,B) \subseteq C^*_\omega(A,B)$, and the identity holds iff for $K = \{k \in \{<j - i\} \mid B \perp k/{}^{\omega}\{<j - k\}\} = \{<j - R_\omega(B) \cup \{i\}\}$ we have $B \perp K/A' \cap {}^{\omega}\{<j - K \cup \{i\}\}$ for each $A' \in \{A, \bar{A}\}$.

Again, the proof essentially requires writing out the appropriate definitions. The theorem says that the identity holds if and only if the variables which are individually independent of *B* given the rest of the actual past of *B* are also collectively independent of *B* given *A* and the rest of the actual past of *B* as well as given $\bar{A}$ and the rest of the actual past of *B*. Both aspects of this condition are easily violated, but it will suffice to exemplify this for the aspect relating to *A* and $\bar{A}$ (and not for the one about collective independence).

Suppose *A* precedes *D, D* precedes *B*, $A \cap D \cap B = \{\omega\}$, $P(B \mid A \cap D) = 0.9$, $P(B \mid A \cap \bar{D}) = 0.9$, $P(B \mid \bar{A} \cap D) = 0.1$, and $P(B \mid \bar{A} \cap \bar{D}) = 0.5$. Here is a dream: there is hardly anything more delicious than red orange juice, but it is not offered in the deli-shops. So I thought that this was a way to become rich (*B*) and started a red orange juice enterprise. But what should I charge? Either \$2.99 (*D*) or \$1.99 ($\bar{D}$) per half a gallon; the prices in between are taboo, and higher or lower prices would be disastrous. In my dream I was lucky; nobody had the same idea (*A*). But then it is quite plausible to assume that it does not matter how I fix the price. If I fix the price to be high, I sell less with a larger profit per unit; otherwise, I sell more with a smaller profit per unit. My prospects of *B* are equally favorable. Thus, according to the numbers and Definition 2, *D* is directly causally irrelevant to *B* in ω. If there

[25] This identity follows from the fact that for $k \in R_\omega(B)$ $\{D \mid D \text{ is a } k\text{-state and } D \xrightarrow[\omega]{\pm} B\} = {}^{\omega}k$.

were competitors ($\bar{A}$), however, the price would of course make a big difference. Now look at the relation between A and B. A is a direct cause of B in ω, and also in $A \cap \bar{D} \cap B$; the fact that I have a monopoly is in any case advantageous to B. What are the circumstances of $A \xrightarrow[\omega]{+} B$? The crucial comparison is that $C_\omega(A,B) = D$, but $C^*_\omega(A,B) = \Omega$. Thus, we face here the strange fact that D is directly causally irrelevant to B, but relevant to $A \xrightarrow[\omega]{+} B$.

This possibility is, I think, responsible for quite some perplexity found in the literature. One may explain it away by resorting to a finer causal analysis in which D turns out to be indirectly causally relevant to B; but it is an open question whether this strategy always works. One may take it as constituting an objection against Definition 2; but this does not invalidate the other reasons for our explication. Maybe there are other ways to deal with the problem, but I think the possibility must be admitted that the two causal roles of D fall apart, i.e. that D's being relevant to the direct causal relation of other facts to B does not coincide with D's itself being directly causally relevant to B. However, if such behaviour is considered an anomaly, I propose to state an assumption excluding it. Then one can study how causal structures behave in general and how much more nicely they behave when this assumption is satisfied. Indeed, this assumption will play an important role later on.

What is the assumption? It was already stated in Theorem 6; it is the identity of $C_\omega(A,B)$ and $C^*_\omega(A,B)$. This explains why I have called $C^*_\omega(A,B)$ the ideal circumstances of A and B in ω; it specifies how the circumstances ideally are, but need not be. Finally this is the deeper reason why $C_\omega(A,B)$ is the preferred explication of the actually relevant circumstances; among all the otherwise equally acceptable explications this is the only one which lends itself to a statement of the assumption of ideal circumstances.

2.5 The Difficulties with Indirect Causation

It is now time to tackle the explication of indirect causation and hence of causation in general which, as the literature shows, is a difficult matter. Why? The general reason is that, even within our parsimonious framework, there is a bewildering plethora of plausible conditions for causation which cannot be simultaneously satisfied. The main purpose of this section is to present and untangle these conditions. Three kinds of conditions will be dealt with extensively and two others mentioned. A secondary goal is to show that the difficulties with these conditions are largely independent of the particular definition of direct causation one adopts. Therefore, little use of Definition 2 is made in this section; the synthesis is undertaken only in the final section.

The first condition is rather a matter of faith: namely that an explication of causation be simple. This sounds quite airy because simplicity ratings often diverge. But it helps to avoid the manifest danger of lapsing into the strategy of trying to solve difficulties by piling up clauses and provisos, each of them plausible, but all together unintelligible.

The second condition is that there must be a good overall fit between an explication and the many more or less problematic examples found in the literature. Obviously the whole story necessary to show that a given explication satisfies this condition is long, indeed too long for this essay. But I have reservations about abbreviating the story. There is some tendency to focus on this or that problematic type of example as the central touchstone of any theory of causation. But this would be too narrow an attitude; there are too many types of examples to be considered, and intuitions about examples are not fixed enough to constitute an unshakable reference point. As I said, a good *overall* fit is to be achieved, even if this standard opens a door to vagueness and subjectivity. Moreover, examples are in a sense theoretically barren. We do not understand them as long as we have no theoretical structure enabling us to integrate them and to explain why they are examples for this or against that; and staring at them probably is bad heuristics for arriving at that structure. This is why I concentrate here on three further kinds of conditions of a theoretical nature.

The third kind of condition consists in structural conditions concerning the formal structure of causal relations. The fourth kind consists in Markovian conditions: there is a strong intuition that causal chains are Markov chains; and of course an indirect cause should be connected to its indirect effect by some causal chain. The fifth kind consists in positive relevance conditions: there is also a strong intuition that a cause is in some sense positively relevant to its effect; it is, indeed, embodied in the basic conception of causation cited in the very first sentence of Section 2.3, and Definition 2 also relies on it.

There are alternative ways of specifying each kind of condition. It will turn out that the most plausible candidates are mutually incompatible. Recognition of this fact is important to the explanation of a number of examples and confusions. Let us look at these conditions in more detail.

Structural conditions: The first structural condition for the relation $\xrightarrow[\omega]{+\ldots+}$ of being a (direct or indirect) cause in ω is trivial, but should be made explicit:

(S0) *Lower bound*: If $A \xrightarrow[\omega]{+} B$, then $A \xrightarrow[\omega]{+\ldots+} B$.

I shall continually use (S0) without mention. The next condition sets an upper bound:

(S1) *Upper bound*: If $A \xrightarrow[\omega]{+\ldots+} B$, then A stands to B in the transitive closure of $\xrightarrow[\omega]{+}$.

This condition is not acceptable for continuous time. But given discrete time, (S1) seems compelling; I cannot imagine how indirect causation could extend farther than what is allowed by direct causal steps. The next all-important condition is

(S2) *Transitivity*: If $A \xrightarrow[\omega]{+\ldots+} B$ and $B \xrightarrow[\omega]{+\ldots+} C$, then $A \xrightarrow[\omega]{+\ldots+} C$.

(S0), (S1), and (S2) entail that $\xrightarrow[\omega]{+\ldots+}$ is the transitive closure of $\xrightarrow[\omega]{+}$. Thus, as mentioned in the introduction, these conditions yield a definition of causation in general. So where is the snag? It lies in the fact that all Markovian and positive relevance conditions violate transitivity. This will become fully clear below. But the gist is easily summed up:

Though transitivity looks very natural, one would expect transitivity to ensue from a general definition of causation. If it is the other way around, naturalness is tarnished. Now, if deterministic causes are defined as sufficient and/or necessary conditions, transitivity follows immediately – at least when such conditions are explained in terms of logical or nomological entailment. This is certainly the strongest source of the intuition of transitivity. But even in the case of deterministic causation the issue is not clear. If such conditions are explained in terms of the subjunctive conditional, transitivity fails because the subjunctive conditional fails to be transitive.[26] Thus, even in this case a conflict arises. Lewis (1973b) resolves it by axiomatically accepting transitivity, at the cost of renouncing the general equation between causation and sufficient and/or necessary conditions and taking transitivity as a primitive property.

In the case of probabilistic causation, the issue is even less clear. Here, a direct causal impact has, so to speak, no necessitating force, but is only weak and imperfect.[27] Hence, it seems plausible that such a weak impact is not preserved over long causal chains, but fades sooner or later. For instance, given our very coarse and only probabilistic meteorological models, each day's weather may be granted to causally influence the next day's weather. But does the weather, say, at the turn of the last century still influence today's weather? It does not seem so; somewhere in between the influence has faded completely, even though it may be difficult to tell precisely when or where. If this is plausible, the intuition of transitivity totters. Indeed, this intuition is not generally respected by theorists of probabilistic causation. For example, Suppes (1970, p. 58), dryly states that all causal relations he has defined fail to be transitive as long as the limiting probabilities 0 and 1 are not involved.[28]

Thus, a profound uncertainty about this issue may be observed, and there is reason for looking for alternatives to transitivity. Here is a possible approach: Certainly, each indirect cause and effect should be connected by a causal chain. Everything then depends on how causal chains are characterized, and they may indeed be characterized in several, apparently nonequivalent ways:

Definition 5:

(a) $\langle A_1,\ldots, A_n\rangle$ is a *weak causal chain in* ω iff $A_1 \xrightarrow[\omega]{+} A_2 \xrightarrow[\omega]{+} \ldots \xrightarrow[\omega]{+} A_n$.
(b) $\langle A_1,\ldots, A_n\rangle$ is a *connected causal chain* in ω iff it is a weak causal chain in ω and, for all r and s with $1 \le r < s \le n$, $A_r \xrightarrow[\omega]{+\ldots+} A_s$.
(c) $\langle A_1,\ldots, A_n\rangle$ is a *strict causal chain in* ω iff it is a connected causal chain in ω and, for no r and s with $r < s$, $A_r \xrightarrow[\omega]{+} A_{s+1}$.

---

[26] Cf. Lewis (1973a, pp. 32–34).

[27] One might well find this idea and thus probabilistic causation unintelligible; many have done so. But in the light of the last remark in the introduction the present discussion should be illuminating for them, too.

[28] Eells and Sober (1983) take up another remark of Suppes on that page and investigate under which special circumstances transitivity of probabilistic causation is preserved.

(d) $\langle A_1,\ldots, A_n\rangle$ is an *effective causal chain* in ω iff it is a weak causal chain in ω and, for all $r > 1$, $A_1 \xrightarrow[\omega]{+\ldots+} A_r$.
(e) $\langle A_1,\ldots, A_n\rangle$ is an *affective causal chain in* ω iff it is a weak causal chain in ω and, for all $s < n$, $A_s \xrightarrow[\omega]{+\ldots+} A_n$.

Of course, (a), (b), and (c) are the more promising definitions; the reason for introducing also (d) and (e) will be clear in due course. Note the difference between (b) and (c): if $A \xrightarrow[\omega]{+} B \xrightarrow[\omega]{+} C$ and also $A \xrightarrow[\omega]{+} C$ – a situation which has in no way been excluded so far – then $\langle A, B, C\rangle$ is a connected, but not a strict causal chain in ω. Correspondingly, there are five structural conditions for $\xrightarrow[\omega]{+\ldots+}$:

(S3) *Structural chain conditions*: Whenever $A \xrightarrow[\omega]{+\ldots+} B$, but not $A \xrightarrow[\omega]{+} B$, then there are $C_1,\ldots, C_n$ $(n \geq 1)$ such that $\langle A, C_1, \ldots, C_n, B\rangle$ is a (a) weak, (b) connected, (c) strict, (d) effective, (e) affective causal chain in ω.

It might be tempting to reverse (S3), in particular part (b), i.e. to take the fact that in a series of states starting with *A* and ending with *B* all causal relations except the one from *A* to *B* obtain to imply that *A* also causes *B*. Formally:

(S3b′) *Reversed chain condition*: if there is a connected causal chain $\langle C_1, \ldots, C_n\rangle$ in ω such that $\langle A, C_1,\ldots, C_n\rangle$ is an effective causal chain in ω and $\langle C_1,\ldots, C_n, B\rangle$ is an affective causal chain in ω, then $A \xrightarrow[\omega]{+\ldots+} B$, i.e. $\langle A, C_1,\ldots, C_n, B\rangle$ is also a connected chain in ω.

Another idea is that: by assuming transitivity the causal relation $\xrightarrow[\omega]{+\ldots+}$ is boosted to its maximal extension within the upper bound (S1). Thus, if transitivity is dropped, the effective range of a state *A* in ω, i.e. the set of its effects in ω, may comprise less than all states which can be reached from *A* via weak causal chains. How much less? It is hard to say. But in any case, it seems impossible that the effective range of *A* extends farther than the effective ranges of all its immediate causal successors:

(S4a) *Local effective maximum*: whenever $A \xrightarrow[\omega]{+\ldots+} B$, but not $A \xrightarrow[\omega]{+} B$, then there is a *C* with $A \xrightarrow[\omega]{+} C \xrightarrow[\omega]{+\ldots+} B$.

The same consideration holds, of course, for the affective range of *A*, i.e. the set of its causes:

(S4b) *Local affective maximum*: whenever $B \xrightarrow[\omega]{+\ldots+} A$, but not $B \xrightarrow[\omega]{+} A$, then there is a *C* with $B \xrightarrow[\omega]{+\ldots+} C \xrightarrow[\omega]{+} A$.

These suggestions demonstrate the ease with which further conditions may be invented. But there is no point in doing so. More interesting is the relation between the conditions stated so far. This is given completely by:

Theorem 7:

(a) Upper bound (S1) and transitivity (S2) are equivalent to the assertion that $\xrightarrow[\omega]{+\ldots+}$ is the transitive closure of $\xrightarrow[\omega]{+}$.
(b) Upper bound (S1) and the reversed chain condition (S3b′) are also equivalent to this assertion.

(c) Given (S1), transitivity (S2) implies the connected chain condition (S3b); but the reverse does not hold.
(d) The connected (S3b) and the strict (S3c) chain condition are equivalent.
(e) The connected chain condition (S3b) implies the effective (S3d) and the affective (S3e) chain condition; but even jointly, (S3d) and (S3e) do not imply (S3b).
(f) Local effective maximum (S4a) is equivalent to the affective chain condition (S3e).
(g) Local affective maximum (S4b) is equivalent to the effective chain condition (S3d).
(h) Each of (S3d) and (S3e) imply the weak chain condition (S3a); but the reverse does not hold.
(i) Upper bound (S1) is equivalent to the weak chain condition (S3a).

Proof:

(a) Is trivial.
(b) For the direction (⇒) suppose that it has been shown for all $r < n$ that we have $C_1 \xrightarrow[\omega]{+\ldots+} C_r$ for each weak causal chain $\langle C_1, \ldots, C_n \rangle$ in ω of length r. Now, let $\langle C_1, \ldots, C_n \rangle$ be a weak causal chain in ω of length n. Because of the supposition the premises of (S3b′) are satisfied, and so we may infer that $C_1 \xrightarrow[\omega]{+\ldots+} C_r$. Hence, the existence of a weak causal chain in ω from A to B already ensures $A \xrightarrow[\omega]{+\ldots+} B$. With upper bound (S1) this implies the desired result. The other direction (⇐) is trivial.
(c) The direction (⇒) is trivial. Concerning the reverse, imagine that $A \xrightarrow[\omega]{+} B \xrightarrow[\omega]{+} C$ but not $A \xrightarrow[\omega]{+\ldots+} C$. This situation satisfies (S3b), but not transitivity (S2).
(d) Each strict causal chain is connected; thus (S3c) implies (S3b). On the other hand, suppose that $\langle A_1, \ldots, A_n \rangle$ is a connected causal chain in ω. Let $A_1 = B_1$; $B_2 = A_r$, where r is the maximal index for which $A_1 \xrightarrow[\omega]{+} A_r$; $B_3 = A_s$, where s is the maximal index for which $A_r \xrightarrow[\omega]{+} A_s$; etc. Thus, for some m $B_m = A_n$. Obviously, $\langle B_1, \ldots, B_m \rangle$ is a strict causal chain in ω. This shows that each connected causal chain has a strict causal subchain with the same start and end. Hence, (S3b) also implies (S3c).
(e) The direction (⇒) is trivial. That the reverse does not hold, may be seen in the following way: Take a weak causal chain $\langle A_1, \ldots, A_5 \rangle$ in ω with five members, and suppose that $A_r \xrightarrow[\omega]{+\ldots+} A_s$ for all $r < s$ with the exception of $r = 2$ and $s = 4$. Moreover, assume states B and C such that both, $\langle A_1, B, A_4 \rangle$ and $\langle A_2, C, A_5 \rangle$, are effective and affective causal chains in ω, but neither $A_1 \xrightarrow[\omega]{+\ldots+} C$ nor $B \xrightarrow[\omega]{+\ldots+} A_5$ holds. In this situation, there is an effective and affective causal chain from each D to each E for which $D \xrightarrow[\omega]{+\ldots+} E$, but there is no connected causal chain from A_1 to A_5.
(f) For the direction (⇒), suppose that $A \xrightarrow[\omega]{+\ldots+} B$,; but not $A \xrightarrow[\omega]{+} B$. According to (S4a), there is a C_1 with $A \xrightarrow[\omega]{+} C_1 \xrightarrow[\omega]{+\ldots+} B$.. Now, if $C_1 \xrightarrow[\omega]{+} B$, we are finished. If not, we again apply (S4a) and find a C_2 with $C_1 \xrightarrow[\omega]{+} C_2 \xrightarrow[\omega]{+\ldots+} B$. And so on. In the end, this process yields an affective causal chain in ω from A to B. The reverse direction (⇐) is trivial.

(g) This is perfectly symmetric to (f).
(h) and (i) are trivial.

The fact that some reverse inferences are not valid is perhaps a little surprising. In any case, Theorem 7 exhibits clearly the differing strengths of the various conditions.

It is not yet the time to decide which structural conditions are the most plausible ones. But one conclusion is quite obvious: if one should give up the transitivity of $\xrightarrow[\omega]{+\ldots+}$ and decide to settle for something weaker, one gets onto a slippery slope, at least from a purely structural point of view. For instance, it is an unpleasant fact that then one has to cope with various non-equivalent concepts of a causal chain. However, if transitivity is assumed, each weak causal chain is also effective, affective, and connected, and, thus, no ambiguity would arise. A further point is that, intuitively, it may be not so clear which chain condition direction to endorse. The reverse direction – stated in (S3b′) in its weakest form – may also seem plausible; but it implies transitivity according to Theorem 7(b). Thus, in the light of structural considerations alone, transitivity (S2) has a clear preponderance over the alternatives. So, let us look more closely at the reasons against transitivity.

Markovian conditions: There is a strong intuition that indirect effects are screened off from their indirect causes by the mediating links, i.e. that the indirect causal efficacy of a state is completely contained in the mediating links, or, in other words, that, if the intermediate members of a causal chain are realized in some way or other, then the past of the chain is irrelevant to its future. This intuition is commonly expressed in the Markovian way; indeed, it is often said that Markov chains have no memory, that they are characterized by the absence of after-effect. The central concept is:

Definition 6: $\langle i_1,\ldots, i_n\rangle$ *is a (finite) Markov chain* iff $i_{r+1} \perp \{i_1, \ldots, i_{r-1}\} / i_r$ for all $r = 2,\ldots, n-1$. Moreover, $\langle i_1,\ldots, i_n\rangle$ is a *causal Markov chain in* ω iff it is a Markov chain and a weak causal chain in ω (where $\langle i_1,\ldots, i_n\rangle$ is a weak causal chain in ω iff $\langle {}^{\omega}i_1, \ldots, {}^{\omega}i_n\rangle$ is).

Unfortunately, there are various choices for rendering precise the Markovian intuition. Is it intuitively a necessary condition for a causal chain to be a causal Markov chain? Or a sufficient condition that it be both a Markov and a strict causal chain? This is not easy to decide. Let us look at one attempt a bit more closely:

(M1) *Markov chain condition*: suppose that $A = {}^{\omega}i$ and $B = {}^{\omega}j$. Then $A \xrightarrow[\omega]{+\ldots+} B$ iff there exist $k_1, \ldots, k_n$ ($n \geq 0$) such that $\langle i, k_1, \ldots, k_n, j\rangle$ is a causal Markov chain in ω.

(M1) is a biconditional and thus bolder than (S3). But it rests on the same basic idea, namely, that indirect causation must be mediated by a causal chain; and it adds a particular explication of causal chains. Indeed, (M1) provides an explicit definition of $\xrightarrow[\omega]{+\ldots+}$ according to which it behaves thus:

Theorem 8: The Markov chain condition (M1) implies the strict chain condition (S3c), but it does not imply transitivity (S2).

Proof: The first part follows from the well-known fact that, if $\langle i_1,\ldots, i_n\rangle$ is a Markov chain, then any subsequence $\langle j_1,\ldots, j_m\rangle$ of $\langle i_1,\ldots, i_n\rangle$ is also a Markov chain.

The failure of transitivity is due to the fact that the serial connection of two or more Markov chains will not generally result in one large Markov chain.

Hence, the Markov chain condition does away with the unwelcome splitting up of the structural characterizations of causal chains. Apparently, it is a serious alternative to the assumption of transitivity. But there are, on the contrary, also clearly disconcerting features.

First, it is easy to see that according to (M1) there may be connected causal chains which are not Markovian. Suppose that $A = {}^{\omega}i$, $B = {}^{\omega}j$, $C = {}^{\omega}k$, and $D = {}^{\omega}l$, and that $A \xrightarrow[\omega]{+} B \xrightarrow[\omega]{+} D$, $A \xrightarrow[\omega]{+} C \xrightarrow[\omega]{+} D$, and $A \xrightarrow[\omega]{+\ldots+} D$. Then both, $\langle A, B, D\rangle$ and $\langle A, C, D\rangle$, are connected causal chains in ω. But in order to satisfy (M1) only one of $\langle i, j, l\rangle$ and $\langle i, k, l\rangle$ needs to be a Markov chain; there is nothing so far to guarantee that the other is so, too. If $\langle i, j, l\rangle$ is the Markov chain, should then the other, $\langle i, k, l\rangle$, be denied to be a genuine causal chain? Or should one stipulate that such a situation does not arise, i.e. that not only some, but all connected causal chains leading from one state to another are Markovian? No; a more trenchant conclusion is called for.

Consider an illustration of the very common abstract situation just described. At a signal of the romantic lover (*A*), a fiddler (*B*) and a mandolin player (*C*) strike up a sweet melody in order to tenderly wake the beloved (*D*). Here we have, as required, two causal chains running from *A* to *D*, one through *B* and the other through *C*. It is plausible in this case, and easily done, to distribute the probabilities in such a way that, given *B* alone, *A* is still probabilistically positively relevant to *D* (via *C*, so to speak), and also given *C* alone; the situation is symmetric with respect to *B* and *C*. This, however, means that, contrary to (M1), no causal chain between *A* and *D* is Markovian and thus that (M1) somehow fails to capture the Markovian intuition.

A well-known move for coping with such problems is to generalize the concept of a Markov chain to that of a Markov field.[29] In these terms, the case exemplifies a Markov field characterized by the conditional independence $l \perp i \,/\, \{j, k\}$ which says that *D* is screened off from *A* only jointly by *B* and *C*. In principle, I fully endorse this strategy,[30] but not at the present stage where it seems to me to overshoot the mark. If one adopts this strategy, the conceptual key role is taken over by the notion of a Markov field and the corresponding causal notion of a causal net, which are more complex notions and more difficult to grasp. The structural and the Markov conditions would then have to be expressed in these more complex terms. And causal chains become derivative entities definable only as certain parts of causal nets. This seems to be the wrong direction of analysis; we should build up complexities from simpler units already understood.

Indeed, a less radical move will do; a slight, though basic conceptual modification will save the old strategy. In explicating direct causes, the positive correlation between direct cause and effect was considered not in isolation, but embedded in the given past

---

[29] Cf., e.g., Lauritzen (1982).

[30] The theory of Markov fields is indeed utterly illuminating for the causal theorist; cf., e.g., Kiiveri et al. (1984) or the rich material presented in Pearl (1988, ch. 3).

course of events. Similarly, the members of a Markov chain should be taken not in isolation, but rather as embedded in a given setting. Thus, I propose:

Definition 7: $\langle i_1,\ldots, i_n\rangle$ is an ω-*Markov chain iff* $i_1 < \ldots < i_n$ and if, for all $r = 2,\ldots, n-1$, $i_{r+1} \perp \{i_1, \ldots, i_{r-1}\}/i_r, {}^{\omega}\{< i_{r+1} - i_1, \ldots, i_r\}$; this means that the conditional independence characteristic of a Markov chain holds only given the rest of the past of i_{r+1} in ω. Moreover, $\langle i_1,\ldots, i_n\rangle$ is a *causal* ω-*Markov chain* iff $\langle i_1,\ldots, i_n\rangle$ is an ω-Markov chain and a weak causal chain in ω.

Let us modify (M1) correspondingly:

(M2) ω-*Markov chain condition*: Suppose that $A = {}^{\omega}i$ and $B = {}^{\omega}j$. Then $A \xrightarrow[\omega]{+\ldots+} B$, iff there exist $k_1,\ldots, k_n$ $(n \geq 0)$ such that $\langle i, k_1,\ldots, k_n, j\rangle$ is a causal ω-Markov chain.

This amendment takes care of the example of the romantic lover; there, $\langle i, j, l\rangle$ and $\langle i, k, l\rangle$ both plausibly are ω-Markov chains. Indeed, I think that (M2) reflects the Markovian intuition better than (M1). Generally, the expectation should be that a more proximate cause screens off the effect from a more remote cause only given the circumstances and not unconditionally. The structural properties have not changed, however:

Theorem 9: The ω-Markov chain condition (M2) implies the strict chain condition (S3c), but it does not imply transitivity (S2).

Proof: It is easily shown that, if $\langle i_1,\ldots, i_n\rangle$ is an ω-Markov chain, then any subsequence $\langle j_1,\ldots, j_m\rangle$ of $\langle i_1,\ldots, i_n\rangle$ is also an ω-Markov chain. This immediately entails the first part. Again, the serial connection of two or more ω-Markov chains will in general not result in one large ω-Markov chain. Thus, transitivity need not hold.

I conclude that (M2) is a viable alternative to transitivity (S2). But the matter is still open, and the evidential basis should be further augmented.

Positive relevance conditions: A third important theoretical constraint is introduced by the conception that a cause is in some sense positively relevant to its effects. The account of direct causation above is based on that conception; and theoretical unity seems to be best preserved by further relying on it. This sets the task to extend Section 2.4 and to determine the circumstances also of indirect causal relations, which is in fact so intricate that it can only be started, but not completed here.

Recall first Cartwright's circularity problem. In the case of direct causation, it could be argued that the whole past of the effect may be taken as obtaining circumstances. But, for indirect causation, obviously this will not do. Precisely because of the Markovian intuition, some causal intermediates must be excluded from the circumstances in this case; and the problem is to say which ones. However, we need not yet worry about the threat of circularity because we are now after plausible conditions only and not after definitions.

A useful perspective on the problem is gained when we think of the relativity of the direct/indirect-distinction to the given descriptive frame. The core of the idea of positive relevance is, I think, the expectation that what is an indirect cause in the given descriptive frame should be a direct cause in some reduced descriptive frame.

The advantage of putting the core idea in this way is that it avoids reference to a specific explication of direct causal relevance (though I shall employ the account of Section 2.3 later on). The problem takes then the form of determining which reduced frame to consider. There are, prima facie, several options:

(P) *Positive relevance conditions*: Let A be an i-state and B a j state. Then $A \xrightarrow[\omega]{+\ldots+} B$ holds relative to the descriptive frame given by I iff $A \xrightarrow[\omega]{+\ldots+} B$ holds relative to the descriptive frame given by $I - J$

(1) for some $J \subseteq (i, j)$,

(2) for some, or (3) for each, $J = \{k_1, \ldots, k_n\} \subseteq (i, j)$ such that $\langle i, k_1, \ldots, k_n, j\rangle$ is (a) a weak, (b) a connected, (c) a strict causal chain in ω, (d) a causal Markov chain in ω, (e) a causal ω-Markov chain,

(4) for $J = \{k \in (i, j) \mid k$ is a member of (a) a weak, (b) a connected, (c) a strict causal chain in ω, (d) a causal Markov chain in ω, (e) a causal ω-Markov chain running from i to $j\}$,

(5) for $J = (i, j)$.

As a sufficient condition for $A \xrightarrow[\omega]{+\ldots+} B$ (P1) is certainly too weak. But taken as a necessary condition, (P1) seems to be the inalienable minimum of the positive relevance idea. However, even this minimum need not be satisfied in the light of the theory proposed below.

(P5) is, as noted above, the version favored by Good (1961–63). The example of the two-person game in Section 2.3 may also be used to cast doubt on (P5). Suppose that the choice A of the first player is negatively relevant to a certain outcome C, given the later, but independent choice B of the second player. On the current account, A is then a counter-cause of C. But it is easily imaginable that averaging over the choice of the second player makes A unconditionally positively relevant to C; just assume that C is sufficiently unlikely given $\bar{A}$ and $\bar{B}$. According to (P5), A would then be a cause of C. This seems inadequate.

(P4) is the best approximation to the position of Cartwright (1979), though which version of (P4) she would prefer is not clear. Is (P4) plausible? If there is only one causal chain running from i to j then (P2), (P3), and (P4) coincide. But if there is more than one chain, the three conditions may diverge almost arbitrarily; only the versions of (P3) are guaranteed to be stronger than the corresponding versions of (P2). In view of this divergence, it is hard to say which condition is preferable.

But we know some things. An important observation is that each version of (P) violates transitivity (S2), at least if Definition 2 is presupposed. One numerical example covers all versions. Suppose that $A \cap B \cap C = \{\omega\}$, $P(B \mid A) = 0.8$, $P(B \mid \bar{A}) = 0.4$, $P(C \mid A \cap B) = 0.6$, $P(C \mid A \cap \bar{B}) = 0.1$, $P(C \mid \bar{A} \cap B) = 0.9$, and $P(C \mid \bar{A} \cap \bar{B}) = 0.4$. Then $A \xrightarrow[\omega]{+} B \xrightarrow[\omega]{+} C$ holds according to Definition 2; but we have $P(C \mid A \cap B) < P(C \mid \bar{A} \cap B)$ and $P(C \mid A) = 0.5 < 0.6 = P(C \mid \bar{A})$. Thus, according to all versions of (P) we cannot have $A \xrightarrow[\omega]{+\ldots+} C$. Generally, structurally good behavior may at most be expected from the (d) and (e) versions of (P2) and (P3) which incorporate Markovian elements.

Of course, it is rather the harmony between Markovian and positive relevance conditions which is hoped for. This hope is based on the following well-known result:

Theorem 10: Let $\langle i_1,\ldots, i_n\rangle$ be a Markov chain of binary variables and $A_r = {}^{\omega}i_r$ for $r = 1,\ldots, n$. Then, if $P(A_{r+1} \mid A_r) - P(A_{r+1} \mid \bar{A}) = x_i$ for $r = 1,\ldots, n-1$, $P(A_n \mid A_1) - P(A_n \mid \bar{A}_1) = x_1 \cdot \ldots \cdot x_{n-1}$. This implies in particular that, if each A_r is positively relevant to A_{r+1}, then A_1 is positively relevant to A_n.[31]

However, Theorem 10 does not achieve the desired harmony. If additional variables are dispersed between $i_1,\ldots, i_n$, then, according to Definition 2, the positive relevancies assumed in Theorem 10 need not indicate direct causal relations. Moreover, the theorem refers only to the Markovian chain condition (M1). If, however, the right strategy is to replace Markov chains by ω-Markov chains, then the theorem does not apply; and there is no corresponding theorem about ω-Markov chains.

So the situation is, in fact, as bad as I have indicated. All Markovian and positive relevance conditions are incompatible with the favourite structural condition of transitivity; and the preferred Markovian condition need not preserve positive relevance in any of the ways considered.

Suppes (1984, pp. 55ff.), devotes a whole section to "conflicting intuitions" concerning causality, and Salmon (1988b) appreciatively adopts this phrase, though he has, in part, different things in mind. It is my experience that, given the current parsimonious framework and no further notions or distinctions bearing on causality, the intuitions and conflicts described in the present section are central to the discussion of probabilistic causation. Is there any way to resolve these conflicts? The final section suggests one such way.

2.6 Causation

For a long time, Suppes' remark (1970, p. 58) that transitivity is not to be expected in the case of probabilistic causation held me in its grip. All proposed plausible explications so clearly failed to yield transitivity that it seemed crazy to cling to that structural property. The task could thus only be to reconcile the other intuitions and conditions; and Theorem 10 seemed to point the way. Eells and Sober (1983) obviously had the same idea in mind when investigating the lucky circumstances under which the causal relation defined on the basis of the positive relevance idea is transitive.

Moreover, when faced with several options, it is always a wise policy to choose the weakest explication possible. The various strengthenings can then be introduced and studied afterwards. Were one, on the contrary, to start with a stronger notion,

[31] For a proof cf., e.g., Good (1980). Eells and Sober (1983, pp. 49ff.), prove a more general result about the propagation of positive relevance in particular Markov nets. Theorem 10 shows, by the way, that the meteorological example for the fading of probabilistic causal influence and thus for the failure of transitivity is not really convincing. If the meteorological models would explain the weather as a Markov process (I doubt that they actually do), then the theorem says that even the weather at the turn of the last century makes a probabilistic difference for today's weather, though an almost infinitesimally small one.

all the weaker ones would simply drop out of theoretical consideration. But Theorem 7 has revealed transitivity to be a particularly strong structural condition; it implies all the other conditions, given the unassailable (S0) and (S1). So, again, it appeared better to ignore transitivity.

However, the consideration does not apply the right measure of strength. What counts is not structural strength, which was seen to be accompanied by weakness concerning other kinds of conditions. What counts is *conceptual* strength. And the fact is that the transitive closure of direct causation is the *weakest possible notion* of causation in general; it yields the causal relation with the widest possible extension, if the upper bound condition (S1) is presupposed; whenever A is a cause of B in any other feasible sense, A is also a cause of B in this sense.[32] So, I shall settle for the minimal notion of causation, even if the price to be paid is what Lewis (1973b) had to pay, too, namely, that transitivity is a primitive property.

Definition 8: Let A be an i-state, B a j-state, $i < j$, and $\omega \in A \cap B$. Then *A is a cause of B in ω*, i.e. $A \xrightarrow[\omega]{+\ldots+} B$, iff A stands to B in the transitive closure of $\xrightarrow[\omega]{+}$. *A is a counter-cause of B in ω*, i.e. $A \xrightarrow[\omega]{-\ldots-} B$, iff $A \xrightarrow[\omega]{-} B$ or for some D $A \xrightarrow[\omega]{+\ldots+} D \xrightarrow[\omega]{-} B$.. *$A$ is causally relevant to B in ω*, i.e. $A \xrightarrow[\omega]{\pm\ldots\pm} B$, iff A stands to B in the transitive closure of $\xrightarrow[\omega]{\pm}$. Finally, *$A$ is causally irrelevant to B in ω*, i.e. $A \xrightarrow[\omega]{0\ldots0} B$, iff not $A \xrightarrow[\omega]{\pm\ldots\pm} B$.

This definition of counter-causation is, I think, the most plausible one. If counter-causation is to be allowed at all, then the counter-causal influence stops with the realization of the counter-effect and does not extend beyond. At least, I would firmly claim this for deterministic causation (where there may be counter-causation, too) and thus also for probabilistic causation, though less firmly.[33] For the unconvinced there is also the concept of causal relevance which comprises causation, counter-causation, and much more.

Definition 8 covers all kinds of weird cases. First, to repeat, causal chains need not be Markov chains. Also, a cause need not be positively relevant to its effect under admissible conditionalization. The numerical example in the foregoing section demonstrating the incompatibility of (P) and (S2) is a case in point; according to Definition 8, $A \xrightarrow[\omega]{+\ldots+} C$ holds in this case. $A \xrightarrow[\omega]{-} C$ holds also; indeed, this is essential to its construction. This means that a state of affairs may at once be a cause and a counter-cause of another state of affairs, if at least one of these causal relations is indirect. Though this may appear counter-intuitive, it seems to be exactly the right thing to say in many cases – for instance in the famous thrombosis example of Hesslow (1976): the woman's taking a contraceptive is a cause as well as a counter-cause of her thrombosis, mediated by different chains. Otte (1985, pp. 122f.), has drawn this conclusion, also.

---

[32] I owe this point to Karel Lambert; it became really clear to me in a long conversation with him.

[33] Good (1961, p. 311, Axiom 10), and Humphreys (1980, pp. 308f.), seem to be guided by the same conception.

The counterpart may also happen: A is a cause of B, and if $\overline{A}$ had obtained, $\overline{A}$ would have been a cause of B, too. This is the case when causal preemption occurs or a back-up system is installed. Consider, e.g., the equally famous case of the desert traveller introduced by Hart and Honoré (1959, pp. 239ff.). One enemy of the traveller pours poison into his water keg; later, but independently, the other drills a hole into the keg. The latter fact is a cause of the traveller's death. But if the hole had not been drilled, the lack of the hole would also have been a cause of his death, because it would have caused the poison to stay in the keg. In other words, the set-up can be such that a variable has only a relay function; it can turn on different chains which all lead to the same effect. Nevertheless, this relay function must be conceived as a causal function. The list of such oddities could easily be extended.

Settling for the weakest notion is only a start. The essential step consists in showing how to build stronger conceptions upon the weak base. Thus, the task is to specify conditions under which Definition 8 does justice to the Markovian and positive relevance intuitions. And it will not do to find just any sufficient conditions; that would presumably be easy. These conditions must be specified solely in causal terms; only then do we know which causal situations satisfy our causal preconceptions. To this task I now and finally turn.

The Markovian part is the easier one. First some terminology. Since Definition 8 entails that the four kinds of causal chains given by Definition 5(a), (b), (d), and (e) coincide, I shall now talk of causal chains *simpliciter*; only strict causal chains in the sense of Definition 5(c) have to be distinguished. Additionally, we need:

Definition 9: *$\langle A_1,\ldots, A_n\rangle$ is a chain of causal relevance in* ω iff $A_1 \xrightarrow[\omega]{\pm} A_2 \xrightarrow[\omega]{\pm} \ldots \xrightarrow[\omega]{\pm} A_n$; and it is a *strict chain of causal relevance in* ω iff it is a chain of causal relevance in ω and, for no r and s with $r+2 \le s \le n$, $A_r \xrightarrow[\omega]{\pm} A_s$. Moreover, $\langle i_1,\ldots, i_n\rangle$ is a (strict) chain of causal relevance in ω iff $\langle {}^{\omega}i_1,\ldots, {}^{\omega}i_n\rangle$ is.

A first welcome result is:

Theorem 11: The following two assertions are equivalent:

(a) For each $\omega \in \Omega$, all chains $\langle i_1,\ldots, i_n\rangle$ of causal relevance in ω are strict.
(b) For each $\omega \in \Omega$, all chains $\langle i_1,\ldots, i_n\rangle$ of causal relevance in ω are ω-Markov chains.

Proof: (a) $\Rightarrow$ (b): Suppose (b) is false, i.e. there are $i_1,\ldots, i_n$, and ω such that $\langle i_1,\ldots, i_n\rangle$ is a chain of causal relevance in ω, but not an ω-Markov chain. Thus, there is an $s \le n$ such that $i_{s+1} \perp \{i_1,\ldots, i_{s-1}\} / i_s, {}^{\omega}\{< i_{s+1} - i_1,\ldots, i_s\}$ does not hold. Several applications of Theorem 2(e) yield that there is an $r < s$ for which $i_{s+1} \perp i_1/\{i_1,\ldots, i_{r-1}, i_{r+1},\ldots, i_s\}, {}^{\omega}\{< i_{s+1} - i_1,\ldots, i_s\}$. is not true. And this means that there is an $\upsilon \in \Omega$ agreeing with ω outside $\{i_1,\ldots, i_s\}$ such that ${}^{\upsilon}i_r \xrightarrow[\upsilon]{\pm} {}^{\upsilon}i_{s+1}$.

(b) $\Rightarrow$ (a): Suppose (a) is false, i.e. there are $i_1,\ldots, i_n$, and ω such that $\langle i_1,\ldots, i_n\rangle$ is a chain, but not a strict chain of causal relevance in ω. Thus, we have ${}^{\omega}i_r \xrightarrow[\omega]{\pm} {}^{\omega}i_{s+1}$. for some r and s with $r < s$. This immediately entails that $\langle i_1,\ldots, i_n\rangle$ is not an ω-Markov chain.

This is a perspicuous theorem. It says that, if causal relevance spreads strictly in stages in all worlds so that in no world there obtain direct as well as indirect causal relations between any states, then all these chains are ω-Markovian, i.e. they have the preferred Markovian property. However, the order of the quantifiers is not the desired one. There is no reason to expect that all possible worlds are causally well-ordered in this way. Hence, what is needed is a universal equivalence rather than an equivalence of universal statements. Here, at last, Section 2.4 comes into play; the assumption that the actual circumstances are ideal will be required in:

Theorem 12: Let $i_1,\ldots, i_n$ be binary variables and $\langle i_1,\ldots, i_n\rangle$ a chain of causal relevance in ω. Assume that for all $r < n$ $C^*_\omega({}^\omega i_r, {}^\omega i_{r+1}) = C_\omega({}^\omega i_r, {}^\omega i_{r+1})$. Then $\langle i_1,\ldots, i_n\rangle$ is an ω-Markov chain iff it is a strict chain of causal relevance in ω.

Proof: (⇒): Suppose that $\langle i_1, \ldots, i_n\rangle$ is not a strict chain of causal relevance in ω, i.e. there exist $r < s$ with ${}^\omega i_r \xrightarrow[\omega]{\pm} {}^\omega i_{s+1}$.. Again, this entails that $\langle i_1,\ldots, i_n\rangle$ is not an ω-Markov chain.

(⇐): Suppose that $\langle i_1, \ldots, i_n\rangle$ is a strict chain of causal relevance in ω and consider any $s < n$. Then for all $r < s$ ${}^\omega i_r \xrightarrow[\omega]{0} {}^\omega i_{s+1}$, i.e., since the variables are binary, $i_{s+1} \perp i_r \,/\, {}^\omega\{< i_{s+1} - i_r\}$. Thus, if $K = \{k \in \{< i_{s+1} - i_r\} \mid i_{s+1} \perp k / {}^\omega\{< i_{s+1} - k\}\}$, we have $\{i_1,\ldots, i_{s-1}\} \subseteq K$. With the assumption about the circumstances and Theorem 6 we also have $i_{s+1} \perp K / i_s, {}^\omega\{< i_{s+1} - K \cup \{i_s\}\}$. And this implies that $\langle i_1,\ldots, i_n\rangle$ is an ω-Markov chain.

Of course, Theorem 12 in particular applies to causal chains. These theorems are mathematically trivial, but conceptually nice; and I do not see how they can be improved upon much. Their content is certainly plausible; intuitively, it is just the existence of direct bypasses to causal chains which violates the Markovian intuition.

Concerning positive relevance, suppose that $A_r = {}^\omega i_r$ ($r = 1, \ldots, n$) and that $(A_1, \ldots, A_n)$ is a causal chain in ω. Somehow, A_1 should then be positively relevant to A_n. But how? If $\langle i_1,\ldots, i_n\rangle$ were a Markov chain, Theorem 10 could be applied. But it has turned out that ω-Markov and not Markov chains are the ones relevant to our enterprise. So if $\langle i_1,\ldots, i_n\rangle$ is assumed to be an ω-Markov chain, the trouble is that the characteristic conditional independencies refer for each i_r to a different condition, and therefore Theorem 10 is not immediately applicable. But perhaps the different conditions can be equalized and the grounds for Theorem 10 thus prepared. This is the basic idea which will be worked out in the sequel.

A very simple example illustrates all of the essential aspects of that idea. Suppose that there are only four binary variables $i < j < k < l$, $A = {}^\omega i$, $B = {}^\omega j$, $C = {}^\omega k$, $D = {}^\omega l$, $A \xrightarrow[\omega]{+} B \xrightarrow[\omega]{+} D$, and $\langle i, j, l\rangle$ is an ω-Markov chain. In probabilistic terms this means that

(1) $P(B \mid A) > P(B \mid \bar{A})$,
(2) $P(D \mid A \cap B \cap C) > P(D \mid A \cap \bar{B} \cap C)$, and
(3) $P(D \mid A \cap B' \cap C) = P(D \mid \bar{A} \cap B' \cap C)$ for each $B' \in \{B, \bar{B}\}$.

One would like to infer that, if B is omitted, A is positively relevant to D. Here, C may be taken as given or not. Thus, there are two alternatives for expressing this positive relevance:

(4a) $P(D \mid A \cap C) > P(D \mid \bar{A} \cap C)$, or
(4b) $P(D \mid A) > P(D \mid \bar{A})$.

Obviously, neither follows from (1)–(3), because (2) and (3) conditionalize on C, but (1) does not; that is the trouble with ω-Markov chains. The idea is to equalize the conditions in (1)–(3). Again, there are two alternatives. One may keep (2) and (3) and assume

(1a) $P(B \mid A \cap C) > P(B \mid \bar{A} \cap C)$

instead of (1); then (4a) can be inferred with the help of Theorem 10. Or one may keep (1) and assume

(2b) $P(D \mid A \cap B) > P(D \mid A \cap \bar{B})$ and
(3b) $P(D \mid A \cap B') = P(D \mid \bar{A} \cap B')$ for each $B' \in \{B, \bar{B}\}$

instead of (2) and (3); then (4b) can be inferred with the help of Theorem 10. However, (1a), (2b), and (3b) do not yet have causal form. The question thus is which causal assumptions allow them to be derived from (1)–(3). The answer differs for the two alternatives.

Look first at (1a). (1a) obviously follows from (1) together with

(5a) $P(B \mid A' \cap C) = P(B \mid A')$ for each $A' \in \{A, \bar{A}\}$;

and since P is strictly positive, this is equivalent to

(6a) $P(C \mid A' \cap B) = P(C \mid A' \cap \bar{B})$ for each $A' \in \{A, \bar{A}\}$.

Thus, (4a) may be derived from (1)–(3) by additionally assuming (6a). And (6a) has causal form; it says that B is directly causally irrelevant to C, whether A obtains or not.

Now consider (2b) and (3b). They obviously follow from (2) and (3), if

(7b) $P(D \mid A' \cap B' \cap C) = P(D \mid A' \cap B' \cap \bar{C})$ for each $A' \in \{A, \bar{A}\}$ and $B' \in \{B, \bar{B}\}$.

And (7b) already has causal form; it says that C is directly causally irrelevant to D, whether A and B obtain or not.

So at least two simple alternative causal conditions are available which guarantee positive relevance of the indirect cause to the indirect effect in this example. When is neither condition satisfied? When and only when B is causally relevant to C in some world and C is causally relevant to D in some world. But in this case there are two paths of causal influence running from A to D (though not necessarily in one world); so it is not surprising that an account of how causal influence is transmitted through single causal chains is inapplicable to such a case.

These observations are valid in general. In the general case we deal with a causal ω-Markov chain $\langle i_1, \ldots, i_n \rangle$, where again $A_r = {}^{\omega}i_r$ $(r = 1, \ldots, n)$. The strategy of equalizing conditionalization then amounts to finding some set M of variables such that for all $r = 2, \ldots, n$ the positive relevance of A_{r-1} to A_r as well as the characteristic ω-Markov independencies hold also conditional on ${}^{\omega}M$, thus enabling the inference of the positive relevance of A_1 to A_n conditional on ${}^{\omega}M$.

Which properties should M be expected to have? Of course, the basic property is that $i_2, \ldots, i_{n-1}$ do not belong to M. Two further properties are suggested by the above example. Alternative (b) makes it clear that, if a variable is directly causally relevant to i_r, then it must not be deleted from M. In other words, ${}^{\omega}M$ has to preserve the circumstances of A_r in some suitable sense (Theorem 14 below will refer to the probabilistically possibly relevant circumstances, Theorem 16 to the ideal ones). The condition that that much information about A_r must be retained is certainly plausible. Now, by omitting i_r from M we delete the most direct information about A_r. But it seems that we must as well delete any indirect information about A_r which exceeds the information provided by its circumstances; if such indirect information were retained, the averaging with respect to i_r – which is needed for calculating $P(A_r \mid A_1 \cap {}^{\omega}M)$ – may be biased in an undesirable way. This is what emerges from alternative (a) in the above example, in particular from (5a).

In sum, for each $r = 2, \ldots, n$, ${}^{\omega}M$ must include the circumstances of A_r and must not contain any further information about A_r. The fact that ${}^{\omega}M$ then suits the desired equalization of conditions is the intuitive content of the theorems we are after. They are supplemented by two auxiliary theorems providing a way of expressing the exclusion of such further information in causal terms.

For this purpose let us define:

Definition 10: The variables i and j *are causally connected in* ω *within* $J \subseteq I$ iff there are $k_1, \ldots, k_n \in J \cup \{i, j\}$ such that $k_1 = i$, $k_n = j$, and, for all $r = 1, \ldots, n-1$, ${}^{\omega}k_r \xrightarrow[\omega]{\pm} {}^{\omega}k_{r+1}$ or ${}^{\omega}k_{r+1} \xrightarrow[\omega]{\pm} {}^{\omega}k_r$.. And if $i < j$, I call i and j *causally connected in* ω iff they are causally connected in ω within (i, j).[34]

Causal unconnectedness in a lot of worlds implies a lot of probabilistic independence, at least if there are no simultaneous variables.[35] This is ascertained by the auxiliary:

Theorem 13: Suppose that I is linearly ordered by $<$ and that i and $j > i$ are causally unconnected in all $\upsilon \in {}^{\omega}\{< i\}$. Then there are disjoint K and L such that $K \cup L = [i, j]$, $i \in K$, $j \in L$, and $K \perp L / {}^{\omega}\{< i\}$.

This will turn out to be a special case of Theorem 15 below. And it leads to a first result concerning the positive relevance of indirect causes:

Theorem 14: Suppose that I is linearly ordered by $<$, that $\langle i_1, \ldots, i_n \rangle$ is an ω-Markov chain of binary variables. $A_r = {}^{\omega}i_r$, $x_r = P(A_{r+1} \mid A_r \cap {}^{\omega}\{< i_{r+1} - i_r\}) - P(A_{r+1} \mid \overline{A}_r \cap {}^{\omega}\{< i_{r+1} - i_r\})$ and that M is a set of variables such that for all $r = 1, \ldots, n$:

---

[34] This restriction of the connecting sequence to the interval between i and j does certainly not conform to the standard usage of "causally connected", if there is any. Note that despite this restriction causal connectedness is still weaker than causal relevance. A chain of causal relevance from i to j is always future-oriented, whereas a sequence causally connecting i and j may arbitrarily change its temporal direction.

[35] This is rather a technical restriction needed in the proofs below. However, it is not obvious how to avoid it.

(a) $i_r \notin M$,
(b) $R(A_r) - \{i_{r-1}\} \subseteq M$,
(c) each $j \in M$ with $j > i_r$ is causally unconnected with i_r in all $\upsilon \in {}^{\omega}\{M \cap \{< i_r\}\}$).

Then $\langle i_1, \ldots, i_n \rangle$ is a Markov chain conditional on ${}^{\omega}M$, and hence $P(A_n \mid A_1 \cap {}^{\omega}M) - P(A_n \mid \overline{A}_1 \cap {}^{\omega}M) = x_1 \cdot \ldots \cdot x_{n-1}$.

The proof is given together with that of Theorem 16. Condition (b) includes variables into M, (c) excludes variables from M. Thus, (b) and (c) work in opposite directions and may be difficult to satisfy. This can be improved upon, but again at the price of assuming circumstances to be ideal. That assumption is used in the auxiliary:

Theorem 15: Suppose that I is linearly ordered by <, that $\{< i\} \subseteq N \subseteq \{< j\}$, that, for all $\upsilon \in {}^{\omega}N$ and $k, l \in \{i, j\}$ with $k < l$, $C^*_\upsilon({}^{\upsilon}k, {}^{\upsilon}l) = C_\upsilon({}^{\upsilon}k, {}^{\upsilon}l)$, and that i and j are causally unconnected in all $\upsilon \in {}^{\omega}N$. Then there are disjoint K and L such that $K \cup L = [i, j]$, $i \in K$, $j \in L$, and $K - N, {}^{\omega}(K \cap N) \perp L - N, {}^{\omega}(L \cap N) / {}^{\omega}\{< i\}$.

Proof: Let $K = \{i\} \cup \{k \in (i, j) \mid i$ and k are causally connected within (i, j) in some $\upsilon \in {}^{\omega}N\}$ and $L = \{j\} \cup (i, j) - K$. Since causal connectedness within a fixed set is transitive, this definition implies that each $k \in K$ is causally unconnected with each $l \in L$ within (i, j) in all $\upsilon \in {}^{\omega}N$. In particular we thus have for all $k \in K$, $l \in L$, and $\upsilon \in {}^{\omega}N$:

(1) If $k < l$, then $k \notin R_\upsilon({}^{\upsilon}l)$; and if $l < k$, then $l \notin R_\upsilon({}^{\upsilon}k)$.

Now we shall inductively work up from i to j. Suppose we have already shown for some $i^* \in (i, j) \cup \{j\}$ that

(2) $\{< i^*\} \cap K - N, {}^{\omega}(\{< i^*\} \cap K \cap N) \perp \{< i^*\} \cap L - N, {}^{\omega}(\{< i^*\} \cap L \cap N) / {}^{\omega}\{< i\}$.

Let's assume that $i^* \in L$ (for $i^* \in K$ the corresponding reasoning applies). (1) and the ideality of the actual circumstances then imply

(3) $\{< i^*\} \cap K \perp i^* / {}^{\omega}\{< i\} \cap {}^{\upsilon}(\{< i^*\} \cap L)$ for all $\upsilon \in {}^{\omega}N$, i.e,
(4) $\{< i^*\} \cap K \perp i^* / {}^{\omega}\{< i\} \cap {}^{\omega}(\{< i^*\} \cap L \cap N), \{< i^*\} \cap L - N$.

(2) and (4) finally yield according to Theorem 1(d)

(5) $\{< i^*\} \cap K - N, {}^{\omega}(\{< i^*\} \cap K \cap N) \perp (\{< i^*\} \cup \{i^*\}) \cap L - N, {}^{\omega}(\{< i^*\} \cap L \cap N) / {}^{\omega}\{< i\}$.

For $i^* = j$ this is the desired result. Note that for $N = \{< i\}$ Theorem 15 reduces to Theorem 13. In this special case, (3) follows from (1) with the help of Theorem 2(e) alone and without the ideality of circumstances.

This leads to the second result concerning positive relevance:

Theorem 16: Suppose that I is linearly ordered by <, that $\langle i_1, \ldots, i_n \rangle$, A_r, and x_r are as in Theorem 14, and that M is a set of variables such that for all $r = 1, \ldots, n$:

(a) $i_r \notin M$,
(b) $R_\omega(A_r) - \{i_{r-1}\} \subseteq M$,

(c) each $j \in M$ with $j > i_r$ is causally unconnected with i_r in all $\upsilon \in {}^{\omega}M$.

Suppose further that for all $\upsilon \in {}^{\omega}M$ and $k, l \in [i_1, i_n]$ with $k < l$ $C^*_\upsilon({}^{\upsilon}K, {}^{\upsilon}l) = C_\upsilon({}^{\upsilon}K, {}^{\upsilon}l)$. Then the same result obtains as in Theorem 14.

Proof: All that must be shown is that for all $r = 2, \ldots, n$, $A' \in \{A_{r-1}, \bar{A}_{r-1}\}$, and $\upsilon \in \Omega$:

(1) $P(A_r \mid A' \cap {}^{\omega}\{< i_r - i_1, \ldots, i_{r-1}\} \cap {}^{\upsilon}\{i_1, \ldots, i_{r-2}\} = P(A_r \mid A' \cap {}^{\omega}M \cap {}^{\upsilon}\{i_1, \ldots, i_{r-2}\} = P(A_r \mid A' \cap {}^{\omega}M)$

(1) says that all the probabilities showing that $\langle i_1, \ldots, i_n \rangle$ is a Markov chain conditional on ${}^{\omega}M$ coincide with the corresponding probabilities showing that $\langle i_1, \ldots, i_n \rangle$ is an ω-Markov chain. A direct application of Theorem 10 then leads us to the desired result. To show (1) suppose that $M - \{< i_r\}$ consists of $j_1 < \ldots < j_q$ and that we have already shown for some $p - 1 < q$ that

(2) $P(A_r \mid A' \cap {}^{\upsilon}\{< i_r - i_{r-1}\} \cap {}^{\omega}\{j_1, \ldots, j_{p-1}\}) = P(A_r \mid A' \cap {}^{\upsilon}\{< i_r - i_{r-1}\})$

holds for all $\upsilon \in {}^{\omega}M$. Since i_r and j_p are causally unconnected in all $\upsilon \in {}^{\omega}M$ we may infer from Theorem 15 and the ideality of the actual circumstances that there is a partition $\langle K, L \rangle$ of $[i_r, j_p]$ such that $i_r \in K$, $j_p \in L$, and for $M' = M \cap \{< j_p\}$

(3) $K - M', {}^{\upsilon}(K \cap M') \perp L - M', {}^{\upsilon}(L \cap M') / {}^{\upsilon}\{< i_r\}$

for all $\upsilon \in {}^{\omega}M$. Alternatively, we get (3) from Theorem 13 and the stronger unconnectedness assumption (c) of Theorem 14 without the ideality of circumstances. (3) implies with Theorems 1(d) and 2(c)

(4) $i_r \perp j_p / {}^{\upsilon}\{< i_r\} \cap {}^{\upsilon}\{j_1, \ldots, j_{p-1}\}$ for all $\upsilon \in {}^{\omega}M$, i.e,
(5) $i_r \perp j_p / A' \cap {}^{\upsilon}\{< i_r - i_{r-1}\} \cap {}^{\omega}\{j_1 \ldots, j_{p-1}\}$ for all $\upsilon \in {}^{\omega}M$.

(2) and (5) together imply that (2) holds for p, too, and thus also for q, i.e. that for all $\upsilon \in \Omega$

(6) $P(A_r \mid A' \cap {}^{\omega}M \cap {}^{\upsilon}\{< i_r - M \cup \{i_{r-1}\}\}) = P(A_r \mid A' \cap {}^{\omega}(M \cap \{< i_r\}) \cap {}^{\upsilon}\{< i_r - M \cup \{i_{r-1}\}\})$.

Next observe that

(7) $P(A_r \mid A' \cap {}^{\omega}(M \cap \{< i_r\}\}) = P(A_r \mid A' \cap {}^{\omega}(M \cap \{< i_r\}) \cap {}^{\upsilon}\{< i_r - M \cup \{i_{r-1}\}\})$

holds for all $A' \in \{A_{r-1}, \bar{A}_{r-1}\}$ and $\upsilon \in \Omega$ because of the assumption (b) about M and the ideality of circumstances – which is not needed, if, alternatively, the assumption (b) of Theorem 14 is used. (7) says that the R.H.S. of (6) does not depend on υ. Thus, the L.H.S. of (6) does not depend on υ as well. This finally yields

(8) $P(A_r \mid A' \cap {}^{\omega}M) = P(A_r \mid A' \cap {}^{\omega}(M \cap \{< i_r\}) \cap {}^{\upsilon}\{< i_r - M \cup \{i_{r-1}\}\})$

for all $\upsilon \in \Omega$, and this is even somewhat stronger than the desired (1).

A careful analysis of the two proofs will show several steps which do not require the full strength of the premises. Thus, there certainly are weaker and maybe nicer conditions under which the consequence of these theorems still holds. But I wonder whether there are much weaker or much nicer conditions subject to the constraint that they be expressed in causal terms.

When does a set M exist as required by Theorem 16(a)–(c)? I have not found an informative and more perspicuous sufficient condition. But there is a simple necessary condition:

Theorem 17: A set M satisfying clauses (a)–(c) of Theorem 16 exists only if in all $\upsilon \in {}^{\omega}M$ $\langle i_1,\ldots, i_n\rangle$ is the only chain of causal relevance leading from i_1 to i_n.

Proof: Suppose that for some $\upsilon \in {}^{\omega}M$ there is another chain of causal relevance from i_1 to i_n which may be assumed to be of the form $\langle i_1, j_1,\ldots, j_p, i_r,\ldots, i_n\rangle$, where $j_p \neq i_{r-1}$. Hence, clause (b) demands that $j_p \in M$. But i_1 and j_p are causally connected in υ. Hence, $j_p \notin M$ and thus a contradiction is entailed by clause (c).

Since there may be causal connections between earlier members and direct causal antecedents of later members of $\langle i_1,\ldots, i_n\rangle$ which are not chains of causal relevance – a situation which again excludes the existence of an appropriate M –, Theorem 17 may not be strengthened to a biconditional.

Theorem 17 limits the scope of Theorem 16 to cases where causal influence is transmitted through a single causal chain. More powerful theorems are therefore required for dealing with the transmission of causal influence through more complex causal nets. Eells and Sober (1983, pp. 49ff.) should be of help here in a similar way as Theorem 10 has guided Theorems 14 and 16.

But it is clear that this is only the beginning of a much fuller theory of causation. For instance, I have not returned to the very first characterization of causation in Section 2.3 and tried to say what the circumstances of an indirect causal relation are. Theorems 14 and 16 suggest that ${}^{\omega}M$, for a minimal M satisfying (a)-(c) of Theorems 14 or 16, provides such circumstances. However, this suggestion is neither general nor worked out.

Still, I hope to have developed the program far enough to justify the impression that the right direction towards a prosperous theory has been found. In particular, Theorems 11, 12, 14, and 16 explain how the three basic intuitions here discussed come to be held, even though they are not generally compatible. This explanation is as plausible as the assumption that the conditions which have been shown to guarantee agreement between the intuitions are taken for granted. And I think this assumption is not too implausible.[36]

[36] In fact, I believe that there is a deeper explanation for the conditions that actual circumstances are ideal, that causal chains are strict, and the like; these conditions are crucial to an objectivization of our causal picture. This conjecture would emerge more clearly in the unwritten deterministic counterpart of this paper; but it certainly cannot be part of this inquiry.

Chapter 3
Causation: An Alternative[†,*]

3.1 Introduction

Counterfactual analyses of causation have run through many epicycles.[1] It is time to look at a genuine alternative, which is in principle available for 20 years.[2]

The logic of counterfactuals, as fully developed in Lewis (1973a), was an indispensible means for such counterfactual analyses. This is why they developed into a constructive research programme only with Lewis (1973b). It has been carried on throughout the years. But it seems it has received greatest attention just recently.[3] Reading all the papers leaves one in bewilderment: too many examples not adequately dealt with or only by increasingly imperspicuous clauses, and too few satisfying theories. The prospects of the programme have clouded. What to do?

First, we must step back from the glaring equation '*A* is a cause of *B*' = 'if *A* had not happened, *B* would not have, either', which has been stated for centuries, but could not be theoretically exploited until the advent of the logic of counterfactuals. We should rather look at the other classic formula that a cause is a necessary and/or sufficient condition for its effect under the obtaining circumstances. I shall develop this thought in Section 3.4. The formula may appear worn out, and all the familiar explications of 'necessary and/or sufficient condition' turned out to be unfit. But

† This paper was originally published in: *The British Journal for the Philosophy Science* 57 (2006) 93–119. It is reprinted here with kind permission by the British Society for the Philosophy Science and Oxford University Press.

* I am indebted to an anonymous referee whose extensive and careful comments led to numerous improvements of the paper.

[1] The major cycles have been produced by David Lewis himself. See Lewis (1973b, 1986d, 2000). Hints to further cycles may be found there.

[2] It is first presented in my (1983a).

[3] See, e.g., the April issue of the *Journal of Philosophy* 97 (2000), or the collection by Collins et al. (2004). See also the many references therein, mostly referring to papers since 1995.

there is a thoroughly epistemological notion of ‘condition’, or ‘reason’, as I shall say, which is based on so-called ranking theory and which *does the job*. This will be elaborated in Section 3.3. Before this, Section 3.2 will briefly address the nature of causal relata as far as needed. An important general lesson of these sections will be that the theory of deterministic causation can be built in *perfect* analogy to the theory of probabilistic causation (as developed from Suppes 1970 up to Spirtes et al. 1993, Shafer 1996, and Pearl 2000).

We will be rewarded in Section 3.5 where we shall see how to account for some of the most stubborn problem cases in an entirely straightforward way. How can this be? The reader will already have sensed the catch. Obviously, I am going to develop a thoroughly epistemological theory of causation, and this is simply not what we want and what the counterfactual analysts were after. This point is set straight in Section 3.6 where I shall suggest how to objectivize my account so far relativized to an epistemic subject. This will double the reward; we can explain then why the problem cases are so stubborn for objective theories of causation. In short, there *is* a workable alternative. Let us see how it works.

3.2 Variables, Propositions, Time

I have to be quite brief concerning the ontology of causal relata. Ordinary language is an unreliable guide in these matters. We often speak of events, or facts, or states, or even changes as causes and effects (for some early linguistic observations cf., e.g., Vendler 1967), and then we might enter philosophical argument to clarify the subtle differences between these entities. However, this argument tends to be endless. As far as I see, we face here a largely tactical choice. Purely philosophical discussions of causation tend to get entangled into the notion of an *event* (as is amply exemplified by the work of David Lewis). By contrast, I observe that *state-space* terminology prevails among authors concerned rather with scientific applications and less with ontological subtleties. This terminological split is unfortunate, since by taking one side one presents oneself in an unfamiliar way to the other side. The split is all the more unfortunate as the approaches are, I think, intertranslatable to a large extent (and the issues where translatability may fail will not become relevant in this paper). I prefer state-space to event terminology (because this is the one I have grown up with and because it facilitates rigorous theorizing). Formally, of course, I shall use just some set-theoretic construction that is largely open to philosophical interpretation. It is useful to explicitly introduce the construction.

It starts with a set U of variables, a *frame*; members of U are denoted by x, y, z, etc., subsets of U by X, Y, Z, etc. All definitions to follow, in particular the notion of causation I am going to explicate, will be relative to this frame. This may be cause for concern which I shall take up in Section 3.6. Each variable can realize in this or that way, i.e., take one of several possible values (and may indeed be conceived simply as the set of its possible values). A *small world*[4] w is a function that tells how each variable realizes, i.e., assigns to each variable one of its possible values. W denotes the set of small worlds.

Typically, a variable consists of an object, a time, and a family of properties (say, color, charge, marital status, income, etc.); and a realization of such a variable consists in the object's having at that time a certain property of that family (a certain color, charge, marital status, income, etc.). This entails that variables are here considered to be specific, not generic.[5] Let me give a slightly extended example. Meteorologists are interested in generic variables like temperature, air pressure, humidity, wind, etc., which can take various values (the latter, for instance, a velocity vector). But these generic variables realize at certain times and places; only then do we have specific variables. Thus, the temperature at noon of January 1, 2004, in Konstanz is a specific variable that may take any value on the Celsius scale and actually took 2 °C. For each of the generic meteorological variables there are hence as many specific variables as there are spatio-temporal locations considered by the meteorologist. A small metereological world, then, is a weather course, that is, a specification of all specific meteorological variables, or of all generic variables for all the locations considered. Those in pursuit of causal laws or correlations tend to consider generic variables. Here, however, we are interested in singular causation which I, as well as counterfactual analyses, take to be primary. Hence, our causal investigation will focus on specific variables.

As usual, *propositions* are sets of small worlds, i.e. subsets of W; I use A, B, C, etc. to denote them. I could also call them *states of affairs*. But this is only to say that the subtle difference between the ontological connotation of 'state of affairs' and the epistemological connotation of 'proposition' is not my topic here, though it hides deep problems for any theory of causation.

Let us say that A is a proposition *about* the set $X \subseteq U$ of variables if it does not say anything about the other variables in $U \setminus X$, i.e., if for any small world w in A all other small worlds agreeing with w within X are also in A. For instance, a proposition about temperatures only consist of small (meteorological) worlds which realize air pressure, humidity, etc., for all locations considered in any way whatsoever. The set of propositions about X is denoted by $\mathbf{P}(X)$. Hence, $\mathbf{P}(U)$ is the set of all propositions considered. $\mathbf{P}(x)$ is short for $\mathbf{P}(\{x\})$. Indeed, propositions about single variables, i.e., in $\mathbf{P}(x)$ for some $x \in U$, are my candidates for causal relata.

---

[4] This allusion to Savage (1954, sect. 5.5) is to emphasize that we are dealing here with restricted well-defined model worlds and not yet with grand Lewisian possible worlds.

[5] This remark is directed to the state-space camp where the point is often unclear. The event camp is not concerned; if one translates event into state-space terminology one automatically ends up with what I call specific variables.

The difference from event ontology is not so large as it may appear. For instance, events in the sense of Kim (1973) are the very same as my causal relata, i.e., my propositions about single variables. What Lewis (2000) ends up with may be translated into the language of variables, too. He is reluctant to decide how fragile (in his sense) events really are. In any case, an event has very fragile versions (which represent the same event), and it has very fragile alternatives, which may be either different or sufficiently similar. In the latter case, they are alterations of the event just as its versions are. Lewis then explains causation between events in terms of counterfactual dependence between their alterations; the details are not yet relevant, though. Now, the versions of an event and all its alternatives make up for a very fine-grained variable in my sense taking very many possible values, and each version or alternative of the event is a possible realization of the variable. The event itself realizes if a value from some subset of the set of all values realizes. How small this subset is depends on how fragile the event is taken to be. The alterations form a somewhat larger subset. In any case, what Lewis considers as causal relata also qualifies as causal relata in my sense.

Since variables are specific, they have a natural temporal order. $x < y$ means that the variable x realizes before the variable y, and if $A \in \mathbf{P}(x)$ and $B \in \mathbf{P}(y)$, $A < B$ is to say that A precedes B. I shall facilitate matters by assuming that the variables are indeed linearly (not only weakly) ordered in time. Thereby I avoid nasty questions about simultaneous causation and neglect the even less perspicuous case of causation among temporally extended variables that possibly overlap one another. These complications are not our concern.

Finally, I will make the simplifying assumption that the frame U and the set W of small worlds it generates are finite; this entails in particular that temporal order is discrete. Loosening this assumption is a foremost mathematical, though not philosophically insignificant task.

3.3 Induction First

Let us turn to causation after these preliminaries, and let us, as announced, start from the classic formula abundantly found in the literature: A is a cause of B iff A and B both occur, if A precedes B, and if A is a necessary and/or sufficient condition for B under the obtaining circumstances.[6]

The requirement of the cause preceding the effect is often doubted in the philosophical literature, for reasons I do not understand well. I take this requirement

[6] This neglects Hume's contiguity condition, which is inexpressible in the framework introduced above, since it leaves out (or implicit) all spatial relations between variables.

simply for granted. The only implicit argument I shall give is that the theory of causation I shall propose would not work at all without it. Hence, I will leave it open whether this is an argument for temporal precedence or against this theory.

What do 'the obtaining circumstances' refer to? Let us postpone this question to the next section. We should first note that a cause must not be a redundant condition for its effect given the circumstances.[7] If *A* is, say, a sufficient condition for *B* given circumstances *C*, this means that *B* is necessary given *A* and *C*, but not given *C* alone, i.e., that, given *C*, *A* raises the modal status of *B* from impossibility or contingency to necessity. Likewise, in case *A* is a necessary condition for *B*.

Hence, my favorite variant of the classic formula is, generally, this: *A is a cause of B iff A and B both occur, if A precedes B, and if A raises the metaphysical or epistemic status of B given the obtaining circumstances.* This makes explicit the relevance of *A*. It also adds the basic ambiguity in the notion of a condition between a metaphysical and an epistemic reading, which will acquire great importance later on. And it is even general enough to cover probabilistic causation as well where the statuses are probabilistic ones.

Note that counterfactual analyses are a special case of this general formula. They take the statuses metaphysically as counterfactual necessity and possibility. The temporal precedence is entailed by the constant reminder that all counterfactuals involved in the analysis must be read in a non-backtracking way. And the reference to the obtaining circumstances is always implicit in the antecedent of a counterfactual. However, they are only a special case; stepping back from them means widening the view and seeing what else might fall under the general formula.

Well, what else might fall under it? The traditional Humean view is that the talk of necessary and/or sufficient conditions should be explained in terms of nomological or lawful implication, where laws in turn are taken as mere regularities. However, I take it that all regularity accounts of causation have failed.[8]

Thus, we are back at Hume's famous question: what more is causal necessity than mere regularity? Hume should not be reduced to the answer: nothing. He was rather peculiarly ambiguous. More prominent in his writings is an associationist theory of causation, according to which the causal relation between two events is constituted by their being associated in our minds. Association, in turn, is explained as the transfer of liveliness and firmness, the marks by which Hume characterizes belief. Thus, we may say in more modern terms that, if *A* precedes *B* (and is contiguous to it), *A* is a cause of *B* for Hume iff *B* may be inductively inferred from *A*

---

[7] This is what Reichenbach's screening-off is about in the probabilistic case and Mackie's INUS conditions in the deterministic case.

[8] Including John Mackie's account in terms of INUS conditions. Indeed, contrary to his views in (1965) he concludes in (1974, p. 86), that conditionality cannot be understood in terms of the regularity theory.

(and vice versa). At the same time, this entails a fundamental subjective relativization of the notion of causation.[9]

Here, I fully endorse this subjectivist turn. I shall not try to adduce principled reasons for doing so. My argument rather lies in the Sections 3.5 and 3.6: this turn is successful where counterfactual and other objectivistic analyses are not, and there is still a way to escape from subjectivism. However, the equivalence of causation and induction was much too quick. We have to underpin the account of causation we are heading for by an elaborate theory of inductive inference. This is the crucial task for the rest of this section.

What might we expect of a theory of inductive inference? The task of induction is to project from the total evidence we have received all our beliefs transcending the evidence. The task of belief dynamics is to tell which posterior belief state to assume on the basis of the prior belief state and the evidence received in between. It is next to obvious that these two tasks are essentially equivalent (for details see my 2000a). Hence, what we expect of a theory of induction is no more and no less than an account of doxastic states which specifies not only their static, but also their dynamic laws (understood as laws of rationality).

The form of these laws depends, of course, on the chosen representation of doxastic states. The best elaborated representation is certainly the probabilistic one, for which we have well-argued static and dynamic laws (cf., e.g., Skyrms 1990, ch. 5). But that would lead us to a theory of probabilistic causation.

In pursuit of deterministic causation, we should hence focus on plain belief or acceptance that admits, as it were, only of three grades: each proposition is held true, undecided, or held false. The obvious idea is to represent plain belief simply by the set of propositions held true, and the obvious static law for such belief sets is that they be consistent and deductively closed.[10] However, there are no general dynamic laws for doxastic states thus represented. Representing plain belief by extremal probabilities is of no avail, since all laws for changing subjective probabilities fail with the extremal ones.[11] Hence, a different representation is needed in order to account for the dynamics of plain belief.

To cut a long story short, I am still convinced that this is best achieved by the theory of ranking functions.[12] This conviction rests on the fact that ranking theory offers a good solution to the problem of iterated belief revision, and thus a *general*

---

[9] I believe that the associationist theory is conceptually more basic in Hume. But regularities shape our associations and explain why our associations run rather this way than that way. In this way, the associationist theory may eventually reduce to the regularity theory. It is obvious, though, that Hume's ambiguity between causation as a philosophical relation (regularity) and as a natural relation (association) has provoked many exegetic efforts.

[10] This is at least what doxastic logic standardly assumes. There are well-known objections, but no standard way at all to meet them. So I prefer to keep within the mainstream.

[11] Popper measures are often thought to overcome the relevant restrictions of standard probability theory. But they do not go far enough; see my (1986) and (1988) [here: ch. 1].

[12] Proposed in my (1988) [here: ch. 1] under the label "ordinal conditional functions". Their first appearance, though, is in my (1983a, ch. 5).

dynamics of plain belief, whereas the discussion of this problem in the belief revision literature has not produced a serious rival in my view (cf. Hansson 1998 or Rott 2003). So, the next thing to do is to briefly introduce and explain this theory of ranking functions.

The basic concept is very simple:

Definition 1: κ is a ranking function iff it is a function from the set W of small worlds into the set of non-negative integers such that $\kappa^{-1}(0) \neq \emptyset$. It is extended to propositions by defining $\kappa(A) = \min \{\kappa(w) \mid w \in A\}$ for $A \neq \emptyset$ and $\kappa(\emptyset) = \infty$.

A ranking function κ is to be interpreted as a ranking of disbelief. If $\kappa(w) = 0$, w is not disbelieved and might be the actual small world according to κ. This is why I require that $\kappa(w) = 0$ for some small world w. If $\kappa(w) = n > 0$, then w is disbelieved with rank n. The rank of a proposition is the minimum of the ranks of its members; thus a proposition is no more and no less disbelieved than the most plausible worlds realizing it. $\kappa(A) = 0$ says that A is not disbelieved, but not that A is believed; rather, belief in A is expressed by disbelief in $\bar{A}$, i.e. $\kappa(\bar{A}) > 0$ or $\kappa^{-1}(0) \subseteq A$. In other words, all and only the supersets of $\kappa^{-1}(0)$ are believed in κ; they thus form a consistent and deductively closed belief set.

If we were only to represent belief, we would have to distinguish only an inner sphere of not disbelieved worlds having rank 0 and an outer shell of the remaining disbelieved worlds having rank > 0. But as we shall immediately see, more shells are needed in order to cope with the dynamics of belief. The picture of shells or spheres reminds of the entrenchment orderings used in belief revision theory or indeed of the similarity spheres used by Lewis for the semantics of counterfactuals. However, in both pictures the spheres or shells are only ordered. Ranks go beyond by numbering the shells; the arithmetics of ranks will turn out to be crucial.

Two simple, but important properties of ranking functions follow immediately: the *law of negation* that for all $A \subseteq W$ either $\kappa(A) = 0$ or $\kappa(\bar{A}) = 0$ or both, and the *law of disjunction* that for all $A, B \subseteq W$ $\kappa(A \cup B) = \min \{\kappa(A), \kappa(B)\}$.

So far, only disbelief comes in degrees. But degrees of disbelief are tantamount to degrees of belief. It is easy to represent both degrees in one notion:

Definition 2: β is the *belief function associated with* the ranking function κ iff for each $A \subseteq W$ $\beta(A) = \kappa(\bar{A}) - \kappa(A)$ (due to the law of negation, at least one of the two terms is 0). β is a *belief function* iff it is associated with some ranking function.

Thus, $\beta(\bar{A}) = -\beta(A)$, and A is believed to be true, false, or neither according to β (or κ) depending on whether $\beta(A) > 0$, < 0, or $= 0$. Belief functions may be the more intuitive notion; therefore I often prefer to use them. However, they are a derived notion; laws and theorems are more easily stated in terms of ranking functions.

The ranks reveal their power when we turn to the dynamics of plain belief. The central notion is given by:

Definition 3: Let κ be a ranking function and $\emptyset \neq A \subseteq W$. Then the rank of $w \in W$ *given* or *conditional on* A is defined as $\kappa(w \mid A) = \begin{cases} \kappa(w) - \kappa(A) \text{ for } w \in A \\ \infty \text{ for } w \notin A \end{cases}$.

Similarly, the rank of $B \subseteq W$ *given* or *conditional on* A is defined as $\kappa(B \mid A) = \min \{\kappa(w \mid A) \mid w \in B\} = \kappa(A \cap B) - \kappa(A)$. I also call the function $\kappa(. \mid A)$ the *A-part* of κ. If β is the belief function associated with κ, we finally set $\beta(B \mid A) = \kappa(\overline{B} \mid A) - \kappa(B \mid A)$.

Definition 3 is tantamount to the *law of conjunction* which states that $\kappa(A \cap B) = \kappa(A) + \kappa(B \mid A)$ for all propositions $A \neq \emptyset$ and B. The definition and the law essentially refer to the arithmetics of ranks; a mere ordering (of ranks, of entrenchment, or of similarity spheres) would not do. Indeed, in the relevant literature one finds quite often the proposal and elaboration of a theoretical structure that is more or less equivalent to the above laws of negation and disjunction. The divergence starts with the law of conjunction, which may thus be viewed as the distinctive feature of ranking theory. It has important consequences:

First, it is obvious that a ranking function κ is uniquely determined by its A-part $\kappa(. \mid A)$. Its $\overline{A}$ -part $\kappa(. \mid \overline{A})$, and the degree $\beta(A)$ of belief in A. This suggests a simple model for doxastic changes: As is well known, probabilistic belief change is modelled on the assumption that the probabilities conditional on the proposition (or its negation) about which one receives information remain unchanged.[13] Similarly, we can assume here that, if the received information directly concerns only the proposition A (and its negation), only the ranks of A and $\overline{A}$ are changed – such that, say, the posterior rank of A is 0 and that of $\overline{A}$ is n so that A becomes believed with degree n –, whereas all the ranks conditional on A and on $\overline{A}$ remain unchanged. Thereby, the doxastic change results in a fully determinate posterior ranking function which one may call the *A,n-conditionalization* of the prior one.

The picture of shells or spheres may again be helpful. If A is not disbelieved in the prior state the effect of A,n-conditionalization is just to add A to the old beliefs (and to draw all logical consequences). This would be absurd, though, if A would be priorly disbelieved. In this case, the effect of A,n-conditionalization is to move to the innermost shell compatible with A; its intersection with A (and all the logical consequences thereof) then constitutes the posterior belief set. In order to allow for a differentiated revision behavior, more than one shell around the inner sphere are needed. So far, all accounts working in this picture agree. However, for a full and iterated belief dynamics one must not only say what the posterior beliefs are, but also how the systems of spheres gets rearranged in revision. This issue is precisely answered by the arithmetical method of A,n-conditionalization, but it presents great difficulties for other approaches.These remarks may suffice for indicating that ranking theory successfully provides a completely general dynamics of belief.[14]

Secondly, this account of conditionalization immediately leads to the crucial notion of doxastic dependence and independence: two propositions are independent iff conditionalization with respect to one does not affect the doxastic status of the

---

[13] This is true of simple conditionalization as well as of generalized conditionalization proposed by Jeffrey (1965, ch. 11).

[14] For more details, see my (1988, sect. 5) [here: sect. 1.5]. The present paper will use only the precise definition of conditional ranks.

other. More generally, two sets of variables are independent iff conditionalization with respect to any proposition about the one set does not affect the doxastic status of any proposition about the other. Or formally:

Definition 4: Let β be the belief function associated with the ranking function κ. Then A and B are *independent* given $C \neq \emptyset$ relative to β (or κ) iff $\beta(B \mid A \cap C) = \beta(B \mid \bar{A} \cap C)$, i.e. iff $\kappa(A' \cap B' \mid C) = \kappa(A' \mid C) + \kappa(B' \mid C)$ for all $A' \in \{A, \bar{A}\}$, $B' \in \{B, \bar{B}\}$; unconditional independence results for $C = W$. Moreover, if $X, Y, Z \subseteq U$ are three sets of variables, X and Y are independent given Z relative to β (or κ) iff for all $A \in \mathbf{P}(X)$, all $B \in \mathbf{P}(Y)$, and all realizations C of Z (or atoms or logically strongest a posteriori propositions in $\mathbf{P}(Z)$) A and B are independent given C w.r.t. β (or κ); unconditional independence results for $Z = \emptyset$.

Unconditional and conditional ranking independence conforms to the same laws as probabilistic independence.[15] This entails in particular that the whole powerful theory of Bayesian nets (cf. Pearl 1988, ch. 3, or, e.g., Jensen 1996), which rests on these laws, can immediately be transferred to ranking functions.[16] Indeed, it may have become clear in the meantime that ranking functions, though their appearance is quite different, behave very much like probability measures.[17] So, in a way, my further procedure is simply to transfer what can be reasonably said about probabilistic causation to deterministic causation with the help of ranking theory.

Before doing so, we have to add a third and final observation: dependence, which negates independence, may obviously take two forms: positive relevance and negative relevance. Intuitively, we would say that a proposition A is a reason for a proposition B (relative to a given doxastic state) if A strengthens the belief in B, i.e., if the belief in B given A is firmer than given $\bar{A}$. This is something deeply rooted in everyday language; we also say that A supports or confirms B, that A speaks for B, etc. All this comes formally to positive relevance. There are even more ways to express negative relevance; this is, for instance, the essential function of 'but' (cf. Merin 1996). Hence, these notions deserve a formal explication:

Definition 5a: Let β be the belief function associated with the ranking function κ. Then A is a *reason for B given C relative to* β (or κ) iff $\beta(B \mid A \cap C) > \beta(B \mid \bar{A} \cap C)$. Again, the unconditional notion results for $C = W$.

---

[15] As I was eager to prove in my (1983a, sect. 5.3), and in my (1988, sect. 6) [here: sect. 1.6]. For a fuller comparison see my (1994).

[16] If one notes, moreover, how tight the relation between Bayesian nets and causation is assumed to be – see my (1978, sect. 3.3), Spirtes et al. (1993), or Pearl (2000) – the bearing of ranking theory on the theory of causation becomes already obvious.

[17] The deeper reason is that ranks may be roughly seen as the orders of magnitude of infinitesimal probabilities in a non-standard probability measure. Thus, by translating the sum of probabilities into the minimum of ranks, the product of probabilities into the sum of ranks, and the quotient of probabilities into the difference of ranks one transforms most theorems of probability theory into ranking theorems. This transition has niceties, though, which are not really clarified; cf. my (1994, pp. 183–185).

According to this definition, being a reason is a symmetric, but not a transitive relation. This is analogous to probabilistic positive relevance, but in sharp contrast to being a deductive reason, which is transitive and not symmetric. However, being a reason thus defined embraces being a deductive reason (which amounts to set inclusion between propositions $\neq \emptyset$, W). Indeed, when I earlier referred to inductive inference, this comes down to the theory of positive relevance or the relation of being a reason.[18] It is also worth mentioning that being a reason does not presuppose the reason to be actually given, i.e. believed. On the contrary, whether A is a reason for B relative to β is independent of the degree $\beta(A)$ of belief in A.

The value 0 has the special role of a dividing line between belief and disbelief. Therefore, different kinds of reasons must be distinguished:

Definition 5b: *Given C, A* is a

$$\begin{Bmatrix} \textit{additional} \\ \textit{sufficient} \\ \textit{necessary} \\ \textit{weak} \end{Bmatrix} \textit{reason for } B \text{ w.r.t. } \beta \text{ iff } \begin{Bmatrix} \beta(B \mid A \cap C) > \beta(B \mid \bar{A} \cap C) > 0 \\ \beta(B \mid A \cap C) > 0 \geq \beta(B \mid \bar{A} \cap C) \\ \beta(B \mid A \cap C) \geq 0 > \beta(B \mid \bar{A} \cap C) \\ 0 > \beta(B \mid A \cap C) > \beta(B \mid \bar{A} \cap C) \end{Bmatrix}$$

Hence, if A is a reason for B, it belongs to at least one of these kinds. There is just one way of belonging to several of these kinds; namely, by being a necessary and sufficient reason. Sufficient and necessary reasons are certainly salient. But additional and weak reasons, which do not show up in plain beliefs and are therefore usually neglected, deserve to be allowed for by Definition 5b.

This presentation of ranking theory suffices as a refined substitute for Hume's rudimentary theory of association. Thus equipped, we may return to causation.

3.4 Causation

We had started with the formula that A is a cause of B if, among other things, A is a necessary and/or sufficient condition for B under the obtaining circumstances, and we have seen that the point is rather that A is a positively relevant condition for B given the circumstances. In all frameworks for deterministic causation I know of and in particular in a regularity as well as in a counterfactual framework, being

[18] Recall that inductive logic and qualitative confirmation theory were considered to be one and the same project. Recall also that there has been a rigorous, although less successful, discussion of qualitative confirmation theory; cf. the survey of Niiniluoto (1972). If my (2005a) makes sense, it is a promising task to revive qualitative confirmation theory in terms of ranking theoretic positive relevance.

positively relevant automatically comes down to being a relevant necessary and/or sufficient condition. However, with the richer conceptual resources of the previous section, we may and should distinguish just as many kinds of causes as there are kinds of reasons. This point will become important.

The only thing so far left for clarification are the obtaining circumstances. The most plausible thing to say is that the circumstances relevant for judging the causal relation from *A* to *B* consists of all the other causes of *B* that are not caused by *A*. But this is obviously circular.[19] However, the circularity dissolves, if only *A*'s being a *direct* cause of *B* is considered. In this case there are no intermediate causes, i.e., no causes of *B* caused by *A*; the relevant circumstances may hence include all the other causes of *B*. Moreover, it seems to do no harm when all irrelevant circumstances are added as well, i.e. all the other facts preceding, but not causing *B*. Thus, we have arrived at conceiving the obtaining circumstances of *A*'s directly causing *B* as consisting of all the facts preceding *B* and differing from *A*.

A slightly more detailed argument (worked out in my 1983a, ch. 3, sect. 6.1) leads to the same result. Given that *A* and *B* are facts about single variables and that *A* precedes *B* (that's always tacitly understood), *A*'s being a reason for *B* according to Definition 5 is obviously the deterministic analogue to *A*'s being a prima facie cause of *B* in the probabilistic sense of Suppes (1970, ch. 2). But the prima facie appearance may change in three ways. First, facts preceding the cause *A* may turn up which render *A* irrelevant and thus only a *spurious* cause for *B*. Think, e.g., of the case of the falling barometer prima facie causing the thunderstorm, but being screened off, of course, by the low air pressure. This case is usually interpreted probabilistically, but has a deterministic reading as well. Secondly, facts realizing between *A* and *B* may turn up which render *A* irrelevant and thus, at most, an *indirect* cause for *B*. Any deterministic causal chain exemplifies this possibility. These two points were already considered by Suppes. Thirdly, however, if *A* is irrelevant to *B* given some condition, further facts preceding the effect *B* may add to the condition such that *A* is again positively relevant, and thus apparently a *hidden* cause for *B*. Suppose you press a switch and, unexpectedly, the light does not go on. You conclude that the switch does not work and that your pressing it had no effect whatsoever. The truth, however, is that someone else accidentally pressed another switch for that light at the very same time. So, given these circumstances, your pressing the switch indeed caused the light *not* to go on.

The three cases entail that every new fact preceding the effect may, in principle, change the assessment of the causal relation from *A* to *B* and suggest, respectively, that *A* is a direct, or an indirect, cause of *B*, or neither. The assessment is guaranteed to settle only when the whole past of the effect *B* has been taken into account.[20]

---

[19] This is basically the fundamental objection which Cartwright (1979) raised against all probabilistic explications of causation.

[20] It should be clear at this point that facts occurring after the effect have no such force. This does not preclude, of course, that, given incomplete knowledge about the past, the future may carry information about the past, and thus about the causal relation between *A* and *B*.

But what is the whole past of the effect? Within the given frame U, this can only mean the well-defined past as far as it is statable in this frame; this is the source of the *frame-relativity* of the theory developed here.

Both this and the previous consideration lead thus to the same explication of direct causation:

Definition 6: Let $A \in \mathbf{P}(x)$, $B \in \mathbf{P}(y)$ for some $x,y \in U$, and $w \in W$. Then A is a *direct cause* or, respectively, an *additional, sufficient, necessary*, or *weak direct cause* of B in the small world w relative to the ranking function κ iff:

(1) $w \in A \cap B$,
(2) $A < B$[21],
(3) A is a reason, or, respectively, an additional, sufficient, necessary, or weak reason for B given $w_{<B, \neq A}$ w.r.t. κ – where $w_{<B, \neq A} = \{w' \mid w'$ agrees with w on $\{z \in U \mid z < y \text{ and } z \neq y\}\}$ denotes the past of B except A as it is in w (which collects, as argued, the obtaining circumstances).

As an illustration, let us look at the cases of a causal chain and of a conjunctive fork, which are hard or impossible to distinguish for a regularity account, but present no problem to counterfactual analyses.

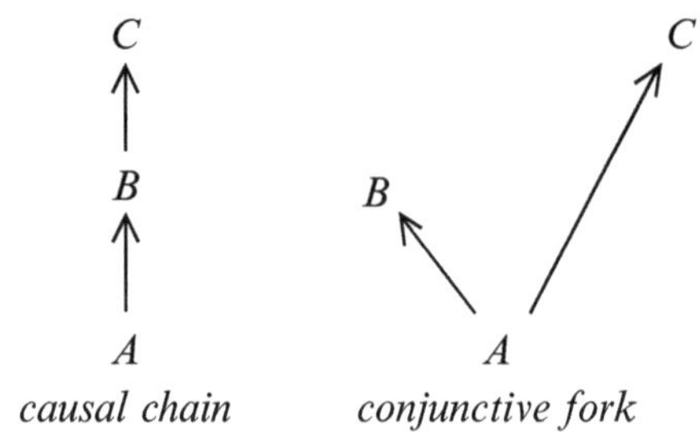

They are easily distinguished also with the help of ranking functions: suppose $\kappa(A) = \kappa(\bar{A}) = 0$, so that we have to specify only the ranks conditional on A and $\bar{A}$. One specification is this:

$\kappa(. \mid A)$	C	$\bar{C}$
B	0	1
$\bar{B}$	2	1

$\kappa(. \mid \bar{A})$	C	$\bar{C}$
B	1	2
$\bar{B}$	1	0

(1) *causal chain*

[21] One may wonder why A is not required to immediately precede B. But clearly, this is inadmissible as long as the frame U may contain any variables whatsoever, and may thus miss variables that are intuitively causal intermediates. I return to this point in Section 3.6.

A is here a direct cause of B in w (where $A \cap B \cap C = \{w\}$), in fact a necessary and sufficient one (because $\kappa(B \mid A) = \kappa(\overline{B} \mid \overline{A}) = 0$ and $\kappa(\overline{B} \mid A) = \kappa(B \mid \overline{A}) = 1$), and indeed the only one due to temporal order. B is a direct cause of C in w, again, a necessary and sufficient one (because $\kappa(C \mid A \cap B) = \kappa(\overline{C} \mid A \cap \overline{B}) = 0$ and $\kappa(\overline{C} \mid A \cap B) = \kappa(C \mid A \cap \overline{B}) = 1$), and indeed the only one because C is independent of A given B as well as given B (i.e., the figures just stated would be the same if A were replaced by $\overline{A}$ – this is what probability theory refers to as the Markov property). So, we have here an example for a causal chain, in fact the simplest one in which the ranks simply count how many times the obtaining causal relations are violated (in the sequence A, $\overline{B}$, and C, for instance, two such violations occur).

Another specification is this:

$\kappa(. \mid A)$	C	$\overline{C}$
B	0	1
$\overline{B}$	1	2

$\kappa(. \mid \overline{A})$	C	$\overline{C}$
B	2	1
$\overline{B}$	1	0

(2) *conjunctive fork*

Again, A is the only direct cause of B in w, in fact a necessary and sufficient one (because $\kappa(B \mid A)$ etc. are the same as in (1)). But now, A is also a necessary and sufficient cause of C (because $\kappa(C \mid A \cap B) = \kappa(\overline{C} \mid \overline{A} \cap B) = 0$ and $\kappa(\overline{C} \mid A \cap B) = \kappa(C \mid \overline{A} \cap B) = 1$), and the only one because C is independent of B given A as well as given $\overline{A}$ (i.e., the figures just stated would be the same for $\overline{B}$ instead of B). We might also say that A screens off B from C. So, we now have the simplest example for a conjunctive fork where ranks again just count the violations of causal relations; the more violations, the more disbelieved.[22] One should note, though, that in both cases the causal relations could be realized by many different distributions of ranks.

One may wonder whether the relevant circumstances are now extremely embracive, much more than intuition requires. The reason is that we have constructed 'relevant' extremely weak. Thus, a lot is relevant. Indeed, all of $w_{< B, \neq A}$ is relevant for the causal relation between A and B, but, as we might say, only potentially relevant on purely temporal grounds. The crucial advantage of this construal is, however, that it is free of any circularity. On this basis we may then search for more restrictive interpretations of 'relevant' which are hopefully provably equivalent to this construal.

The search is indeed successful. We may distinguish five narrower senses of relevant circumstances (for details see my 1990a, sect. 4, [here: sect. 2.4]), where I

[22] The term 'conjunctive fork' has been introduced by Salmon (1980), in distinction to what he calls 'interactive forks'. It is still a matter of debate whether the latter can and should be explained away; cf., e.g., Martel (2003). My framework, in any case, cannot represent interactive forks as intended by Salmon.

have investigated the issue in relation to probabilistic causation). Three of them are provably equivalent to the richest sense above. According to the fourth, the circumstances for the direct causal relation from *A* to *B* consist just of all the other direct causes of *B*, as suggested at the beginning of this section. This is equivalent to the other ones only under special conditions. The fifth sense, finally, is provably equivalent only in the case of necessary and/or sufficient causes. 'Equivalent' means here that the cause's raising of the probability or the rank of the effect is exactly the same given the narrower circumstances as given the richest circumstances. These are essentially satisfying results completing in my view the refutation of the circularity objection of Cartwright (1979), and they perfectly carry over to deterministic causation as explained here.

So far, we have dealt only with direct causation. How are we to extend our account to causation in general? I agree with Lewis (1973b, 2000) and many others that we should respect our structural intuition that causation is transitive. It goes without saying that direct causes are causes. And, clearly, causal relations should not further extend than direct causal relations – at least as long as we consider only discrete time. The three assumptions entail in fact that causation is the transitive closure of direct causation.

Definition 7: *A* is a *cause* of *B* in *w* relative to κ (or β) if and only if there are $A_1,\ldots, A_n$ $(n \geq 2)$ such that $A_1 = A$, $A_n = B$, and, for all $i = 1,\ldots, n\text{-}1$, A_i is a direct cause of A_{i+1} in *w* relative to κ (or β).

This allows for a lot of causes. Intuitively, though, we speak of much less. This is no cause for worry, however, as has been often observed. Intuitively, we speak of surprising or important causes, of the crucial or most informative cause, etc. But all this belongs to the pragmatics of causal talk. And if there is any hope of doing justice to the pragmatics, it is certainly only by first developing a systematic theory of causation that abstracts from pragmatic considerations and then trying to introduce the relevant distinctions. My interest is the former, not the latter.

However, there are various real prices to pay for this definition, and this is why we find so much uncertainty in the literature about this issue, mainly, but not only in the camp of probabilistic causation. One important price is that we thereby decide against the deeply entrenched intuition that causal chains should be something like Markov chains; the relevant conditional independences are not guaranteed by Definition 7. Another important price is that the basic idea that a cause is positively relevant to its effect under the obtaining circumstances, though useful for explicating direct causation, does not generally hold; an indirect cause may well be even negatively relevant to its effect.[23] These prices may well seem too high.[24]

The predicament should not be solved by merely pondering about which intuition is weightier or fits the examples better. A more theoretical solution is called

[23] The mutual incompatibilities of the three intuitions are thoroughly explained in my (1990a), sect. 5 [here: sect. 2.5]. What I say there for the probabilistic case again applies just as well to the deterministic case.

[24] Hall (2000) thoroughly discusses similar conflicts and draws more complicated conclusions. See also Lewis (2000, pp. 191ff.).

for. In my view the following theoretical maxim is decisive: Whenever there are several plausible explications of some notion we are interested in, the theoretically most enlightening procedure is to look for the weakest of these explications; only thereby we can gain theoretical insight about the conditions under which the stronger explications apply as well.

Concerning causation, it is obviously the transitive closure of direct causation that yields the weakest or widest permissible causal relation (within discrete time). The other intuitions, by contrast, would lead to stricter causal relations permitting only shorter causal chains. Thus, the maxim just stated speaks in favor of Definition 7.

Satisfaction of the maxim further demands, then, investigating the conditions under which causal chains as specified in Definition 7 have the desired stronger properties. Section 6 of my (1990a) [here: sect. 2.6] contains such an investigation in probabilistic terms. But again, the results obtained there fully carry over to the deterministic case. They do justice to our intuitions to an arguably sufficient extent.

3.5 Redundant Causation

We have already seen that counterfactual analyses and my ranking theoretic account of causation do equally well in distinguishing between causal chains and forks. Let us therefore look at more discriminatory cases. Counterfactual analyses always had a hard time with the various forms of redundant causation. Hence it is interesting to see how these can be handled by the account proposed here.

A and *B* *redundantly* cause *C* iff it holds: if neither *A* nor *B* had realized, *C* would not have occurred; but if one of *A* or *B* had not realized, in the presence of the other *C* would still have occurred. The following ranking (in terms of belief functions), which restrict themselves to necessary and/or sufficient causes, may be instructive:

$\beta(C \mid .)$	B	$\overline{B}$
A	1	−1
$\overline{A}$	−1	−1

(3) *joint necessary and sufficient causes*

$\beta(C \mid .)$	B	$\overline{B}$
A	1	0
$\overline{A}$	0	−1

(4) *joint sufficient, but not necessary causes*

$\beta(C \mid .)$	B	$\overline{B}$
A	1	1
$\overline{A}$	1	−1

(5) *redundant causes*

One problem with redundant causation is that according to a naive counterfactual analysis neither *A* nor *B* is a cause of *C*. The deeper problem, though, is that redundant causation comes in various forms. In cases of symmetric overdetermination we tend to say that both, *A* and *B*, cause *C*, whereas in cases of asymmetric preemption we want to deny that the preempted cause is a cause. However, as long as we present things as in figure (5) there is no way to give *A* and *B* different roles.

In his (1986d, pp. 193–212) Lewis paradigmatically discusses various strategies to settle the issues. One strategy is fine-graining of events. The prince plays the mandoline and simultaneously sings a love song to wake the princess. One musical part would have sufficed for waking. Thus this is a case of overdetermination. But hearing both, the princess wakes up in a slightly different way, which is hence *jointly* caused by the simultaneous performances. Likewise, the poison in the famous desert traveler's keg would have killed him, but it is preempted by the hole in the keg. He rather dies of thirst, and that's a different death not producible by the poison. Lewis (1986d, pp. 197–199) explains why he does not want to fully rely on this strategy, and I agree.

He goes on to discuss the other strategy, fine-graining of causal chains.[25] There he arrives at quite complicated conclusions. Let us consider cases of *overdetermination* first. In (1973b), footnote 12, Lewis declares such cases as useless as test cases because of lack of firm naive opinions about them, something he almost literally repeats in (2000, p. 182). In his (1986d, pp. 207ff.) he is more optimistic and agrees with Bunzl (1979) that fine-graining of causal chains shows most alleged cases of overdetermination to reduce either to ordinary joint causation or to preemption via an intermediate Bunzl event, as he calls it. The remaining cases, if there are any, might then be resolved by his doctrine of quasi-dependence. In (2000) he repudiates this doctrine. However, the uncertainty has no weight, because the remaining cases are 'spoils to the victor', anyway.

Thus, overdetermination puts a lot of strain on counterfactual analyses. This is in strange disharmony to the great ease with which at least prima facie cases of overdetermination can be produced; they abound in everyday life. Moreover, I do not believe in the uncertainty of pure intuition concerning these cases; uncertain intuitions are already tinged by uncertain theory. My intuition (or my theory) is quite determined. Why not take the prima facie cases at face value? I find it desirable to have a simple account of a simple phenomenon. And there is one. We need neither fine-graining of events nor fine-graining of causal chains, ranks offer a third method for dealing with problem cases. Definition 5b allowed for additional reasons, Definition 6 similarly allowed for additional causes, and this is exactly what overdetermining causes are. This is displayed in the following table:

$\beta(C \mid .)$	B	$\overline{B}$
A	2	1
$\overline{A}$	1	−1

(6) *overdetermining causes*

[25] Salmon (1980) already concludes that fine-graining of events and fine-graining of causal chains or, in his terms, 'the method of more detailed specification of events' and 'the method of interpolated causal links' are the two main strategies for dealing with problematic examples.

According to this table, each of *A* and *B* would have been a necessary and sufficient cause of *C* in the absence of the other; in the presence of the other each is still positively relevant to, i.e., a cause of *C*, but then each can only be an additional cause. This scheme, I find, fits naturally all the intuitive cases of causal overdetermination: Usually, if a sufficient cause occurs it is unbelievable that the effect does not occur; this applies to the cases (3) and (4) above. If in a case of overdetermination the effect does not occur, for some or no reason, at least two things appear to have gone wrong at once; and this is at least doubly unbelievable, as represented in (6).

The case of *preemption* appears even more complicated from the point of view of counterfactual analyses. Lewis discusses it already in his (1973b) where he considers normal cases, as it were, in which fine-graining of causal chains does the trick. The hole in the desert traveler's keg causes him to be thirsty, and the thirst eventually causes his death, but the thirst is in no way caused by the preempted poison. The poison would rather have caused a heart attack leading to death. But this causal chain never went to completion, it was cut off by the hole in the keg. This solution works in counterfactual as well as in ranking terms.

In (1986d, pp. 200ff.), Lewis calls the easy case *early* preemption and distinguishes it from *late* preemption where the causal chain from the preempting cause to the effect is somehow empty from the preempting action of the preempting cause onwards. Thus, in late preemption one does not find an event like the above traveler's thirst, and hence the easy solution does not work. According to Lewis (1986d), late preemption can even take three different forms. However, we do not need to discuss them here. Lewis himself takes the first two possibilities to be too far-fetched to worry about and rejects his (1986d) solution of the third possibility in his (2000).

Hall and Paul (2003) are not happy with Lewis' presentation of late preemption. For them, the mark of late preemption is that "at no point in the sequence of events leading from cause to effect does there fail to exist a backup process sufficient to bring about that effect" (p. 111). And they take this to be an obvious possibility. Their example is Lewis':

> Suzy and Billy, both throw rocks at a bottle. Suzy is quicker, and consequently it is her rock, and not Billy's, that breaks the bottle. But Billy, though not as fast, is just as accurate. Had Suzy not thrown, or had her rock somehow been interrupted mid-flight, Billy's rock would have broken the bottle moments later. (Hall and Paul 2003, p. 110)

Now the obvious asymmetry between Suzy and Billy is the temporal one. Lewis (2000) argues that it does not matter whether Suzy's and Billy's breaking the bottle are taken as two versions of the same event or as two alternative events. In any case, one must look at the fine-grained alterations, and then the case is not different from early preemption. Suzy's rock touches the bottle, whereas Billy's does not, and thus the causal chain from Suzy to the bottle goes to completion, whereas the one from Billy is cut.

However, Hall and Paul declare the temporal asymmetry to be inessential. They continue:

> It is perfectly easy to construct late preemption examples in which, had the cause not occurred – or indeed, had any of the events connecting the cause to the effect not occurred

– the effect would have occurred at exactly the same time, and in exactly the same manner.… for example, suppose that the signal from *C* [= the preempting cause] exerts a slight retarding force on the signal from *A* [= the preempted cause]. Pick any point before this signal from *C* reaches *E* [= the effect], and ask what would have happened if, at that time, the signal had been absent. Answer: the signal from *A* would have accelerated, and we can stipulate that it would have accelerated enough to reach *E* at exactly the time at which the signal from *C* in fact reaches *E*. (Hall and Paul 2003, pp. 112f., brackets added by me)

This is not perfectly easy, it is highly contrived. What is worse, the retarding effect of the chain from *C* to *E* on the chain from *A* to *E* must become smaller and smaller and converge to 0. Otherwise, there must be a time at which it is too late for the chain from *A* to arrive at *E* at the same time as the chain from *C*. Hence, despite the retarding force of *C*, the chain from *A* arrives at *E* at exactly the same time as the chain from *C*. And then there is no reason to take the chain from *A* to *E* as preempted; the case rather seems to be one of symmetric overdetermination.

One may wonder, hence, whether there are any convincing cases of late preemption which, by definition, have to be such that fine-graining of causal chains does not reveal an asymmetry. Yes, there are. So far, we have considered only cases of preemption which turned out to be cases of *cutting* (possibly after fine-graining) and thus not of late preemption. However, Schaffer (2000) has forcefully argued that there is also preemption by *trumping*. In Lewis' words:

The sergeant and the major are shouting orders at the soldiers. The soldiers know that in the case of conflict, they must obey the superior officer. But as it happens, there is no conflict. Sergeant and major simultaneously shout 'Advance!'; the soldiers hear them both; the soldiers advance. Their advancing is redundantly caused. … But the redundancy is asymmetrical: since the soldiers obey the superior officer, they advance because the major orders them to, not because the sergeant does. The major preempts the sergeant in causing them to advance. The major's order *trumps* the sergeant's. (Lewis 2000, p. 183)

Schaffer insists that his examples should be taken at face value; they are 'intuitively clear' and 'empirically and pretheoretically plausible'.[26] And Lewis concurs:

We can speculate that this might be a case of cutting. Maybe when a soldier hears the major giving orders, this places a block somewhere in his brain, so that the signal coming from the sergeant gets stopped before it gets as far as it would have if the major had been silent and the sergeant had been obeyed. Maybe so. Or maybe not. *We do not know one way or the other. It is epistemically possible*, and hence it is possible simpliciter, that this is a case of preemption without cutting. (Lewis 2000, p. 183, my italics)

Schaffer shows that four variants of the counterfactual analysis founder at trumping. In response Lewis proposes a fifth which returns again to the strategy of fine-graining events (or rather alterations). In my terminology, he proposes not to look

[26] In my (1983a, ch. 3), I have discussed a structurally similar example. It was a case in which it is unclear who trumps whom. Schaffer's main example is one with Merlin and later on Morgana casting a spell to turn the prince into a frog and a wizard's law to the effect that the *first* spell cast on a given day match the enchantment that midnight. In that case Merlin trumps Morgana. As Schaffer argues such a law is even compatible with Lewis' best-system analysis of laws. But suppose we live in a world in which each day when two wizards cast a spell they, perhaps accidentally, cast the same spell. In this case it is not clear who trumps whom. I was unsure what to think about such a case.

at binary variables ('whether-on-whether dependence'), but rather at more-than-two-valued variables ('how-when-whether-on-how-when-whether dependence'). This may be successful with the major and the sergeant. However, Suppes (1970, ch. 5), was the first to attempt to specify causal relations between multi-valued variables. This attempt was heroic, but not well received. Indeed, the probabilistic camp prefers to talk only about causal dependence between variables and to be silent on causation between events (cf., e.g., Spirtes et al. 1993). Lewis now also favors talking about causal dependence between variables. Insofar I agree with the criticism of Collins (2000, sect. IV), that Lewis changes the topic. Moreover, why should there be no trumping with respect to binary variables? No reason; Lewis would have to argue that the fragility of events (in his sense) entails that there are no binary variables (in my sense).

We seem to be on the wrong track. Fine-graining of causal chains is disallowed by definition, fine-graining of variables or events helps in some cases, but not necessarily in all. However, ranks again offer a straightforward account of trumping. Look at the following table:

$\beta(C \mid .)$	B	$\overline{B}$
A	2	2
$\overline{A}$	1	−1

(7) *trumping*

Here, A is a cause of C w.r.t. β independently of B, though only an additional one in the presence of B, whereas B is no cause of C in the presence of A, but a sufficient cause in the absence of A. This matches well the story of the major and the sergeant. The soldiers' disobedience to the major's orders is more incredible than their disobedience to the sergeant's orders.

The reason why these simple accounts of overdetermination and trumping are available to me, but not to any counterfactual theory is obvious: ranking functions specify varying degrees of disbelief and thus also of positive belief, whereas it does not make sense at all, in counterfactual theories or elsewhere, to speak of varying degrees of positive truth; nothing can be truer than true. Hence, nothing corresponding to the schemes (6) and (7) is available to counterfactual theories. In fact, ranking functions have so many more degrees of freedom that I am confident that they are able to account for all kinds of recalcitrant examples. Still, I would like to emphasize that I am not just playing around with numbers. Ranking functions have a perfectly clear epistemological interpretation,[27] and in all formal representations of examples the ranks must be specified in a way which is at least plausible.

[27] They can even be measured on a ratio scale by multiple contractions, as Matthias Hild has first shown; cf. my (1999a). [Or see now Hild and Spohn (2008).]

3.6 Objectivization

Ranks other than 0 and 1 as they appear in the cases (6) and (7) are not gratuitous, however; I do not want to deny that there is something puzzling about these cases. So let us finally consider the costs; this will help us to explain the puzzle without thereby depreciating the simple account given so far. The costs should have been clear all along; they consist in subjectively relativizing causation to an observer or epistemic subject. We did not get little in return, I think, indeed, things unattainable to others. There are many philosophers, though, who find the price too high. Therefore, I finally want to indicate at least that there are ways to reestablish objective causation on the subjective basis presented here.

Let me first emphasize, though, that this subjective relativization is not arbitrary. It is a response, indeed, as mentioned in Section 3.3, Hume's response, to a deep philosophical problem Hume has raised: namely, what is the nature of nomic and causal necessity? Long is the list of great philosophers who indulged in Hume's subjectivistic turn, equally long the list of those standing firmly objectivistic, and many tried to take some middle course, the most prominent perhaps being projectivism famously elaborated by Kant's transcendental idealism. Clearly, the issue is anything but settled. Hence, unintuitive as subjectivism certainly appears, it is not philosophically disreputable.

Lewis' response to the deep problem is his doctrine of Humean supervenience. Thereby, he hopes to achieve to give an objectivist account of the problematic ilk of laws, counterfactuals, and causation (and even objective probability) instead of just postulating that ilk (as does Armstrong 1983 concerning lawhood and Tooley 1987 concerning causation). This is not the place to discuss that doctrine (cf., however, Spohn forthcoming a). I only want to mention that it is not entirely clear how well Lewis succeeds in keeping his enterprise on the objectivistic side. His account of causation is as objective as his account of counterfactuals. The latter again turns on the objectivity of the similarity between possible worlds. There he admits at least that 'plenty of unresolved vagueness remains' (Lewis 1979a, p. 472).[28] Moreover, similarity essentially refers to laws the objectivity of which he tries to save by his 'best-system analysis of laws'. However, Lewis himself acknowledges that 'best' is quite a human category, and he consequently tries to dissolve subjectivistic implications (cf. Lewis 1994b, pp. 478ff.). So, there is at least cause for concern.

On the other hand, if one starts right on the subjectivistic side as I do, one should at least attempt to oblige to objectivistic intuitions as far as possible (but it is up to the objectivist to decide whether he is satisfied by the offers). I mentioned in Section 3.3 how I think Hume backed up his associationist by a regularity account of causation. If association is replaced by ranks, a more complicated story must be told. Indeed, the objectivization of the account of causation given so far has two aspects which I can only indicate here.

---

[28] At this point I was more attracted then by the 'epistemic approach to conditionals' which Peter Gärdenfors developed since 1978 (see his 1988) and from which ranking functions descend.

The first is to eliminate the frame-relativity of the account. This may be done by appealing to the universal frame consisting of all variables whatsoever, though this appeal is doubtlessly obscure. The somewhat homelier method is to relate small worlds not to indescribably grand, but just to larger worlds, i.e., to conjecture that the causal relations obtaining relative to a small frame are maintained in the extensions of that frame. It should be a fruitful task, then, to investigate under which conditions the relations within a coarse frame are indicative of those in the refined frame.[29]

The second and main aspect of objectivization, however, pertains to the ranking functions. In my account, they played a role corresponding to that of regularities in the regularity theory of causation or to that of the similarity ordering of worlds in Lewis' counterfactual analysis, and they played it more successfully. However, the only interpretation I have offered for them is as subjective doxastic states. So, what we are seeking is a way of viewing them more objectively. Is there such a way?

For some of them, yes.[30] The basic idea is this: We may assume that the propositions generated by the given frame have unproblematic objective truth conditions. Ranking functions, however, usually don't have them. A ranking function may be said to be true or false according to whether the beliefs embodied in it are true or false. But this refers only to ranks being 0 or larger than 0, it does not confer objectivity to varying distributions of ranks larger than 0. Generally, though, we might say that ranking functions are objectivizable to the extent we succeed in uniquely associating them with unquestionably objective propositions.

There is such an association answering our present needs. First observe that a *causal law* L may be associated with each ranking function κ: define L as a big conjunction of material implications, of all implications of the form 'if A and $w_{<B,\neq A}$, then B', whenever A is a sufficient direct cause of B in w relative to κ, and all implications of the form 'if $\bar{A}$ and $w_{<B,\neq A}$, then $\bar{B}$', whenever A is a necessary direct cause of B in w relative to κ. So, L is the conjunction of all causal conditionals obtaining according to κ, reduced to material implications. Hence, L is simply a true or false proposition generated by the given frame.

The crucial question is whether a ranking function can be reconstructed from its associated causal law; our objectivization strategy works to the extent in which this is feasible. The reconstructibility is limited, of course; there are always many ranking functions with which the same causal law is associated. But there is a narrow class of ranking functions which uniquely correspond to their causal laws and may thus be assigned the same truth values as their associated laws. We may call them *fault counting functions*: for a given law L simply define κ_L such that for each $w \in W$ $\kappa_L(w)$ is the number of times the law L is violated in w. In figures (1)–(5) above, I have used such fault counting functions.

---

[29] Spirtes et al. (1993, chs. 6, 7, and 10) have a lot to say about the probabilistic side of this issue.

[30] In the following, I give a very rough sketch of what is worked out in formal detail in my (1993a) [here: ch. 5].

However, even then the unique reconstructibility, and thus the objectivization of ranking functions through causal laws, works only under two conditions: (i) a certain principle of causality is required to hold, and (ii) each direct cause must immediately precede its direct effect (cf. my 1993a, pp. 243–246) [here: pp. 129–132]. These conditions certainly invite further scrutiny and evaluation. Here, I shall confine myself to three concluding remarks:

First, it would have been natural to wonder why Definition 6, explicating direct causation, does *not* require the direct cause to immediately precede the direct effect. Generally, this requirement would have been clearly unreasonable. As long as we do not put any constraints on the frame to be chosen, the frame considered may well omit all intermediate members of a causal chain, and thus represent the causal relation between two temporally quite distant events as a direct one. Hence, it is interesting to see that the temporal immediacy returns in condition (ii) via the objectivizability of causal relations; it is objectivization which requires frames to be rich enough to always provide immediately preceding direct causes.

The second remark is that additional causes (and weak causes) cannot be objectivized according to this theory. The reason is that, if A is an additional cause of B in w relative to κ, the corresponding causal law contains only the material implication 'if $w_{<B,\neq A}$, then B'; this is what is believed in κ. Then, however, we cannot read off from the law whether or not A is positively relevant to B given $w_{<B,\neq A}$. This entails that it is impossible to objectivize my treatment of overdetermination and trumping in the schemes (6) and (7) above which crucially relied on additional causes.[31] Objectively, these cases *must* be explained away, as Bunzl and Lewis have succeeded to a large extent in the case of overdetermination. Lewis refuses to take the same route in the case of trumping, with the surprising hint at epistemic possibilities, which I have italicized in the last of the longer quotations above in Section 3.5. I suspect here a confusion of epistemic and metaphysical possibilities. Scheme (7) well represents the epistemic possibilities. Objectively, though, one has to inquire how trumping works; and then a more detailed story, about the brains of the soldiers or whatever, has to be told. Thus, there is trouble with overdetermination and trumping also according to my account. The point is, however, that I have both, a straightforward account of these cases as well as an explanation of our urge to explain them away.

The final remark is that all this entails a certain view of causal laws. The objectivization just sketched yields two things. On the one hand, it delivers the objectively true or false proposition L. Thus, causal laws reduce to mere regularities, as the regularity theorist always pleaded. On the other hand, it produces the objectivizable ranking functions uniquely corresponding to these propositions. This accounts for the modal (inductive, explanatory, or counterfactual) force of causal laws. It does not do so by simply postulating this modal force, as proposed by Armstrong

[31] (7) does clearly not represent a fault counting function. One may be tempted to think that (6) does; there seem to be two faults when an overdetermined effect does not occur. Objectively, though, there is only one fault in this case, only one event not occurring as expected.

(1983) thus provoking bewilderment as to how to distinguish presence from absence of the modal force. It rather gives a Humean explanation of that modal force via ranking theory and the appertaining theory of objectivization by uniquely associating with a causal law a characteristic inductive behavior encoded in the corresponding ranking function.[32] All this would have been entirely out of reach, however, without a general theory of inductive behavior or of doxastic dynamics applying to plain belief, as I have presented it in Section 3.3.

[32] I have elaborated on this view of laws in my (2002) [here: ch. 6], arguing that this characteristic inductive behavior is indeed the mark of lawlikeness.

Chapter 4
Bayesian Nets Are All There Is to Causal Dependence[†1]

4.1 Introduction

There are too many theories of causation to get into the focus of a small paper. But there are two in which I have a natural interest since they look almost the same: namely the theory of Clark Glymour, Peter Spirtes, and Richard Scheines, so vigorously developed since 1983[1] and most richly stated in Spirtes et al. (1993) (whence I shall refer to it as the SGS theory), and my own theory, published since 1978 in a somewhat irregular way. They look almost the same, but the underlying conceptions turn out to be quite dissimilar. Hence, the original idea for this paper was a modest one: simply to compare the philosophical basics of the two theories. However, no paper without a thesis! Therefore I have sharpened my comparison to the thesis written right into the title.

The plan of the paper is simple. Section 4.2 sets out the formal theory of Bayesian nets in an almost informal way, and Section 4.3 analyses the philosophical differences hidden in the common grounds. Section 4.4 briefly extends the comparison to the treatment of actions or interventions.

4.2 Causal Graphs and Bayesian Nets

Whenever we want to conduct a causal analysis in a given empirical field, we have to start by conceptually structuring this field. This is usually done by specifying a frame or a set U of variables characterizing the field. Each variable $A \in U$ can take

[†1] This paper was originally published in: M.C. Galavotti, P. Suppes, D. Costantini (eds.), *Stochastic Causality*, Stanford: CSLI Publications, 2001, pp. 157–172. It is reprinted here with kind permission of CSLI Publications.

[1] The acknowledgments of Glymour et al. (1987) report that the work on that book took about 4 years.

some value from the set of its possible values. Thus, by specifying a value for each variable in U we specify some possible small world, some way how the empirical field characterized by the frame U may realize.

Variables should be conceived here as specific and not as generic variables. A generic variable would be something like social status or annual income which may take different values for different persons at different times. However, it is hard to find any causal order among generic variables. One then finds causal circles – high social status tends to generate high annual income, and vice versa – and one even finds apparent self-causation – social status tends to reproduce itself.

By contrast, a specific variable is something like my social status today or my annual income in 1998, not conceived as it actually is, which is given by some particular figure, but conceived as something which may take any value, say, between 0 and 1 billion Euros. There is a proper causal order among specific variables. For instance, there is no self-causation. If my social status today is high, it tends to be high tomorrow as well (though there is no guarantee, see the sudden fall of politicians), but this is a causal relation between two different specific variables.

Indeed, the causal structure within the frame U of specific variables is neatly captured by a causal graph over U which is nothing but a *DAG*, a *directed acyclic graph* $\langle U,E \rangle$ with U being its set of nodes and E being its set of edges. That the graph is directed means that its edges are directed, i.e. that E is an asymmetric relation over U, and that it is acyclic means that the directed edges don't form circles, i.e. that even the transitive closure of E is asymmetric.

Let me give a standard example (used by Pearl 1998 and elsewhere): U consists of five variables:

A_1: season of a given year (spring, summer, fall, winter)
A_2: rain fall during the season (yes, no)
A_3: sprinkler during season (on, off)
A_4: wet pavement (yes, no)
A_5: slippery pavement (yes, no)

which we might plausibly arrange into the following DAG (if the variables refer to some place in Southern California).

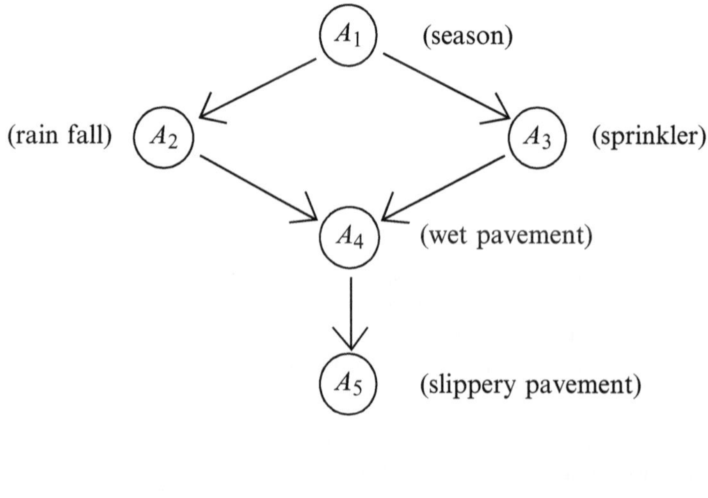

The DAG $\langle U,E\rangle$ becomes a *causal graph*, if the edges in E are given a causal interpretation, i.e. if an edge $A \rightarrow B$ is interpreted as stating that A is directly influencing B, or that B is directly causally dependent on A, within the given frame U. Thus, so far the DAGs simply express the formal properties of direct causal dependence.

Specific variables have a specific temporal location. Hence, the variables in U are temporally ordered. So I shall add the natural constraint that in any edge $A \rightarrow B$ of a causal graph A temporally precedes B. Some philosophers oppose, but this is not the place to discus their worries.

The next and crucial step is to introduce probabilities. The frame U generates, as mentioned, a space of possible small worlds the subsets of which may take probabilities according to some probability measure P. In particular, each event of the form $\{A = a\}$, stating that the variable A takes the value a, gets a probability. Accordingly, there is probabilistic dependence and independence among variables. More explicitly, we may define the sets X and $Y \subseteq U$ of variables to be *probabilistically independent* given or *conditional on* the set $Z \subseteq U$, i.e. $X \perp_P Y/Z$, iff for all x,y,z $P(X = x \mid Y = y, Z = z) = P(X = x \mid Z = z)$, i.e. iff, given any realization z of Z, any event about X is probabilistically independent of any event about Y.

Following SGS, we can state two conditions concerning a DAG $\langle U,E\rangle$ and a measure P for U, in which $Pa(A)$ denotes the set of parents or immediate predecessors of the node A, $Nd(A)$ denotes the set of non-descendants of A, and $Pr(A)$ denotes the set of nodes temporally preceding A.

There is, first, the *Markov condition* (cf. Spirtes et al. 1993, pp. 53ff.) stating that for each $A \in U$ $A \perp_P Nd(A)/Pa(A)$, i.e. that each variable is independent from all its non-descendants given its parents. If the DAG agrees with the given temporal order this condition is equivalent to the apparently weaker condition that for each $A \in U$ $A \perp_P Pr(A)/Pa(A)$. This condition is also equivalent to the decomposability of P:

$$P(U = u) = \prod_{A \in U} P(A = a \mid Pa(A) = x),$$

where a and x, respectively, are the realizations of A and $Pa(A)$ according to the realization u of U. This decomposability harbors enormous computational advantages so ingeniously exploited by Pearl (1988) and others.

For instance, the above example satisfies the Markov condition iff

$$A_3 \perp_P A_2 \,/\, A_1,$$

$$A_4 \perp_P A_1 \,/\, \{A_2, A_3\}, \text{and}$$

$$A_5 \perp_P \{A_1, A_2, A_3\} \,/\, A_4,$$

or iff, for all $a_1, \ldots, a_5$ realizing $A_1, \ldots, A_5$

$$P(a_1, \ldots, a_5) = P(a_1)\ P(a_2 \mid a_1)\ P(a_3 \mid a_1)\ P(a_4 \mid a_2, a_3)\ P(a_5 \mid a_4).$$

There is, second, the *minimality condition* (cf. Spirtes et al. 1993, pp. 53f.) stating that no proper subgraph of the DAG $\langle U,E\rangle$ satisfies the Markov condition.

Following Pearl (1988, p. 119) a DAG satisfying the Markov and the minimality condition is called a *Bayesian net(work)*. In a Bayesian net, the parents of a node thus form the smallest set of variables for which the relevant conditional independence holds.

For instance, the above example satisfies the minimality condition iff none of the following independencies holds:

$$A_2 \perp_P A_1$$

$$A_3 \perp_P A_1$$

$$A_4 \perp_P A_2 \,/\, A_3 \text{ and } A_4 \perp_P A_3 \,/\, A_2 \text{ and}$$

$$A_5 \perp_P A_4$$

SGS further introduce a third condition, *the faithfulness condition* (cf. Spirtes et al. 1993, pp. 56), which is, however, more complicated and slightly less important so that I shall neglect it in the sequel.

So far, I have only introduced two distinct graph-theoretical representations: one of causal dependence between variables and one of conditional probabilistic dependence. However, the core observation of each probabilistic theory of causation is that there is a close connection between causal and probabilistic dependence, that the two representations indeed coincide, i.e. that *each causal graph is a Bayesian net*. Thereby, the Markov and the minimality condition turn into the *causal* Markov and the *causal* minimality condition. This means, to repeat, that the set of variables on which *A directly causally depends* within the frame *U* is *the* smallest set conditional on which *A* is probabilistically independent from all its other non-effects or, equivalently, from all other temporally preceding variables.[2] This assertion may indeed be used to *define* direct causal dependency within the frame *U*. At least I proposed to do so in Spohn (1976/78, sect. 3.3, in particular pp.117f.). The definitional equivalence also follows from the assumptions made by SGS.

So far there is perfect agreement between SGS and me. However, there are also differences: first, concerning the development of causal theory, and second, concerning the understanding of the basic theory thus laid out. I shall dwell on the second point, but let me briefly mention the main differences of the first kind.

In my work, I did not use, and did not even think of, any graph-theoretical methods. These methods, graph-theoretic representations of independence relations, so-called d-separation, etc., were essentially introduced and pushed forward by Judea Pearl and his group after around 1985 (cf. Pearl 1988, pp. 132ff.). I am enthusiastic about these methods. They add powerfully to the strength, beauty, and vividness of the theory. Of course they are richly used by SGS. What I did have, however, in Spohn (1976/78, sects. 3.2, 3.3), with some variations translated in Spohn (1980), was the above-mentioned probabilistic definition of direct causal dependence and the full theory of conditional probabilistic independence on which

[2] That there is exactly one such set is a consequence of the properties of conditional probabilistic independence.

this definition and the graph-theoretic methods rest, i.e. the graphoid and the semi-graphoid axioms, including the conjecture of their completeness (refuted by now) and the weaker conjecture of the completeness of the properties of direct causal dependence entailed by them (proved by now).[3]

Naturally, I wondered how the above account of causal dependence between variables may be founded on an account of causal relations between events or states of affairs or singular propositions. This is obviously philosophically important, but of little use in scientific and statistical methodology, and thus of no concern to SGS. The foundation seemed straightforward: the event $\{A = a\}$ is a *direct cause* of the event $\{B = b\}$ in the possible small world u if and only if both events occur in u, if $\{A = a\}$ precedes $\{B = b\}$, and if $\{A = a\}$ is positively relevant to $\{B = b\}$ according to P under the obtaining circumstances C, which are best identified with the event that all the variables preceding B (and differing from A, of course) take the values they take in u. Thus, the variable B directly causally depends on the variable A iff some event about A is a direct cause of some event about B in some possible small world. For a long time, I was under the influence of the view of Suppes (1970, p. 58) that probabilistic causation cannot be transitive. In Spohn (1990a) [here: ch. 2]. I changed my mind and started to prefer defining (direct or indirect) causation as the transitive closure of direct causation, though, as explained there, the issue is quite intricate.

Finally, in Spohn (1983a, chs. 5 and 6; see also 1988 [here: ch. 1]). I have proposed the theory of ranking functions, as they are called nowadays, which yield a perfect deterministic analogue to probability theory, to conditional probabilistic dependence and independence, to the theory of Bayesian nets, and thus to the above account of probabilistic causation, and I have suggested there that this is how deterministic causation should be analyzed.[4]

So I have always moved within the philosophical confines. By contrast, Judea Pearl and his collaborators have done impressive work developing and utilizing the whole theoretical field for the purposes of artificial intelligence in a most detailed and fruitful way. And SGS have done impressive work developing sound statistical methodology on a sound philosophical basis, a different and in many respects much more difficult endeavor which starts to be successful in the big statistical community. Though all this work is addressed, to a large extent, to other departments, it contains a lot of high philosophical interest. But there is no place to further expand on this.

4.3 About the Causal Import of Bayesian Nets

Let me turn, then, to the interpretational differences between SGS and me which are my main concern. For this purpose, let us look again at the proposed definition: the variable A directly causally depends, within the frame U, on all and only the

[3] For the conjectures see Spohn (1976/78, pp. 105, 119). For the positive and negative results see, e.g., the overview in Spohn (1994).

[4] A suggestion which I have coherently explained in English only in Spohn (2000b).

members of the smallest set of variables in *U* preceding *A* conditional on which *A* is probabilistically independent from all other variables in *U* preceding *A*. This definition hides two relativizations which deserve closer scrutiny.

First, direct causal dependence is obviously frame-relative according to this definition. The relativization would be acceptable, if it concerned only the direct/indirect distinction: what appears to be a direct causal dependency within a coarse-grained frame may well unfold into a longer causal chain within a more fine-grained frame. In this sense the frame-relativity is also accepted by SGS (cf. Spirtes et al. 1993, pp. 42f.). It's worse, however. The whole notion of causal dependence is frame-relative according to this definition: where there appears to be a direct or an indirect causal dependency within a coarse-grained frame, there may be none within a more fine-grained frame, and vice versa. This consequence seems harder to swallow.

The second relativization is better hidden. The talk of conditional independence refers, of course, to an underlying probability measure. Where does it come from?

It might come from reality, so to speak. This raises the question, of course, how to conceive of objective probabilities – a large question which I want to cut short by simply saying that they should best be understood as chances or propensities. This, however, is obscure enough. I have three reservations about using chances in the present context.

The first reservation is that chances are hard to find. But we want, and do, apply the probabilistic theory of causation almost everywhere, and in particular to fields where it is very unclear whether genuine chances exist. Almost all examples of SGS are from social sciences, medical sciences, etc. Maybe, if basic physics is chancy, everything else in the universe is chancy, too. But if so, we suffer from a complete lack of understanding of the chances, say, in economics or medicine, and whatever the probabilities are we are considering in these fields, they are certainly not suchlike chances.

A further reservation is that I find it very awkward in the meantime to talk of chancy events being caused (as has been most forcefully argued by Railton 1978). The idea behind genuine chances is that of partial determination without further determinability, and the idea behind causation is that of full determination. So, it's rather only the chances of events which are fully determined or caused and not the chancy events themselves. I certainly agree with Papineau (1989, pp. 308, 320) that we need a probabilistic theory of causality in any case and that it is then largely a matter of terminology whether we should say that something that has raised the chance of an occurring event is among the causes of that event or only among the causes of the chance of that event. Still, my terminological preference is clear.

Mainly, however, my reservation is due to the fact that the above theory would be doomed as an analysis of causation if it starts with the notion of chance. The philosophical point of the enterprise is to elucidate the obscure notion of causal necessitation or full determination, and then the notion of chance or partial determination is presumably part of the package to be elucidated. To analyze the one in terms of the other does not seem helpful. I rather hoped to get a grip somehow on both notions together, on causation and chance.

If objective probabilities are thus to be avoided in the above definition of causal dependence, the only alternative is to use subjective probabilities. This is certainly an option, indeed the one I always preferred. However, it clearly amounts to a further relativization of causation to an epistemic subject or to its epistemic state. The above definition then says not what causal dependence *is*, but only how it is *conceived* by some epistemic subject.

This relativization is certainly in good Humean spirit. But even Hume who maintained it so bravely, was ambiguous and denied it at other places. Likewise, I have never been happy with these relativizations, but I did not get clear about how to get rid of them and what else to say about causation.

For instance, I could not see that the manipulability account of causation is of any help. Whether to explain the notion of something influencing something else by the notion of myself influencing something else or the other way around does not seem to make much of a difference. Moreover, actions, goals, etc. always deemed to me extraneous to the topic of causation. I found no help in the process theory of causation of Salmon (1984). Rich and illuminating as it is, its fundamental distinction between processes and pseudoprocesses leads in a large circle back to counterfactuals. So why not immediately engage into a counterfactual analysis of causation? Alluding to mechanisms is unhelpful since mechanisms seem to be nothing but suitably refined causal chains. The idea of energy transfer seems entirely beside the point when it comes to causation in the social sciences. Postulating a second-order universal of causal necessitation adds little in itself. And so forth.

So, the crucial question persisted: what else to say about causation? Only slowly it dawned upon me that I might, and indeed should, turn the inability to say more into a positive thesis. In a sense which I shall explain below there is nothing more to say about causation then I already did!

By contrast, these relativizations are plainly unacceptable to SGS, and this is, I admit, only common-sensical. They do not want, and do not pretend, to give an analysis of causation. They rather want to develop a theory over some undefined notion of causation, just as statistics is a big theory over some undefined notion of probability. So, in effect, they develop a theory jointly about causation and probability (cf. Spirtes et al. 1993, pp. 5ff., 41ff.).

Their attitude, then, is this. Causal dependence, whatever it is, is ubiquitous. However, we are able to model only small parts of empirical reality by tentatively describing them by causal graphs and statistical hypotheses. The basic axiom of this model building is that these causal graphs are Bayesian nets, i.e. satisfy the Markov and the minimality condition introduced above (and also the faithfulness condition). The frame-relative definition of direct causal dependence is thus only an equivalence following from their axiom and has no explicative status. This shows clearly that their underlying conception is quite different from mine.

The natural follow-up question is: why should the axiom hold? SGS do not claim universal validity. The Einstein-Podolsky-Rosen paradox and quantum entanglement in general seem to provide a noticeable exception on which, however, I would like to be silent as well. But this does not diminish the success of the axiom elsewhere. They summarize their defense of the axiom in the following way:

> The basis for the Causal Markov Condition is, first, that it is necessarily true of populations of structurally alike pseudo-indeterministic systems whose exogenous variables are distributed independently, and second, it is supported by almost all of our experience with systems that can be put through repetitive processes and whose fundamental propensities can be tested. (Spirtes et al. 1993, p. 64)

I am not quite satisfied by this. The first defense points to an interesting and important fact, but defers the issue to deterministic causation. And the second defense shows that we have a lot of intuitive skills and scientific knowledge in order to select appropriate sections of reality. But they continue the summary of their defense:

> Any persuasive case against the condition would have to exhibit macroscopic systems for which it fails and give some powerful reason why we should think the macroscopic natural and social systems for which we wish causal explanations also fail to satisfy the condition. It seems that no such case has been made.

Indeed, it is interesting how they argue about specific putative counter-examples. Their strategy is always the same: whenever there is a causal graph which is not a Bayesian net, there exists a suitable causal refinement of the original graph which is a Bayesian net. In the specific cases they discuss I find their argument convincing, for instance, when they reject the interactive forks of Salmon (1984, pp. 168ff.).[5]

But why should this strategy always work (with the disturbing exception already noticed)? Two possible explanations come to my mind. One possibility is that we have an *independent* notion of causation, and using that notion we generally happen to find suitable refined causal graphs which are Bayesian nets. But surely it is incredible that we merely happen to find these refinements. There should be a general reason for this success. Here one might continue in the following way.

Basically, causation is deterministic, and then, given a specific conception of deterministic causation, we can specify very general conditions under which such causal relationships get displayed in Bayesian nets. This is the strategy pursued by Papineau (1985). It is also the strategy behind SGS' theorem that (linear) pseudo-indeterministic systems, i.e. systems with a suitable (linear) deterministic extension in which the exogenous variables are independently distributed, satisfy at least the Markov condition (cf. Spirtes et al. 1993, pp. 58ff.).

This strategy is very illuminating as far as it goes. But I doubt that it works in the end. My reason for my doubt is that I don't believe that we have a workable theory of deterministic causation which could play this independent role. Rather I believe, as already indicated, that all our problems and arguments about probabilistic causation turn up all over again when deterministic causation is at issue.[6]

---

[5] This rejection is of vital importance to their and my enterprise. If interactive forks were not only an apparently unavoidable, strange exception, as in the EPR paradox, but a perfectly normal and unsurprising phenomenon, as Cartwright (2001) argues again, then Bayesian nets would lose much of their interest, and my title thesis would simply be wrong.

[6] See Spohn (2000b) for some substantiation of this claim [or here: ch. 3].

Hence, I don't think that the strategy presently envisaged works on the basis of deterministic causation. And I do not see any other independent notion of causation for which it has been, or could be, argued that it generally exhibits itself in Bayesian nets. So I am indeed skeptical of the whole approach.

How else might we explain that there always are suitably refined causal graphs which are Bayesian nets? The only other possibility which comes to my mind is to say that there is *no* independent notion of causation to be alluded to, that this *is* our understanding of causation. In other words: it is the structure of suitably refined Bayesian nets which decides about how the causal dependencies run. *We cannot regard B to be causally dependent on A unless we find a sequence of arrows or directed edges running from A to B in a suitably refined Bayesian net and unless, of course, this stays to be so in further refinements.* The last clause shows that the talk of suitable refinements is unnecessarily vague. In the final analysis it is the all-embracive Bayesian net representing the whole of reality which decides about how the causal dependencies actually are.

Of course, we are bound to have only a partial grasp of this all-embracive Bayesian net. Therefore it is important to have theorems telling under which conditions and to which extent our partial grasp is indicative of the final picture, that is, under which conditions the causal relations in a fine-grained Bayesian net are maintained in coarsenings. The theorem of SGS about pseudo-indeterministic systems is a good example. Clearly, however, the conditions to be specified in such theorems cannot be but assumptions about the shape of the final picture.

These remarks indicate how I propose to get rid of the two relativizations of causal dependence explained above. If the notion of causal dependence is prima facie frame-relative, we can eliminate this relativity only by moving into the all-embracive frame containing all variables needed for a complete description of empirical reality. The all-embracive Bayesian net, then, does not distribute subjective probabilities over this frame in some arbitrary way. Rather, full information about the maximal frame should be accompanied by full information about the facts, so that subjective probabilities are optimally informed and thus objective at least in the sense proposed by Jeffrey (1965, ch. 12). In this way, the relativization of causal dependence to an epistemic state is eliminated as well.[7]

I am well aware that by referring to the all-embracive frame and to objective probabilities in this sense I am referring to entirely ill-defined and speculative entities. It is clear, moreover, that all causal theory can only deal with specific frames and specific Bayesian nets and their relations. Still, I find it philosophically inevitable to refer to such ill-defined entities, and the philosophical task is to try to strip them at least of some of their obscurity.

This finally explains my claim that in a sense there is no more to causal dependence than the above definition: this definition *with* its relativizations does all the theoretical work, and the move just proposed to eliminate these relativizations

[7] Or at least reduced. My vague formulations do not allow conclusions concerning the uniqueness of the objective probabilities thus understood.

and thus to say what causal dependence really is only a philosophical appendix adding no substantial theoretical content.

This needs two qualifying remarks. The first remark is that, even in the sense intended here, it is not wholly true that Bayesian nets exhaust all there is to the notion of causal dependence. I have hardly addressed the relation between time and causation and not at all the relation between space and causation, and both add considerably to the notion of causal dependence, i.e., to how the all-embracive Bayesian net has to look in the final analysis. By contrast, I have already expressed my doubts that such notions as action, mechanism, energy transfer, or process further enrich the notion of causal dependence. Anyway, whatever the further aspects of the notion of causal dependence, the theory of Bayesian nets covers its central conceptual content.

The second remark is that one must be very clear about the status of my claim that unrelativized, i.e. actual causal dependence is relativized causal dependence relative to the all-embracive frame and Jeffreyan objective probabilities. This is very much like the claim of Putnam (1980) that the ideal theory cannot be false. Both assertions are *a priori true*. Something is a priori true iff it cannot *turn out* to be otherwise. By contrast, something is necessarily true iff it cannot *be* otherwise. Hence, there is nothing metaphysically necessary about the truth of the ideal theory. The world could easily be different from what the ideal theory says even given the truth of the ideally complete evidence on which it relies. But the world cannot turn out to be different from what the ideal theory says because this theory exhausts all factual and counterfactual means of evidence.

Similarly, causal dependence cannot turn out to be different from what it is in the all-embracive Bayesian net. But again this is only an epistemological claim, slightly more contentful than Putnam's claim, which has nothing to do with the metaphysics of causation. Indeed, I was completely silent on the latter. If I had wanted to say something about the metaphysics, I should have entered the whole of science, and then, of course, much more could be said.

Let me emphasize once more that I believe exactly the same story to apply to deterministic causation. There, again, Bayesian nets form the conceptual core of causal dependence, the only difference being that Bayesian nets are now constructed not in terms of probability measures, but in terms of ranking functions, their deterministic analogue.

4.4 Actions and Interventions

When I started to write about causation in Spohn (1976/78), my real interest was decision theory. Therefore action variables were part of my picture from the outset. More precisely, I considered not only a set U of occurrence variables, as I called them for want of a better term, but also a set V of action variables. Thus the frame considered was always $U \cup V$. In decision contexts the task is to find the optimal action, action sequence, or strategy, and once one has found it, one starts executing it (unless weakness of will interferes). Hence, it does not make sense to assume the decision maker to have a probabilistic assessment of his *own* possible actions. For this reason I postulated that a decision model must not explicitly or implicitly contain any probabilities for the action variables in V (and thus took opposition to Jeffrey 1965).[8] So instead of considering one probability measure P over $U \cup V$ I followed Fishburn (1964, pp. 36ff.), and assumed a family $\{P_v\}$ of probability measures over U, parametrized by the possible action sequences v realizing V, which were to express probabilities of events over U conditional on v. It is straightforward then to extend the notion of conditional dependence and independence to such a family $\{P_v\}$, with the effect that relativized causal dependence can be explained relative to the frame $U \cup V$ in the way sketched above and that a causal graph over $U \cup V$ can be constructed which is a Bayesian net (in a slightly generalized sense). Consequently, all action variables are exogenous variables in that graph (but there may be more), and they introduce an asymmetry into the independence relation since occurrence variables can be (conditionally) independent from action variables, whereas the question whether an action variable is independent from another variable cannot arise simply because no probabilities are assigned to actions.[9]

A natural application of this account is Newcomb's problem, of course, which is basically a problem about the relation between probability and causality. As I observed in Spohn (1978, sect. 5.1), the account just sketched entails that among the four combinations of probabilistic dependence on and independence from action variables on the one hand and causal dependence on and independence from action variables on the other exactly one is impossible, namely the case that something is probabilistically dependent on, but causally independent from the action variables. But this, and only this, was the case Nozick (1969) worried about. Accordingly, there is no Newcomb problem, and two-boxing emerges as the only rational option. I still think that this observation is basically sound.[10,†2]

When studying causation more closely later on, I neglected action variables for the sake of simplicity. But one can observe a growing interest in the explicit consideration of action variables in the theory of causation and the surrounding statistical and AI literature which certainly relates also to the triumph of the graph-theoretical methods. Thus, a theory of intervention or manipulation has become also a central part of the SGS theory.

Their picture is this (cf. Spirtes et al. 1993, pp. 75ff.). They start with an unmanipulated graph, as they call it, over a frame U. Then they consider one or several

[8] See also our exchange in Spohn (1977) and Jeffrey (1977). I still believe my principle "no probabilities for one's own options" to be correct and full of important consequences. It expresses, for instance, the most basic aspect of the freedom of the will since it exempts the will, i.e. willful actions, from causes, at least in the eyes of the agent. Cf. Spohn (1978, p. 193).

[9] For all this see Spohn (1976/78, sects. 3.1, 3.2).

[10] Of course, I have become aware that this observation does not exhaust the problem. It is a rich problem indeed, and at least in the iterated Newcomb problem I have converted to a one-boxer. Cf. Spohn (2000c).

[†2] In fact, I have converted to a one-boxer even in the single-shot case; cf. Spohn (2003c).

manipulations which they represent through a set *V* of variables enriching the original frame *U* in such a way that they are exogenous variables in the enriched or combined graph and directly manipulate or act on some variables in *U*. These intervention variables in *V* have a zero state which says: "Don't interfere!" or "Let it go!" If they take this state, the original unmanipulated graph stays in force. But if they take another state they enforce a new distribution on the directly manipulated variables irrespective, and thus breaking the force, of the ancestors of the directly manipulated variables in the unmanipulated graph. In the simplest case the new distribution will outright dictate a certain value to the directly manipulated variables. Their so-called manipulation theorem says then how to compute all the probabilities of the manipulated graph from the unmanipulated graph and the new distributions of the directly manipulated variables. All this provides also a nice and precise explanation of the epistemological difference between observing a variable to take a certain value and making it to take that value[11] which entail two quite different belief revisions (cf. also Meek and Glymour 1994, pp. 1007ff.).

However, the SGS theory of manipulation strikes me as being essentially equivalent with my old proposal just sketched. I did not distinguish a particular unmanipulated graph or, what comes to the same, a special zero state of the intervention variables, because there is not always a natural zero state – in the Newcomb situation you have to take one or two boxes, you cannot just let it go – and because non-interference or refraining seemed to me to be an action as well. One could, however, distinguish some values of action variables as such zero states in my framework and thus define the unmanipulated graph in the sense of SGS as the subgraph determined by these action variables taking their zero states. Their manipulation theorem then simply states the recursive decomposition of probabilities characteristic of Bayesian nets and their slight generalization to a probability family $\{Pv\}$.[12]

Again, a crucial difference lies in the fact that SGS build a very detailed statistical theory of prediction (of the effects of intervention) on their basic definitions.[13] Our basic agreement, however, is also displayed in our treatment of Newcomb's problem, where Meek and Glymour (1994, p. 1015) reach the same conclusion as the one I have sketched above.

---

[11] A distinction which has been observed also by Kyburg (1980).

[12] The comparison extends to Pearl (1998, sect. 4) which summarizes his work on the role of actions in Bayesian networks. His procedure superficially differs from SGS's. Instead of expanding the original to a manipulated graph he includes action variables in the original graph (which, however, may merely be observed, from outside, as it were), and for representing actions as choices enforcing a certain value of the action variables he mutilates the original graph by cutting out all edges ending in actions variables. The mutilation also leads to a changed probability distribution, the same as the one described by SGS in their manipulation theorem. In Spohn (1978, sect. 5.2) I considered the very same problem – how to turn a theoretically detached view of a set of variables which does not give action variables a special role into a practically relevant view which does respect the special role of actions for the agent? – and I arrived at the very same cutting procedure.

[13] This remark applies mutatis mutandis to the work of Judea Pearl.

To sum up: There is a large agreement between SGS and me in the formal basics of a probabilistic theory of causal dependence, including even the extension to actions or interventions. The main difference is that they abstain from any bold statement about what causation is, wisely so for their purposes, whereas I have advanced and argued for the, positive or negative, thesis that from an epistemological point of view the theory of Bayesian nets exhaust, with the caveats mentioned, the theory of causal dependence.

Chapter 5
Causal Laws are Objectifications of Inductive Schemes†1

And this paper is an attempt to say precisely how, thus addressing a philosophical problem which is commonly taken to be a serious one. It does so, however, in quite an idiosyncratic way. It is based on the account of inductive schemes I have given in (1988, 1990b) and on the conception of causation I have presented in (1980, 1983a, 1990a), and it intends to fill one of many gaps which have been left by these papers.

Still, I have tried to make this paper self-contained. Section 5.1 explains the philosophical question this paper is about; in more general terms it asks what might be meant by objectifying epistemic states or features of them and to which extent epistemic states can be objectified. The next sections introduce the basis I rely on with formal precision and some explanation; Section 5.2 deals with induction and Section 5.3 with causation. Within these confines, Section 5.4 attempts to give an explication of the relevant sense of objectification and Section 5.5 investigates the extent to which various features of epistemic states are objectifiable. The two most salient results are roughly that the relation "*A* is a reason for *B*" cannot be objectified at all and that the relation "*A* is a cause of *B*" can be objectified only under substantial, though reasonable restrictions.

What has all of this to do with probability? A lot. The paper trades on a pervasive duality between probabilistic and deterministic epistemology, between a probabilistic representation of epistemic states together with a theory of probabilistic causation and another representation of epistemic states which I call deterministic because it lends itself, in a perfectly parallel fashion, to a theory of deterministic causation.[1] Here I explicitly deal only with the deterministic side, but the duality should pave the way for further conclusions concerning objective probabilities and statistical laws. This outlook is briefly expanded in the final Section 5.6.

†1 Published in Jacques-Paul Dubucs (ed.), Philosophy of Probability, Kluwer, Dordrecht 1993, pp. 223–255.

[1] I have more fully presented this duality in (1983a) and (1988, sect. 7).

5.1 Is Causation Objective?

Objectivity has many different facets which call for many different explanations. One facet is truth. We think that what is true is objectively true, independent from any subjective point of view. In this sense it is an open issue whether causation is objective, whether causal statements are (objectively) true or false. The common intuition is affirmative, but it's not easy to philosophically account for it.

The issue initiates with David Hume. Indeed, it hides right in his two definitions of causation as what he calls a philosophical and a natural relation.[2] Causation as a philosophical relation is constituted by precedence, contiguity, and regularity; it is objective because precedence, contiguity, and the existence of a suitable regularity are objective matters. Whereas causation as a natural relation is constituted by precedence, contiguity, and association (of the effect with the cause in the mind of some epistemic subject); it is not objective because on this view causal statements as such are neither true nor false, but depend on the epistemic state of the subject.[3] It is an intricate exegetical issue precisely how Hume understands the relation between his two definitions.[4] The most plausible view is, roughly, that the associationist theory is conceptually more basic and is provided with an explanation by the regularity theory because it is the regularities which, to a large extent, shape our associations.[5] However, Hume is not free from ambiguity; in his response to the charge of an imagined realist that his notion of causation is not objective he quickly resorts from the associationist to the regularity theory.[6]

Since then the problem stays with us; and the ways sought to get out of it are too numerous to be counted here. I mention only some of them.

One may deny the problem by giving an outright objectivist account of causation. One may conceive of causation as a kind of physical ingredient of the world, e.g. as energy transfer, as is often thought.[7] Or one may conceive of it as an objective structural feature of the world constituted by laws of nature (this is the most popular view[8]), by a relation of counterfactuality (as has been urged in our

---

[2] Cf. Hume (1739, pp. 170ff.).

[3] Thus, a statement of the form "*A* is a cause of *B* relative to the subject *X*" may well be objective and objectively true; relativization yields objectification (cf. Mühlhölzer 1988). But of course, we are interested in the truth of the unrelativized statement, and it would certainly be inappropriate to get rid of the relativization simply by existential quantification.

[4] Cf. Mackie (1974, ch. 1), and Beauchamp and Rosenberg (1981, ch. 1), for two thorough discussions.

[5] That is the line of thought Beauchamp and Rosenberg (1981, ch. 1), end up with – plausibly in my view.

[6] Cf. Hume (1739, pp.167–169).

[7] Cf., e.g., Aronson (1971) and Fair (1979).

[8] To be found in many places; see, e.g., Hempel (1965, pp. 348ff.), and Carnap (1966, ch. 19). Of course, this popular view runs into the well-known difficulty of characterizing laws of nature, i.e. of specifying a criterion of lawlikeness, without recourse to causation.

days in particular by Lewis 1973a), or as a certain second order universal (an Australian proposal[9]). But I remain skeptical: because there is a need to explain the most prominent and peculiar epistemological role of the notion of causation rightly emphasized by Hume, because it seems that this explanation cannot simply be given in terms of the subject's grasp of how causation objectively is, and because it is hard to see which other kind of explanation is available to purely objectivistic accounts of causation – though this is not the place to argue this point.[10]

Or one may deny the problem by acquiescing in an epistemologically relativized notion of causation and talking us out of our realistic intuition. This line is most prominently pursued today by Putnam (cf., e.g., his 1983b) and, in quite a different way, also by van Fraassen.[11] But this subjectivistic strategy can at most succeed, if it does not only try to make us believe that the realistic intuition concerning causation is a confusion or an illusion, but offers us a plausible account or a convincing substitute for it.

So, there is no way of avoiding to face the problem. Facing the problem means trying to integrate the two one-sided positions, that is, to give both an objectivistic and subjectivistic account of causation and to specify their relation. If it is true that this relation does not simply consist in the subject's grasp of objective causation, then the direction of analysis should presumably be reversed, i.e. the objectivistic account should be understood as some kind of objectification of the subjectivistic one.

There are not so many models for doing this. One may indulge into Kant's complicated doctrine of transcendental idealism in his *Kritik der reinen Vernunft* in which the present objectification problem is meshed with other and, in the Kantian context, more salient ones concerning space, time, the self, and other objects. In modern times the awareness that the subjectivist and the objectivist side need to be mediated is still lively; and Salmon (1984) is certainly one of the most forceful attempts to meet this need, i.e. to defend an, as he calls it, ontic conception of causation without losing the virtues of an epistemic conception. However, I am not sure how to categorize this and other recent attempts as objectifications of a subjectivistic account.

In a way, Hume himself may be said to have offered a solution of the problem. As already mentioned, one may take causation as basically non-objective as specified in his associationist theory of causation, and one may then objectify it to the extent to which our associations can be explained or supported by existing regularities; insofar our associations do not have such an objective basis, causation is not objectifiable.

Isn't this good enough a solution? No, because the associationist theory isn't good enough. There are various well-known problems in the logic of causation it cannot cope with. Among them, the basic problem is that it cannot distinguish

[9] Cf. in particular Tooley (1987, sect. 8.3).

[10] The most extensive recent criticism of objectivistic or realistic approaches to laws of nature and related things may be found in van Fraassen (1989, part I).

[11] Cf. his (1980a, pp. 112ff.), where he argues the theory of causation to be almost wholly absorbed by a theory of explanation which can be understood only in a subjectively or pragmatically relativized way; only an empty objective characterization of "the causal net = whatever structure of relations science describes" (p.124) remains.

between the causes of an effect and mere symptoms or indicators preceding it.[12] I propose a simple remedy: improve the associationist theory and then adapt the account of objectification.

The improvement consists of two steps. Since for Hume induction is more or less synonymous to association – the inductively inferred beliefs are those associated with other beliefs – the first step is to give a general and precise account of a subject's inductions or associations; this is the intent of the theory of inductive schemes explicated as so-called natural conditional functions (NCFs) in the next section. The second step, then, is to reconstruct a Humean theory of causation on that improved subjective basis; this will be undertaken also in Section 5.3 as far as it will be required.

Afterwards, we can turn to the question how this subjectively relativized theory of causation can be objectified. In fact, I propose to investigate a more general question: An inductive scheme, or a NCF, characterizes the epistemic state of some subject. Such an epistemic state has various features. It includes a specific causal picture, for instance, or it contains specific beliefs. These features are sometimes a matter of truth and falsity and sometimes not. For example, beliefs can certainly be true or false, whereas a subjective probability for some contingent proposition cannot sensibly be called true or false; it can only be well-advised or ill-guided.[13] So we need a general explanation of what it means to make such a feature a matter of truth and falsity. This allows us to pose the question of objectification for each feature of an epistemic state, namely as the question to which extent that feature can be made a matter of truth and falsity. Section 5.4 attempts to give that explanation and Section 5.5 attempts to answer the question of objectification for some features of an epistemic state, among them its causal picture as explicated in Section 5.3.

5.2 Induction

What might a theory of induction be expected to yield? No more and no less, I think, than a dynamic account of epistemic states which specifies not only their static laws, but also their laws of change – where these laws are most plausibly understood as laws of rationality.[14] The forms these laws take depend, of course, on how epistemic states are represented. The axioms of mathematical probability are the static laws of a probabilistic representation, and the principle of maximizing relative entropy as well as various rules of conditionalization are its most plausible candidates for dynamic laws. *Plain belief* which affirms or denies a proposition or does neither and thus admits only of three grades[15] is most easily represented by a

---

[12] Cf., e.g., Mackie (1974, pp. 81ff.).

[13] Even if the subjective probability matches the objective probability of that proposition, should it have one, it would be inappropriate to call it true.

[14] In my (1993b) I have tried to characterize the role of laws of rationality.

[15] This is how I want "plain belief" to be understood. It is the kind of belief all epistemic logics are about.

set of propositions, namely those held to be true. The most plausible static laws are that such a set be consistent and deductively closed. However, there is no general dynamic account of epistemic states represented in this way. Even if one returns to the probabilistic representation and equates plain belief with subjective probability 1, one does not arrive at a general dynamic theory of plain belief because all standard probabilistic laws of change do not allow what must be allowed, namely to retract from probability 1 (whatever has probability 1, keeps it according to these laws) and thus to give up plain beliefs. Hence, a different representation of epistemic states and a different theory is required in order to account for the dynamics of plain belief. In (1988) I have presented such a theory and explained its details and the drawbacks of rival theories.[16] Here, I have to restrict myself to briefly presenting its formal structure.

Throughout, I shall make the convenient assumption that propositions construed as set of possible worlds serve as objects of belief and as objects of causation as well. This is problematic in various ways[17]; but I shall not bother with these problems because they do not essentially affect the present issue. Thus, let Ω denote a set of *possible worlds*, as philosophers say, or a sample space, as probability theorists say, i.e. just an exhaustive set of mutually exclusive possibilities; there is no need of further clarifying the nature of these possibilities. Elements of Ω will be denoted by ω, υ, α, etc. Not worrying about questions of measurability, we take each subset of Ω to represent a *proposition*; propositions are denoted by *A, B, C, D, E*, etc. The basic concept, then, is formally very simple; it is given by:

Definition 1: κ is a *natural conditional function* (a *NCF*) iff κ is a function from Ω into the set of natural numbers such that $\kappa^{-1}(0) \neq \varnothing$. A NCF κ is extended to propositions by defining $\kappa(A) = \min \{\kappa(\omega) \mid \omega \in A\}$ for each $A \neq \varnothing$ and $\kappa(\varnothing) = \infty$.[18]

A NCF κ is to be interpreted as a *grading of disbelief.* If $\kappa(\omega) = 0$, then ω is not disbelieved, i.e. ω might be the actual world according to κ. Because not every world can be denied to be the actual one, Definition 1 requires that $\kappa(\omega) = 0$ for some $\omega \in \Omega$. If $\kappa(\omega) = n > 0$, then ω is disbelieved with degree *n*. A proposition is then assigned the minimal degree of disbelief of its members. Thus, if $\kappa(A) = n > 0$, then *A* is disbelieved with degree *n*. And if $\kappa(A) = 0$, then *A* is not disbelieved, i.e. *A* might be true

---

[16] In particular, I there explain in which way the theory of NCFs generalizes the theory by which it was most heavily influenced, namely the somewhat restricted account of belief change developed by Gärdenfors and his collaborators (and most extensively presented in Gärdenfors 1988).

[17] For example, it seems that one must take sets of centered instead of uncentered possible worlds in order to account for indexical belief. Also, one might argue that the identification of the objects of belief and those of the causal relation is guilty of a confusion of metaphysical and epistemic modality. And these are only two of many problems.

[18] In my (1988) [here: ch. 1], I have defined so-called ordinal conditional functions which are a bit more general (and a bit more awkward) in taking ordinal numbers as values. This generality will not be required here.

according to κ. However, $\kappa(A) = 0$ does not mean that A is believed according to κ. Belief in A is rather expressed by disbelief in $\sim A$,[19] i.e. by $\kappa(\sim A) > 0$ or $\kappa^{-1}(0) \subseteq A$. I call $\kappa^{-1}(0)$ the *net content* of the epistemic state κ. Thus, all and only the supersets of the net content of κ are plainly believed in κ, i.e. held to be true. This implies that plain belief is consistent and deductively closed; but these features go hand in hand with the assumption that the objects of beliefs are propositions.

Two simple, but important properties of NCFs immediately follow: *the law of negation* that for each proposition A either $\kappa(A) = 0$ or $\kappa(\sim A) = 0$ or both, and *the law of disjunction* that for all propositions A and B, $\kappa(A \cup B) = \min(\kappa(A), \kappa(B))$.

According to a NCF κ, propositions are believed in various degrees. It is useful to explicitly introduce the function expressing these degrees, because it is more vivid than the above disbelief talk:

Definition 2: β is *the belief function associated with* the NCF κ iff, for each subset A of Ω, $\beta(A) = \kappa(\sim A) - \kappa(A)$; and β is a *belief function* iff it is associated with some NCF.[20]

Thus, $\beta(\sim A) = -\beta(A)$, and A is believed true or false or neither according to β (or κ) depending on whether $\beta(A) > 0$ or < 0 or $= 0$. However, there is no simple law for disjunctions in terms of belief functions; this is why NCFs are preferable on the formal score.

So far, the various degrees of belief did not really play a theoretical role. But they are crucial for a dynamic account of plain belief. The central notion is specified in:

Definition 3: Let κ be a NCF and A a non-empty proposition. Then *the A-part of* κ is the function $\kappa(.\mid A)$ defined on A by $\kappa(\omega \mid A) = \kappa(\omega) - \kappa(A)$ for each $\omega \in A$. Again, this function is extended to all propositions by setting $\kappa(B \mid A) = \min\{\kappa(\omega \mid A) \mid \omega \in A \cap B\} = \kappa(A \cap B) - \kappa(A)$ for each $B \subseteq \Omega$. Finally, if β is the belief function associated with κ, we define, as in Definition 2, $\beta(B \mid A) = \kappa(\sim B \mid A) - \kappa(B \mid A)$.

Definition 3 is tantamount to *the law of conjunction* that $\kappa(A \cap B) = \kappa(A) + \kappa(B \mid A)$ for all propositions A and B with $A \neq \varnothing$.

The A-part $\kappa(.\mid A)$ of κ can be viewed as a NCF with respect to the restricted possibility space A and thus as a grading of disbelief *conditional on* A. Accordingly, $\beta(.\mid A)$ expresses degrees of belief conditional on A.

It is obvious that a NCF κ is uniquely determined by its A-part $\kappa(.\mid A)$, its $\sim A$-part $\kappa(.\mid \sim A)$, and the degree $\beta(A)$ of belief in A. This marks an important point of difference to various other approaches which in effect operate only with orderings and not with gradings of disbelief. Within these approaches one may perhaps also define A-parts, i.e. orderings of disbelief conditional on A. But there is no way to uniquely combine various parts of an ordering, and therefore the following ideas cannot be carried over to orderings of disbelief.

---

[19] "$\sim$" is used here to denote the set-theoretical complement (with respect to Ω).

[20] Shenoy (1991) has convinced me of the usefulness of explicitly introducing this concept, and the brief definition is due to Bernard Walliser.

This uniqueness suggests a simple model of belief revision for NCFs. If a piece of information consists only in the proposition A, then it is plausible to assume that only the old degree $\beta(A)$ of belief in A gets changed to some new degree $\beta'(A) = n$, whereas the A-part and the $\sim A$-part of the old NCF κ are left unchanged; n, $\kappa(. \mid A)$, and $\kappa(. \mid \sim A)$ then determine a new NCF κ' (and a new belief function β'), which I call the *A,n-conditionalization of* κ. There are also more complicated models in which the information need not concern a single proposition. But these remarks already indicate that a full dynamics of plain belief and thus a full theory of induction can be stated in terms of NCFs.

An account of conditionalization immediately yields the epistemologically important notions of dependence and independence. Two propositions are independent iff conditionalization with respect to the one does not affect the epistemic status of the other. Formally:

Definition 4: Let β be the belief function associated with the NCF κ, and A, B, and C three non-empty propositions. Then A and B are *independent with respect to* β (or κ) iff $\beta(B \mid A) = \beta(B \mid \sim A)$, i.e. iff $\kappa(A' \cap B') = \kappa(A') + \kappa(B')$ for each $A' \in \{A, \sim A\}$, $B' \in \{B, \sim B\}$; and they are *independent conditional on C w.r.t.* β (or κ) iff $\beta(B \mid A \cap C) = \beta(B \mid \sim A \cap C)$.

Of course, (conditional) independence may be generalized to a relation between whole algebras of propositions, and so forth. Indeed, the fact that conditional dependence and independence with respect to belief functions behave precisely like their probabilistic counterparts is the technical reason why NCFs will form a suitable base for a parallel theory of deterministic causation.

A closely related and equally important notion is the concept of a reason. Being a reason is always relative to an epistemic state, and given such a state a reason strengthens the belief in, or, in other words, is positively relevant to, what it is a reason for. Formally:

Definition 5: Let β be the belief function associated with the NCF κ, and A, B, and C three propositions. Then A is a *reason for B relative to* β or κ iff $\beta(B \mid A) > \beta(B \mid \sim A)$; and A is a *reason for B conditional on C relative to* β or κ iff $\beta(B \mid A \cap C) > \beta(B \mid \sim A \cap C)$.

According to this definition, the relation of being a reason is *symmetric, but not transitive*, in analogy to probabilistic positive relevance, but in sharp contrast to the narrower relation of being a deductive reason (which is just set inclusion between contingent propositions).[21] Moreover, being a reason does not presuppose that the reason is actually given, i.e. believed; on the contrary, whether A is a reason for B relative to β is independent of the degree $\beta(A)$ of belief in A.

Since the value 0 has the special role of a dividing line between belief and disbelief, different kinds of reasons can be distinguished:

[21] This structural fact most clearly shows that deductive logic may, in a way, have been misleading as a model of human reasoning.

Definition 6:

$$A \text{ is a} \begin{Bmatrix} additional \\ sufficient \\ necessary \\ weak \end{Bmatrix} \textit{reason for } B \textit{ w.r.t. } \beta \textit{ or } k \textit{ iff} \begin{Bmatrix} \beta(B \mid A) > \beta(B \mid {\sim} A) > 0 \\ \beta(B \mid A) > 0 \geq \beta(A \mid {\sim} A) \\ \beta(B \mid A) \geq 0 > \beta(A \mid {\sim} A) \\ 0 > \beta(B \mid A) > \beta(A \mid {\sim} A) \end{Bmatrix}.$$

Conditional reasons of the various kinds are defined similarly. If A is a reason for B, it belongs at least to one of these four kinds; and there is just one way of belonging to several of these kinds, namely by being a necessary and sufficient reason. Sufficient and necessary reasons are the more important ones; but additional and weak reasons, which do not show up in plain belief and are therefore usually neglected, well deserve to be allowed for by Definition 6.

No further development of the theory of NCFs as a substitute of Hume's theory of association will be needed here. So, the next step in amending Hume is to give an account of causation relative to the NCFs.

5.3 Causation

This account will be quite brief because I have more thoroughly dealt with the matter in the papers referred to.

The first thing to do is to give possible worlds a temporal structure: Let I be a non-empty set of factors or variables; we may assume I to be finite in order to avoid all the technical problems related to infinity. Each variable $i \in I$ is associated with a set Ω_i containing at least two members; Ω_i is the set of values i may take. The set Ω of possible worlds is then represented as the Cartesian product of all the Ω_i ($i \in I$); hence, each $\omega \in \Omega$ is a possible course of events, a function assigning to each variable $i \in I$ the value $\omega(i)$ taken by i in the possible world ω.

Since metric properties of time are irrelevant, time may then simply be represented by a weak, i.e. transitive and connected order $\leq$on the set I of variables; $<$ denotes the corresponding irreflexive order on I. and $i \approx j$ is to say that the variables I and j are simultaneous. Discreteness of time is implied by the finiteness of I. But even if I were infinite, one should assume that time is discrete and treat continuous time as the limit of ever finer discrete time (as it is done in the theory of stochastic processes).

Time should be associated not only with variables, but, if possible, also with propositions. Therefore a proposition A is defined to be a *J-measurable* or, in short, a *J-proposition* for a set $J \subseteq I$ of variables iff for all $\upsilon, \omega \in \Omega$ agreeing on J, i.e. with $\upsilon(i) = \omega(i)$ for all $i \in J$, $\upsilon \in A$ iff $\omega \in A$; intuitively one might say that a J-proposition is only about the variables in J and does not say anything about other variables. There are many contingent propositions which are about single variables, and the temporal order of variables is easily carried over to them. Indeed, I see no loss in restricting causes and effects to be such, so to speak, logically simple propositions which are about one variable only.

The general idea of the explication of causation is now a common one: *A is a cause of B in* the world ω iff A and B obtain in ω, A precedes B, and A raises the epistemic or metaphysical status of B under the circumstances obtaining in ω. This characterization fits many prominent conceptions of causation the main difference of which lies in their account of what I have called the epistemic or metaphysical status and which have a minor difference over whether they should take precedence strictly on loosely so as to include simultaneity. Here, the condition that A and B *obtain in* ω may be expressed by $\omega \in A \cap B$. The condition that *A precedes B* in the strict sense translates into the condition that for some $i, j \in \omega$ A is an i-proposition, B a j-proposition, and $i < j$; however, I shall only require that $i \leq j$, that is, I take precedence in the loose sense including simultaneity.[22] The *epistemic status* of B is specified, of course, by a belief function β, which is *raised* by A iff it is higher conditional on A than conditional on $\sim A$. Finally, it may be argued that in the case of *direct* causation, *the circumstances obtaining in* ω may be identified with the whole past of the direct effect in ω with the exception of the direct cause – where a loose sense of precedence including simultaneity goes hand in hand with a loose sense of past including presence. Thus I define: If A is an i-proposition, B a j-proposition, and $i \leq j$, then the circumstances of A's directly causing B in ω are defined as $C_{A,B,\omega} = \{\upsilon \mid \upsilon(k) = \omega(k)$ for all $k \in I$ with $k \leq j$ and $k \neq i,j\}$, i.e. as the past and presence of B in ω with the exception of A and B. Thus we get:

Definition 7: Let $\omega \in \Omega$, $i, j \in I$, A be a i-proposition, and B a j-proposition. Then *A is a direct cause of B in* ω *relative to* the belief function β or the associated NCF κ iff $\omega \in A \cap B$, $i \leq j$, and $\beta(B \mid A \cap C_{A,B,\omega}) > \beta(B \mid \sim A \cap C_{A,B,\omega})$, i.e. A is a reason for B conditional on $C_{A,B,\omega}$ relative to β or κ. And A is called *an additional, sufficient, necessary, or weak direct cause of B in* ω according to whether A is an additional, sufficient, necessary, or weak reason for B conditional on $C_{A,B,\omega}$.

In my (1980, 1983b, 1990a) I have more fully argued for the adequacy of that definition.[23] And in my (1990a) [here: ch. 2], I have also argued that causation in general (which is direct *or* indirect) should be defined as the transitive closure of direct causation, as seems quite natural, though it is not unproblematic, and as many have assumed, though equally many have rejected it in the case of probabilistic causation.

---

[22] My opinion on this issue vacillates. In my (1980), I allowed for simultaneous causes, and in my other papers I didn't, or rather I avoided the issue. I have no philosophical principle for deciding it; therefore I make the stipulation which best serves my purpose at hand. So I do here, too; the reasons will become clear in Section 5.5 below.

[23] In the latter papers I did so only when simultaneity is excluded. The consequence of Definition 7 that in the case where A and B are simultaneous direct causation is symmetric is certainly forbidding. But in that case one might just read the definiendum in a different way, namely as saying that A and B are causally related without deciding which one is the cause and which one the effect. (Of course, nothing seems to able to decide this, and this is a most powerful argument for taking precedence strictly. But this issue is not our present concern.) And if one should find this last move unacceptable, then one should read this paper as being only about the restricted case where simultaneous variables are excluded.

Definition 7 modifies Hume's associationist theory of causation in three respects. First, it rests on a formally specified account of induction. Second, the circumstances in which causal relations are situated are explicitly taken into account. This is done, at least implicitly, also by Hume, e.g. when he slips into counterfactual explanations of causation,[24] but he does not make a point of it.[25] Third, the contiguity condition,[26] which would be reasonable only for direct causation, anyway, is dropped. As I explain more fully in (1990a), p. 129 [here, p. 52], the contiguity condition would force an unwantedly orderly behavior on subjectively relativized causation; we shall see that it comes into play only in the objectification of causation.[27]

I think that Definition 7 (and the just mentioned extension to indirect causation) is on the whole better able than other accounts to deal with intricate causal situations discussed in the literature. One problem to which we shall return is presented by cases of (symmetric) causal overdetermination which are usually exemplified by the firing squad (where several soldiers shoot the victim at the same time). Such cases are apparently possible. But they are a great mystery, if not an impossibility for all realistic accounts of causation; it seems that the only thing the realist can do is to explain them away (cf. Lewis 1986d, pp.193–212). For Definition 7, by contrast, there is no mystery at all. The two overdetermining causes may be simply conceived as additional causes; each of the two is an additional cause of the effect in the presence of the other one.

There is a snag, however. The notion of an additional cause seems to make sense only within such an epistemically relativized framework and not within a realistic framework, because something true under some conditions cannot be more true under other conditions. On the one hand, the snag would explain and justify the difficulties of the realists in handling overdetermination. On the other hand, it accentuates the problem of objectification. Some causal relations seem to be objectifiable and others do not. But which ones are and which ones are not? And what, in the first place, could all that talk of objectification mean?

5.4 An Explication of Objectification

The having of a certain plain belief is a subjective affair; it applies to some epistemic subjects and does not apply to others. But the plain belief itself can be true or false, and its truth is an objective matter; it is as objective as truth is and in no way to be subjectively relativized. Formally: Let $\alpha \in \Omega$ be the *actual* world. Then the plain belief in the proposition A is true iff the proposition A itself is true, i.e. iff

---

[24] As he does in (1777, sect. VII, part 2).

[25] Of course, the point has soon been acknowledged, e.g. by Mill (1843, book III, ch. V).

[26] That is, the temporal part of it; spatial relations are here out of consideration.

[27] Thus, the role Hume assigns to the contiguity condition indicates to me that he does not clearly separate subjective and objective aspects of causation.

$\alpha \in A$. Thus, plain beliefs have objective truth conditions; the truth condition of the plain belief in A is just the proposition A itself. These are trivial observations, deriving from the obvious one-one-correspondence between plain beliefs and the propositions which they are about. In the light of a lot of literature about the content of beliefs, this is at least implausible.[28] But it is a direct consequence of my not worrying about these problems and burdening propositions with the double role as truth conditions and as contents of beliefs.

My objectification problem now is to which extent these observations apply also to other epistemic states and in particular to NCFs, to which extent other epistemic states and in particular NCFs may be said to be objectively true or false. This is less trivial, as is easy to see.

At first, one might say that NCFs have truth conditions as well. A NCF κ has a net content, and the net content tells which propositions are plainly believed in κ. Thus, κ may be said to be true iff all its plain beliefs are true, i.e. iff its net content is true, i.e. iff $\alpha \in \kappa^{-1}(0)$; and hence its truth condition is just its net content. However, this offers only very partial objectification. The correspondence between NCFs and their truth conditions in this sense is not one-one, but badly many-one, because a NCF's grading of disbelief below the net content may vary arbitrarily without affecting its truth condition. Thus, though objective truth has some selective power concerning NCFs, it affords no distinction among all the NCFs having the same truth condition, but differing wildly in their inductive behavior and other aspects.

These further observations teach us two things. First, when trying to objectify epistemic states, one has to refer to a certain feature or aspect of these states, and it is this feature on which objectification concentrates. In the last paragraph, the feature considered was the set of plain beliefs according to a NCF. But one may, and we shall, consider other features as well, e.g. the relation of direct causation according to a NCF.

Secondly, objectification may be only partial. It would be complete, if one had a one-one-correspondence between *all* the epistemic states considered and something objective like truth conditions. This may be impossible to reach, however. Usually, only epistemic states of a particular kind will yield to such a one-one-correspondence, namely only those epistemic states which behave uniformly with respect to the feature considered. For example, if only belief functions β are considered which behave uniformly with respect to plain belief, e.g. for which $\beta(A) = 1$ for all A with $\beta(A) > 0$, these are easily put into a one-one-correpsondence with propositions. But then only such belief functions may be called objectifiable with respect to plain belief; and in this sense belief functions are only partially objectifiable with respect to plain belief.

Therefore, I propose the following explication of what it means to objectify NCFs. The *first* thing to do is to specify a natural association of propositions with features of NCFs. Here, a feature of NCFs is just any n-place relation ($n \geq 1$) obtaining relative

[28] Indeed, the important insight of Kripke (1972) that a priori and (metaphysically) necessary truths are just two different kinds of things immediately implies that contents of beliefs and truth conditions also are two different kinds of things.

to a NCF (or, alternatively, any $n + 1$-place relation the $n + 1$st place of which is taken by a NCF). The most natural association I can think of is then given by:

Definition 8: Let R be a NCF-relative n-place relation, and $x_1,\ldots, x_n$ be any objects in the domain of R. Then *the proposition* $E_R(x_1,\ldots, x_n)$ *associated with* $R(x_1,\ldots, x_n)$ is to be the strongest (i.e. smallest) proposition A for which $R(x_1,\ldots, x_n)$ w.r.t. a NCF κ implies that A is plainly believed in κ, i.e. which is such that for each NCF κ, $\kappa^{-1}(0) \subseteq A$ if $R(x_1,\ldots, x_n)$ w.r.t. κ.

The significance of this definition is best shown by examples; their claims may easily be seen to be correct:

(1a) If $R(A)$ obtains relative to κ iff A is plainly believed in κ, then $E_R(A) = A$.
(1b) If $R(A, B)$ obtains relative to κ iff B is plainly believed in κ conditional on A, i.e. iff $\kappa({\sim}B \mid A) > 0$, then $E_R(A, B) = {\sim}A \cup B := A \rightarrow B$, whereby $\rightarrow$ is defined as the *set theoretical* operation representing material implication for propositions.[29]
(2a) If $R(A, B)$ obtains relative to κ iff A is an additional reason for B relative to κ, then $E_R(A, B) = B$.
(2b) If $R(A, B)$ obtains relative to κ iff A is a sufficient reason for B relative to κ, then $E_R(A, B) = A \rightarrow B$.
(2c) If $R(A, B)$ obtains relative to κ iff A is a necessary reason for B relative to κ, then $E_R(A, B) = {\sim}A \rightarrow {\sim}B$.
(2d) If $R(A, B)$ obtains relative to κ iff A is a weak reason for B relative to κ, then $E_R(A, B) = {\sim}B$.
(2e) If $R(A, B)$ obtains relative to κ iff A is a reason for B relative to κ, then $E_R(A, B) = \Omega$.
(3a) If $R(A, B, \omega)$ obtains relative to κ iff the i-proposition A is an additional direct cause of the j-proposition B in ω relative to κ, then $E_R(A, B, \omega) = C_{A,B,\omega} \rightarrow B$.
(3b) If $R(A, B, \omega)$ obtains relative to κ iff A is a sufficient direct cause of B in ω relative to κ, then $E_R(A, B, \omega) = C_{A,B,\omega} \cap A \rightarrow B$.
(3c) If $R(A, B, \omega)$ obtains relative to κ iff A is a necessary direct cause of B in ω relative to κ, then $E_R(A, B, \omega) = C_{A,B,\omega} \cap {\sim}A \rightarrow {\sim} B$.
(3d) If $R(A, B, \omega)$ obtains relative to κ iff A is a weak direct cause of B in ω relative to κ, then $E_R(A, B, \omega) = C_{A,B,\omega} \rightarrow {\sim}B$.
(3e) If $R(A, B, \omega)$ obtains relative to κ iff A is a direct cause of B in ω relative to κ, then $E_R(A, B, \omega) = \Omega$.

So far, we have associated a proposition with a NCF-feature R only as applied to certain items $x_1,\ldots, x_n$. But this is immediately extended to an association of a proposition with the NCF-feature itself:

[29] I hope this notation is less misleading than helpful.

Definition 9: Let R be a NCF-relative n-place relation. Then *the proposition* $E_R(\kappa)$ *associated with* R *in relation to* the NCF κ is defined as the proposition $\bigcap\{E_R(x_1,\ldots, x_n) \mid R(x_1,\ldots, x_n)$ obtains relative to $\kappa\}$.

Together with Definition 8 this implies that $E_R(\kappa)$ is plainly believed in κ; indeed $E_R(\kappa)$ is the strongest proposition which is plainly believed in κ in virtue of how the feature R is realized in κ.

The *second* thing to do is to reverse the procedure and to inquire whether a NCF κ may be uniquely reconstructed from the proposition $E_R(\kappa)$ associated with the feature R; this allows, so to speak, to transfer the objectivity of $E_R(\kappa)$ to κ itself. This unique reconstructibility seems also to be required for calling κ to be objectifiable w.r.t. R. However, it seems feasible only when the feature R is realized in κ in some uniform way; I cannot imagine any way of encoding into $E_R(\kappa)$ different ways of realizing R. Therefore we need to refer to such a uniform specification of R. Moreover, we shall see that the reconstruction may work only under certain conditions; therefore we also need to refer to such conditions. Now I am finally prepared to offer my explication of objectifiability:

Definition 10: Let S be any specification of the feature R; this means that, for each NCF κ, $R(x_1,\ldots, x_n)$ obtains relative to κ, if $S(x_1,\ldots, x_n)$ obtains relative to κ; and let F be any condition on the items in the field of R. Then a NCF κ *is objectifiable with respect to R* (or, κ *is an objectification of R*) *under the specification S given condition F* iff, given $E = E_R(\kappa)$, κ is the only NCF such that the following holds:

(a) For all $x_1,\ldots, x_n$, $R(x_1,\ldots, x_n)$ obtains relative to κ if and only if $S(x_1,\ldots, x_n)$ obtains relative to κ,
(b) for all $x_1,\ldots, x_n$, $S(x_1,\ldots, x_n)$ obtains relative to κ if and only if $x_1,\ldots, x_n$ satisfy the condition F and $E \subseteq E_R(x_1,\ldots, x_n)$.

Moreover, we omit reference to the condition F iff F is empty, and we omit reference to the specification S by existential quantification.

Thus I slip into two ways of talking. Sometimes I say that a NCF is objectifiable w.r.t to some feature, and sometimes I say that that feature itself is objectifiable. Both ways of talking seem appropriate, though they may be a bit confusing.

Clause (a) expresses the uniform realization of the feature R in the NCF κ; it requires R to be realized in κ only in the way S. Clause (b) says that the relation R, as it obtains relative to κ, is determined by the condition F and by $E_R(\kappa)$. And the uniqueness clause guarantees that κ may be uniquely reconstructed from the information specified in (a) and (b).

Unconditional objectifiability means that one can infer from $E_R(\kappa)$ alone for which $x_1,\ldots,x_n$ the feature R holds relative to κ. But it may be that the condition F is needed for this inference. Of course, one may find conditions which trivialize conditional objectifiability; how objective conditional objectification really is depends on how objective the condition F is.

Again, all this is too abstract to assess its significance. So let's return to the examples (1)–(3) already introduced and investigate the extent to which they allow objectification.

5.5 The Objectification of Induction and Causation

(1a) Let $R(A)$ obtain relative to κ iff $A \neq \Omega$ and A is plainly believed in κ, and for some positive integer m, let $S(A)$ obtain relative to κ iff $A \neq \Omega$ and $\beta(A) = \kappa(\sim A) = m$.[30] Then κ is an objectification of R under S iff, for all $\omega \in \Omega$, $\kappa(\omega) = m$, if $\kappa(\omega) > 0$. In short, *plain belief is unconditionally objectifiable*.

This is also intuitively plausible; a NCF is uniquely reconstructible from what is plainly believed in it only if it does not differentiate among the disbelieved worlds. Note, however, that such NCFs are unfit epistemic states. Whenever you are in such a state and accept a piece of information contradicting your plain beliefs (this is quite common), you then believe only the information accepted and nothing more; and this is a devastatingly cautious inductive behavior.

(1b) Let $R(A,B)$ obtain relative to κ iff A does not logically imply B and $\kappa(\sim B \mid A) > 0$, i.e. B is plainly believed in κ conditional on A; and for some integer m, let $S(A, B)$ obtain relative to κ iff A is not a subset of B and $\kappa(\sim B \mid A) = m$. Then R is objectified under S by the very same NCF as in (1a).[31] There are thus only poor unconditional objectifications of conditional plain belief because the objectifying NCFs of (1a) have only one genuine level of conditionality. By this I mean that conditional on something disbelieved only that condition and its logical consequences are believed according to these NCFs. Of course, this is tantamount to their unfit inductive behavior just mentioned.

(2) Here the results are even more negative:

(2a, d) If $R(A, B)$ means that A is an additional reason for B w.r.t. κ, there is no unconditional objectification for R, for the simple reason that, since $E_R(A, B) = B$, there is no way of telling from $E_R(\kappa)$ whether A or $\sim A$ is to be an additional reason for B. The same applies to weak reasons, because A is a weak reason for B iff $\sim A$ is an additional reason for $\sim B$.

(2e) A fortiori, there is no objectification at all for the relation of being a reason simpliciter.

(2b, c) If $R(A, B)$ means that A is a sufficient reason for B, R can still not be objectified unconditionally. In this case we have $E_R(A, B) = A \rightarrow B$. So, if κ is to be objectifiable w.r.t. R, A would have to be a sufficient reason for B according to κ if and only if $E_R(\kappa) \subseteq A \rightarrow B$. The set of sufficient reasons for B according to κ would therefore have to be a (complete) ideal, i.e. it would have to be non-empty, it would have to be closed under (arbitrary, not only finite) union, and it would have to contain all the subsets of its members.[32] But usually there are many propositions B for which this set is not an ideal.

---

[30] $A = \Omega$ must be excepted because $\beta(\Omega) = \kappa(\varnothing) = \infty \neq m$ for any m.

[31] For proof consider a NCF κ taking more than the two values 0 and m. Then it is easily seen that κ contains conditional beliefs of differing strength and cannot uniformly realize R as S.

[32] Equivalently, a complete ideal of propositions is just the set of all the negations of the members of a deductively closed set of propositions.

Let's be more precise. First, the special role of $\varnothing$ must be observed. $\varnothing$ is never a reason for B, but $E_R(\kappa) \subseteq \varnothing \to B = \Omega$ always holds. However, there is no problem in making special provisos concerning $\varnothing$. The problem is rather this: Suppose that A is a sufficient reason for B w.r.t. κ. It is easily seen that this is the case if and only if $\kappa(A \cap B) < \kappa(A \cap \sim B)$ and $\kappa(\sim A \cap B) \geq \kappa(\sim A \cap \sim B)$. Thus a subset A' of A which is not a sufficient reason for B may be constructed simply by deleting from $A \cap B$ sufficiently many worlds with low κ-values and putting them into $\sim A' \cap B$. If this construction works, the set of sufficient reasons for B is not an ideal; the construction works for at least some reasons A whenever for each $\upsilon \in \sim B$ there is a $\omega \in B$ such that $\kappa(\upsilon) \leq \kappa(\omega)$; and there are such propositions B whenever there are at least two worlds receiving (not necessarily different) non-zero κ-values.

Since A is a necessary reason for B iff $\sim A$ is a sufficient reason for $\sim B$, the very same observations hold also for the relation of being a necessary reason.

Compare this with the case (1b) of conditional plain belief. There, objectification also requires that the set of conditions under which a proposition B is plainly believed is a (complete) ideal. This is indeed the case for the NCFs specified there, i.e. for the NCFs taking only two values (though for all other NCFs conditional belief is non-monotonous in the sense that strengthening the condition need not preserve conditional belief). However, this tiny success does not carry over to sufficient reasons, because being a sufficient reason consists in conditional belief *plus* no belief under the contrary condition.

The general problem is to reconstruct the (sufficient) reason relation R from all the material implications entailed by $E_R(\kappa)$; and it seems that there is no general solution. Perhaps the device of conditional objectification helps. But what should the condition F be which selects the sufficient reasons from all these material implications? There is a trivial answer: let F itself be the sufficient reason relation according to κ. It was this answer which I had in mind when emphasizing that the objectivity reached by conditional objectification depends on the objectivity of the condition itself. This trivial condition is as subjective as the NCF κ referred to; and I do not see any general, more objective criterion which affords a successful selection.

(3) Matters look much better, however, in the case of causal relations because in this case the relevant material implications are severely restricted in form. So let us return to our initial topic, the objectification of causation. The first assertion is negative: Additional direct causes (3a), weak direct causes (3d), and thus direct causes in general (3e) can obviously be as little objectified as additional and weak reasons. Still, this observation explains why additional and weak causes have been totally neglected in the literature and in particular why causal overdetermination is such a problem for the realist and cannot be accounted for within a realistic setting in the way indicated at the end of Section 5.2. There are more positive news, however, in the remaining cases:

(3b, c) Let $R(A, B, \omega)$ obtain relative to κ iff A is a sufficient direct cause of B in ω relative to κ. $E_R(\kappa)$ may then be called *the causal law of* κ.

Note that necessary causation may be defined by sufficient causation: A is a necessary direct cause of B in ω iff $\sim A$ is a sufficient direct cause of B in a suitable variation ω' of ω concerning the variables A and B are about. Thus, whatever the objectification of sufficient direct causation, it is also an objectification of necessary direct causation to the very same extent; in particular, starting from necessary direct causation we would have arrived at the very same causal law. This is the reason why the following considerations may be restricted to sufficient causation.

When trying to objectify R, we must again refer to some uniform specification S of R. There is obviously only one way of doing so: for some positive integer m, let $S(A, B, \omega)$ obtain relative to κ iff A is an i-proposition, B a j-proposition, $i \leq j$, $\omega \in \Omega$, and $\kappa(\sim B \mid A \cap C_{A,B,\omega}) = m$ and $\kappa(\sim B \mid \sim A \cap C_{A,B,\omega}) = 0$.

In order to objectify sufficient causation we have to rediscover appropriate NCFs from causal laws in some unique way. Thus, the first step is to find a general association of potential causal laws, which, objectively, are just true or false propositions, with NCFs giving these propositions a causal interpretation. The only feasible association I have found is this:

As a piece of notation, define first, for each world ω and each $J \subseteq I$, $\omega_J = \{\upsilon \mid \upsilon(i) = \omega(i)$ for all $i \in J\}$ as the proposition that the variables in J behave as they do in ω; in particular define $\omega_j = \omega_{\{j\}}$ as the proposition that j takes the value in ω, $\omega_{<j} = \{\upsilon \mid \upsilon(i) = \omega(i)$ for all $i < j\}$ as the proposition that the past of j is as it is in ω, and similarly $\omega_{\leq j}$ and $\omega_{\approx j}$; moreover define $I_{<j} = \{i \in I \mid i < j\}$, and similarly $I_{\leq j}$ and $I_{\approx j}$.

With respect to any proposition L, C is called a *L-sufficient condition of* the j-proposition B iff C is a $I_{\leq j}$-$\{j\}$-proposition (i.e. about the past and presence of B) and $L \subseteq C \rightarrow B$; *the L-sufficient condition* $SC_L(B)$ *of* B is defined as the union of all its L-sufficient conditions (i.e. as the largest or weakest $I_{\leq j}$-$\{j\}$-proposition C such that L entails $C \rightarrow B$). We define a proposition L to be a *law of succession* iff for each variable $j \in I$ and j-proposition $B \neq \Omega$ $SC_L(B)$ is a $I_{<j}$-proposition (only about the past of B) and there is an $I_{\approx i}$-proposition $A \neq \Omega$, i being a variable immediately preceding j (if j is the temporally first variable; $I_{\approx i} = \emptyset$), such that $SC_L(B) = A \cup SC_L(A)$. In that case, there is exactly one such $I_{\approx i}$-proposition A, which will be called *the immediate L-sufficient condition of B* and denoted $ISC_L(B)$. Of course, we usually have $ISC_L(B) \neq SC_L(B)$ because whatever is sufficient for $ISC_L(B)$ is also sufficient for B. Intuitively, a law of succession is just a conjunction of material implications of the form "if now A, then next B" which, however, does not entail any categorical proposition about a single variable (but see the qualification in condition (II′) below).

Let us say that L is *violated* by the variable j in the world ω iff the L-sufficient condition of $\sim\omega_j$ obtains in ω, i.e. iff $\omega \in ISC_L(\sim\omega_j)$. Now we can finally specify the NCF κ_L associated with the law L of succession by defining $\kappa_L(\omega) = r \cdot m$, where r is the number of variables by which L is violated in ω (r is finite because I was assumed to be finite).

The κ_L's so defined do not have a wholly satisfying inductive behavior, but they are much more reasonable than those objectifying (conditional) plain belief. The reason is that there may occur as many violations of L as there are variables and that κ_L may thus take as many different values. The predictive and retrodictive behavior of these NCFs under various conditionalizations may therefore be complex; but it is characterized by two simple properties: First, simultaneous variables are always independent conditional on their past (i.e. on the set of variables preceding them), as is easily verified. Secondly, any j-proposition is expected to obtain whenever its immediate L-sufficient condition is satisfied. Hence, such a κ_L has a peculiar inductive rigidity: However often L is violated, κ_L is not deterred from inductive inferences, but expects invariably that no further violations of L will occur.

The κ_L's are my only candidates for objectifying sufficient causation. Are they suited for that purpose? Not generally; there are two problems entailing substantial, though instructive restrictions.

The first problem is that L need not be the causal law of κ_L. We can show that $L \subseteq E_R(\kappa_L)$,[33] but the converse does not necessarily hold. Suppose, e.g., that there are just three variables $i \approx j < k$, that for some ω $A = \omega_i$, $B = \omega_j$, and $C = \omega_k$, and that $L = A \cup B \rightarrow C$. There are then no sufficient causes of C in ω w.r.t. κ_L (even though $\omega_{<k}$ is a L-sufficient condition of C); in fact, there are only two ways for R to obtain relative to κ_L: we have $R(A, C, \omega')$, where $\{\omega'\} = A \cap \sim B \cap C$, and $R(B, C, \omega')$, where $\{\omega'\} = \sim A \cap B \cap C$. Hence, $E_R(\kappa_L) = (A \cap \sim B) \cup (\sim A \cap B) \rightarrow C \neq L$.

But surely, the identity of L and $E_R(\kappa_L)$ is required for a unique association of NCFs with causal laws. So, when does this identity hold? This is answered by a theorem:

$E_R(\kappa_L) = L$ holds if and only if the following two conditions are satisfied:

(I) Each j-proposition $B \neq \Omega$ ($j \in I$) has at least one sufficient direct cause in each $\omega \in ISC_L(B)$ w.r.t. κ_L and

(II) $L = \bigcap\{SC_L(B) \rightarrow B \mid B$ is a j-proposition for some $j \in I\} = \bigcap\{ISC_L(B) \rightarrow B \mid B$ is a j-proposition for some $j \in I\}$

where these conditions are, respectively, equivalent to:

(I′) For each $j \in I$, j-proposition B, and $\omega \in ISC_L(B)$ the intersection of all subsets J of $I_{\approx i}$ with $\omega_J \subseteq ISC_L(B)$ (where i is some variable immediately proceeding j) is non-empty and

(II′) if D_{ω_j} is defined as the smallest or strongest $I_{\approx j}$-proposition such that $L \subseteq \omega_{<j} \rightarrow D_{\omega_j}$ then D_{ω_j} is purely conjunctive for each $\omega \in \Omega$ and $j \in I$, i.e. if $I_{\approx j} = \{i_1, \ldots, i_s\}$, there are i_r-propositions A_r ($r = 1, \ldots, s$) such that $D_{\omega_j} = A_1 \cap \ldots \cap A_s$.

Proof (and remarks): The second equation of (II) always holds for laws of succession; and the equivalence of the first equation of (II) with (II′) is easily proved. I have mentioned (II′) only because it makes the content of (II) more perspicuos.

Concerning the equivalence of (I) and (I′) note that whenever the i-proposition A is a sufficient cause of B in ω w.r.t. κ_L, so is ω_i. Then it is easily seen that ω_i is a sufficient cause of B in ω w.r.t. κ_L if and only if i is a member of each subset J of $J_{\approx i}$ with $\omega_J \subseteq ISC_L(B)$. (I) is, of course, a version of the principle of causality. It is at least interesting that it is required at this stage of the argument (though I am not sure whether the reason for this lies only in my formalization). The point of the equivalent condition (I′) is to show that (I), though it seems to be about causation

[33] *Proof*: Let A be a sufficient cause of the j-proposition B in ω w.r.t. κ_L. This implies that L is violated by j in all worlds υ with $\upsilon \in A$, $C_{A,B,\upsilon} = C_{A,B,\omega}$, and $\upsilon \in \sim B$, i.e. that $A \cap C_{A,B,\omega}$ is a L-sufficient condition of B. Since this holds for all A, B, and ω, we have $L \subseteq E_R(\kappa_L)$.

and thus about κ_L, is in fact a condition solely about the logical form of the law of succession L, and clearly, (II) is so, too.

Now to the main point of the theorem. In view of the observations that, if there is a sufficient direct cause of B in ω, then for some i ω_i is a sufficient direct cause of B in ω and that B can have sufficient direct causes only in the worlds in $ISC_L(B)$, (I) implies that $E_R(\kappa_L) = \bigcap\{A \cap C_{A,B,\omega} \to B \mid R(A, B, \omega)$ obtains w.r.t. $\kappa_L\} = \bigcap(ISC_L(B) \to B \mid$ is a j-proposition for some $j \in I\}$ and (II) ensures that the latter term is equal to L. Conversely, suppose that (I) holds, but (II) doesn't. Thus, some D_{ω_j} of (II′) are not of the conjunctive form described there, but the D_{ω_j} defined w.r.t. $E_R(\kappa_L) = \bigcap\{ISC_L(B) \to B \mid B$ is a j-proposition for some $j \in I\}$ always are. Hence, $E_R(\kappa_L) \neq L$ in this case. Or suppose that (I) does not hold for a certain $\omega \in ISC_L(B)$ and j-proposition B. Then $E_R(\kappa_L) \subseteq \omega_{<j} \to B$ does not hold in virtue of ω, and there is no other causal relationship bringing about this inclusion. But $L \subseteq \omega_{<j} \to B$, and hence $E_R(\kappa_L) \neq L$ also in this case.

This theorem provides a satisfying solution of the first problem of objectification. But there is a second and more substantial problem, namely that the causal relations obtaining relative to NCF κ can generally not be rediscovered from its causal law. The reason is that the causal law is unable to distinguish the following two causal situations:

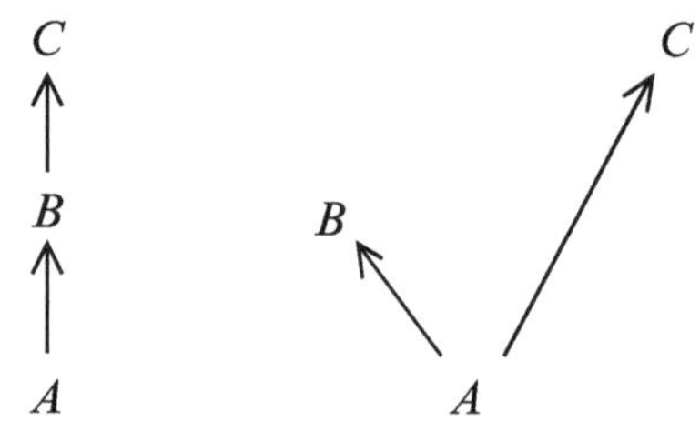

where A precedes B, B precedes C, and A is a necessary and sufficient direct cause of B and where there is a causal chain from A over B to C in the one case and a causal bifurcation in the other.[34] This is a grave and basic problem for all realistic theories of deterministic causation.[35] The only natural way I see to get rid of this problem is to assume that direct causes *immediately* precede their direct effects.[36] This assumption makes the case of a causal bifurcation impossible when B precedes C and allows for it only when B and C are simultaneous; and it does away with the unclear relations between simultaneous variables. Thus it enables us to infer the

[34] It is instructive to construct two NCFs generating the different causal situations and to see the difference vanish in the causal law which is one and the same for the two NCFs.

[35] Cf., e.g., Mackie (1974, pp. 81ff.).

[36] Recall that time was assumed to be discrete. Thus it makes sense to talk of immediate temporal precedence.

causal relations from a causal law. Intuitively it is clear that I have already built in this assumption into our account by deriving the κ_L's from laws L of succession in the way it did.

Note that this assumption to which objectification will be conditioned is a perfectly objective one about how causation relates to time; it does not refer to epistemic states or other subjective matters. The assumption is not really necessary for objectification. Presumably, any other assumption fixing the relation between time and causation (e.g. that each variable directly affects only the second next variable) would do as well. But this would be utterly artificial; contiguity is certainly the most natural thing to assume. Note further that, if I am right, it is at this point of objectification where the (temporal) contiguity condition comes into play and not already in the definition of causation, as Hume thought.

So, let's bring home the point in a precise way. First, there is a little problem with propositions about temporally first variables. Since these propositions do not have causes, causal relations cannot fix any belief values for them, in fact, these belief values may vary arbitrarily in a NCF without changing the causal relations concerning later variables. For the same reason, however, it is desirable that a NCF κ objectifying sufficient causation does not have any opinion on these propositions, i.e. that $\kappa(B) = 0$ for each non-empty proposition B about the first variables. In order to attain this desired result, I propose to apply a technical trick and slightly extend the relations R and S so that they cover also this problematic case: Define that $R'(A, B, \omega)$ obtains relative to κ iff $R(A, B, \omega)$ obtains relative to κ or $\omega \in A \cap B$, B is about a temporally first variable, $A = \Omega$, and $\kappa(\sim B) > 0$, and that $S'(A, B, \omega)$ obtains relative to κ iff $S(A, B, \omega)$ obtains relative to κ or $\omega \in A \cap B$, B is about a first variable, $A = \Omega$ and $\kappa(\sim B) = m$.

Now let's state a condition on the relata of direct causation:

(III) A, B, and ω are such that A is a i-proposition for some $i \in I$ immediately preceding j and B a j-proposition

Then we have the desired theorem: The NCF κ is an objectification of R' under the specification S' given condition (III) if and only if $\kappa = \kappa_L$ for some law L of succession satisfying (I) and (II).

Proof (and remarks): Let L be a law of succession. Trivially, κ_L satisfies clause (a) of Definition 10. It is also clear from the definition of laws of succession and from the construction of κ_L that κ_L satisfies clause (b) of Definition 10, provided L conforms to (I) and (II). Note that we could not do without condition (III) at this point. For, we may have $L \subseteq A \rightarrow B$, $L \subseteq B \rightarrow C$ and thus also $L \subseteq A \rightarrow C$. But the latter material implication is not to represent a direct causal relationship; and this is excluded by condition (III).

In order to see that κ_L is the only NCF satisfying (a) and (b), note first the simultaneous variables are bound to be independent conditional on their past w.r.t. to any NCF κ satisfying (b); this follows from the fact that according to (b) $R(A, B, \omega)$ is not to obtain w.r.t. κ for all $\omega \in \Omega$ and all simultaneous A and B. Next observe that for any NCF κ satisfying (a) and (b) the value $\kappa(\omega_j \mid \omega_{<j})$ is thereby determined for each $\omega \in \Omega$ and $j \in I$ to be equal to $\kappa_L(\omega_j \mid \omega_{<j})$ (= m or 0, depending on whether

$\omega_{<j} \subseteq ISC_L(\sim\omega_j)$ or not). But κ is uniquely determined by all these values since, given the independence of simultaneous variables conditional on their past, $\kappa(\omega) = \sum_{j\in I} \kappa(\omega_j \mid \omega_{<j})$. Thus κ is κ_L.

It is in this reasoning, by the way, where the liberal version of Definition 7 admitting simultaneous causation is technically required. If we would have ruled out simultaneous causation by definition, then no condition whatsoever on causal relations could say anything about simultaneous variables and could force on them the conditional independence needed for uniquely reconstructing κ_L from the causal relationship obtaining relative to it.

Now suppose conversely that κ is the unique NCF satisfying clause (a) and (b) of Definition 10. It is easily seen that (b) entails entails that $E_{R'}(\kappa)$ is a law of succession satisfying (I) and (II). And then it is clear that (a) forces κ to be of the special form κ_L, where $L = E_{R'}(\kappa)$.

Thus, we have finally arrived at an answer to the question to which extent sufficient and/or necessary direct causation is objectifiable. I do not know whether there are other plausible answers to that question within the framework I have been using. Within a richer framework, however, including, e.g., spatial considerations richer results would certainly be forthcoming.

To resume, what is a causal law? The answer is quite trivial. It is a proposition which can be true or false and which has a special form, namely that of a law of succession. This is certainly no news. So why deduce this familiar view in such a complicated fashion? In order to answer the question how such a proposition becomes a causal law. For, where does the causal role of such a proposition come from? From the above objectification, from uniquely associating with such a proposition a particular inductive behavior which is encoded in the corresponding objectifiable NCF and which interprets certain material implications entailed by that proposition as causal relations. This is my construal of the claim that causal laws are objectifications of inductive schemes.

Let me add a philosophical comment. The results obtained may be used as a very partial response to what might be called Quine's challenge. Quine's challenge is to explain how we can scientifically entertain our familiar intensional and intentional talk of beliefs, meanings, dispositions, causes,[37] etc.; in his terms, it is to reconstruct the second grade vernacular within the austere canonical notation of science which is, in effect, extensional first order predicate logic. This is a challenge, of course, because Quine has so forcefully argued over and over again that it cannot be met. And it seems that he is right; if we take the strict standpoint, the loose talk is lost forever.

But then it seems wrong to insist on the strict standpoint from which the rest is just unacceptable. A more fruitful strategy is to start from broader grounds including all kinds of non-extensional phrases, to characterize the virtues of extensional scientific language, and to describe the conditions under which these virtues can be realized. This is more fruitful because it allows us to get a theoretical hold on the

[37] Though Quine rather discusses dispositions, he explicitly assigns causation the same low status; cf., e.g., Quine (1969b, pp. 132f.).

relation between extensional and intensional talk, between our broader picture imbued with subjectivity and our attempts to reach scientific objectivity. In a way, the point of this paper is to show in one particular corner how such a theoretical hold may be constructively gained by specifying the conditions under which intensional causal conditionals reduce to extensional material implication, i.e. under which causation is objectifiable.

Putnam has also been impressed by Quine's arguments; but he draws a different conclusion, namely that the austere standpoint as Quine describes it is not an ideal, but an unattainable chimera.[38] But this conclusion seems to overshoot the mark. A theory of objectification as it is envisaged here should show the extent to which the ideal is attainable. And it partially confirms Putnam. As we have seen, the notion of (deductive *or* inductive) reason is not objectifiable; and if this holds for our narrower and more technical notion, it holds all the more for Putnam's much broader and more sweeping notion of reason. But it also contradicts him. What is true of reason, need not, and does not, carry over to causation.

5.6 Outlook

As I have mentioned in the introduction, there is a far-reaching analogy between the deterministic case with which I have dealt here and the probabilistic case which is more familiar in some respects. It would therefore be most appealing to extend this analogy also to the topic of objectification and to look which grip one thereby gets on the notion of objective probability. Intuitively, the analogy is quite obvious. Sufficient causation is tantamount to the idea of determination, and similarly objective probability is tantamount to the idea of partial determination; the fact that some possible event has, immediately before its time of occurrence, some degree of chance means that it is only partially determined *and* that it is in no way further determinable. We arrived at laws of succession when objectifying sufficient causation; thus Markov processes, which obviously are the probabilistic counterparts of laws of succession and which always were the favorites for probabilistic models of causal processes, may similarly be crucial for understanding objective probability. And so on. But however suggestive such hints are, they clearly need to be worked out in detail. It would then be particularly interesting to compare the resulting account of the relation between subjective and objective probability with other accounts of which I have found Lewis (1980a) and Skyrms (1984, ch. 3), to be most illuminating; I suspect that the similarities by far outweigh the differences.

The considerations presented make only a first step. As I have mentioned, the NCF's which are objectifiable in some respect are not very recommendable inductive schemes (and some are worse than others). Likewise on the probabilistic side,

[38] Cf., e.g., Putnam (1983c).

Bernoulli measures, for instance, according to which the variables considered are independent (and identically distributed) are bad inductive guides. Thus, the necessary second step is to follow de Finetti's strategy, i.e. to inquire the following questions. Which epistemic states are mixtures of objectifiable epistemic states? Which epistemic states can be uniquely represented as mixtures of objectifiable epistemic states? And which of these mixtures converge with increasing evidence to some objectifiable epistemic state?

There are powerful positive results in the probabilistic case: de Finetti's original theorem, a de Finetti-type theorem for Markov chains (proved in Diaconis and Freedman 1980), and, as the strongest of all, the ergodic theorem (cf. Skyrms 1984, ch. 3). The concept of a mixture makes sense also for NCFs; it is easily specified how to mix the NCFs in a given set by a NCF defined on that set so that the mixture is again a NCF. But, of course, there do not exist answers to the questions in terms of NCFs.[†2]

[†2] In chapter 7 I have finally started giving these answers.

Part III
Laws

Chapter 6
Laws, Ceteris Paribus Conditions, and the Dynamics of Belief[†1]

6.1 Preparations[1]

Laws are true lawlike sentences. But what is lawlikeness? Much effort went into investigating the issue, but the richer the concert of opinions became, the more apparent their deficiencies became, too, and with it the profound importance of the issue for epistemology and philosophy of science.

The most widely agreed prime features are that laws, in contrast to accidental generalizations, support counterfactuals, have explanatory power, and are projectible from, or confirmed by, their instances. These characteristics have long been recognized. However, the three topics they refer to – counterfactuals, explanation, and induction – were little elaborated in the beginning and are strongly contested nowadays. Moreover, the interrelations between these subjects were quite obscure. Hence, these features did not point to a clear view of lawlikeness, either. In this paper, I try to advance the issue. We shall see that the advance naturally extends to *ceteris paribus* laws, the general topic of this collection. Let me start with three straight decisions.

The first decision takes a stance on the priority of the prime features. I am convinced that it is the inductive behavior associated with laws which is the most basic one, and that it somehow entails the other prime features. I cannot justify

[†1] This paper was originally published in: *Erkenntnis* 57 (2002) 373–394, a special issue on ceteris paribus laws edited by J. Earman, C. Glymour, and S. Mitchell. It is this issue I refer to when speaking of "this collection" in this essay.

[1] I am deeply indebted to Christopher von Bülow, Ludwig Fahrbach, Volker Halbach, Kevin Kelly, Manfred Kupffer, Arthur Merin, and Eric Olsson for a great lot of valuable comments. I have taken up many of them, but I fear I have dismissed the more important ones, which showed me how many issues would need to be clarified and substantiated, and which would thus require a much longer paper. I am also indebted to Ekkehard Thoman for advice in Latin.

Nancy Cartwright remarks, at the very end of the introduction of her book (1989), that my views on causation are closest to hers. The closeness, though, may not be easy to discover. This is the first paper for several years in which I continue on our peculiar harmony. I dedicate it to her.

this stance in a few lines. Suffice it to say that my study of causation (1983a) led me from Lewis' (1973a) theory of counterfactuals over Gärdenfors' epistemic account of counterfactuals (cf., e.g., Gärdenfors 1981) ever deeper into the theory of induction where I finally thought I had reached firm ground. In Spohn (1991) [here: ch. 9]. I explained my view on the relation of induction to causation and thus to explanation. However, I did not return to counterfactuals (because I always felt that this subject is overlaid by many linguistic intricacies that are quite confusing). My decision finds strong support in Lange (2000) who starts investigating the relation between laws and counterfactuals and also arrives at induction as the most basic issue.

The second decision concerns the relation between laws and their prime features. When inquiring into lawlikeness the idea often was to search for something which *allows* us to use laws in induction, explanation and counterfactuals in the way we do. That is, given that induction is really the most basic aspect, lawlikeness should be something that *justifies* the role of laws in induction. This idea issued in perplexity; no good candidate could be found providing this justification.

There is an alternative idea, namely that lawlikeness is *nothing but* the role of laws in induction. In view of the history of inductive scepticism from Hume to Goodman – which made us despair of finding a deeper justification of induction and taught us rather to describe our inductive behavior and to inquire what is rational about it while being aware that this inquiry may produce only partial justification – this idea seems to be the wiser one. I do not mean to suggest that the lessons of inductive scepticism have been neglected; for instance, Lange (2000) endorses these lessons when explaining what he calls the root commitment concerning the inductive strategies associated with laws. But it is important to be fully aware of these lessons, and hence I shall pursue here the second idea and foreswear the search for deeper justifications. We shall see that we can still say quite a lot about rational induction.

We are thus to study the inductive properties of laws. This presupposes some account of induction or confirmation within which to carry out the study. This is what my third decision is about. I think that on this matter philosophy of science went entirely wrong in the last 25 years. Bayesianism was always strong, and rightly so. In the 1950s and 1960s much effort also went into the elaboration of a qualitative confirmation theory. However, this project was abandoned in the 1970s. The main reason was certainly that the efforts were not successful at all. Niiniluoto (1972) gives an excellent survey that displays the incoherencies of the various attempts. An additional reason may be the rise and success of the theory of counterfactuals, which answered many problems in philosophy of science (though not problems of induction) and thus attracted a lot of the motivation originally directed to an account of induction.

In any case, the effect was that Bayesianism was more or less the only remaining well-elaborated alternative. This hindered progress, because deterministic laws and probability do not fit together well. Deterministic laws are not simply the limiting case of probabilistic laws, just as deterministic causation is not the limiting case of probabilistic causation. It is, for instance, widely agreed that the entire issue of *ceteris paribus* laws, to which we shall turn below, cannot find an adequate probabilistic explication. We find a parallel in the disparity between belief, or acceptance-as true, and subjective probability, which was highlighted by the lottery paradox and has as yet not found a convincing reconciliation. My conclusion is, though I have hardly argued for it, that Bayesianism is of little help in advancing the issue of lawlikeness.

Philosophical logic was very active since around 1975 in producing alternatives, though not under the labels "induction" or "confirmation". However, these activities were hardly recognized in philosophy of science. Instead, they radiated to AI where they were rather successful. It is precisely in this area where we shall find help. Let me explain.

What should we expect an account of induction to achieve? I take the view (cf. Spohn 2000a) that it is equivalent to a theory of belief revision or, more generally, to an account of the dynamics of doxastic states. This is why the topic is so inexhaustible. Everybody, from the neurophysiologist to the historian of ideas, can contribute to it, and one can deal with it from a descriptive as well as a normative perspective.

Philosophers, I assume, would like to come up with a very general normative account. Bayesianism provides such an account that is almost complete. There, rational doxastic states are described by probability measures, and their rational dynamics is described by various conditionalization rules. As mentioned, however, in order to connect up with deterministic laws, we should proceed with an account of doxastic states which represents plain belief or acceptance-as-true. Doxastic logic is sufficient for the statics, but it does not provide any dynamics. Probability < 1 cannot represent belief, because it does not license the inference from the beliefs in two conjuncts to the belief in their conjunction. Probability 1 cannot do it, either, because we would like to be able to update with respect to information previously disbelieved, because disbelieved propositions would have probability 0 according to this approach, and because Bayesian dynamics does not provide an account of conditionalization with respect to null propositions (that is why I called Bayesianism almost complete). Hence, Bayesianism is unhelpful. Belief revision theory (cf., e.g., Gärdenfors 1988) was deviced to fill the gap. Unfortunately, the dynamics it provides turned out to be incomplete as well (cf. Spohn 1988, sect. 3) [here: sect. 1.3]. There have been several attempts to plug the holes (cf., e.g., Nayak 1994 and Halpern 2001), but I still think that ranking theory, proposed in Spohn (1983a, sect. 5.3 and 1988, [here: ch. 1]), though under a different name, offers the most convincing account for a full dynamics of plain belief.

In any case, this is my third decision: to carry out my study of the inductive behavior of laws strictly in terms of the theory of ranking functions. This framework may be unfamiliar, but the study will not be difficult, since ranking theory is a very simple theory. Still, there will be little space for broader discussion. Some of my results may appear trivial and some strange. On the whole, though, the study seems to me to be illuminating. But see and judge by yourself!

The plan of the paper is now almost obvious. In Section 6.2 I shall introduce the theory of ranking functions as far as needed. Section 6.3 explicates lawlikeness, i.e., the difference between laws and accidental generalizations insofar as it can be expressed in ranking terms. We shall see that this explication naturally leads to an inquiry of the role of *ceteris paribus* conditions and the like, a task taken up in Section 6.4. Since Section 6.3 analyzes belief in a law not as a belief in a regularity or some more sophisticated proposition, but rather as a certain inductive attitude, the immediate question arises how a law, i.e., such an inductive attitude, can be confirmed. This crucial question is addressed in Section 6.5. Section 6.6 will close with a few comparative remarks.

I thus focus entirely on the epistemological aspects of laws. I do not deny, but only neglect that laws have important metaphysical aspects as well. I have been less negligent in Spohn (1993a) [here: ch. 5], where I tried to understand causal laws as objectifications of inductive schemes, and in Spohn (1997c) [here: ch. 12], where I discussed both aspects of reduction sentences, the laws associated with disposition predicates. The two papers thus partially precede and partially transcend the present paper, and the unity of the three papers is less than perfect.

6.2 Ranking Functions

Let us start with a set W of possible worlds, small rather than large worlds, as we shall see soon. Each subset of W is a truth condition or *proposition*. I assume propositions to be the objects of doxastic attitudes. Thus I take these attitudes to be intensional. We know well that this is problematic, and we scarcely know what to do about the problem. Hence, my assumption is just an act of front alignment.

The assumption also entails that we need not distinguish between propositions and sentences expressing them. Hence, I shall often use first-order sentences to represent or denote propositions and shall not distinguish between logically equivalent sentences, since they express the same proposition.

That is all we need to introduce our basic notion: κ is a *ranking function* (for W) iff κ is a function from W into $\mathbf{N}$ (the set of non-negative integers) such that $\kappa(w) = 0$ for some $w \in W$. For each proposition $A \subseteq W$ the *rank* $\kappa(A)$ of A is defined by $\kappa(A) = \min \{\kappa(w) \mid w \in A\}$ and $\kappa(\emptyset) = \infty$. For $A, B \subseteq W$ the (*conditional*) *rank* $\kappa(B \mid A)$ of B *given* A is defined by $\kappa(B \mid A) = \kappa(A \cap B) - \kappa(A)$. Since singletons of worlds are propositions as well, the point and the set function are interdefinable. The point function is simpler, but auxiliary, the set function is the one to be interpreted as a doxastic state.

Indeed, ranks are best interpreted as *grades of disbelief*. $\kappa(A) = 0$ says that A is not disbelieved at all. It does not say that A is believed; this is rather expressed by

$\kappa(\bar{A}) > 0$,[2] i.e., that non-A is disbelieved (to some degree).[3] The clause that $\kappa(w) = 0$ for some $w \in W$ is thus a *consistency* requirement. It guarantees that at least some proposition, and in particular W itself, is not disbelieved. This entails the *law of negation*: for each $A \subseteq W$, either $\kappa(A) = 0$ or $\kappa(\bar{A}) = 0$ or both.

The set $C_\kappa = \{w \mid \kappa(w) = 0\}$ is called the *core* of κ (or of the doxastic state represented by κ). C_κ is the strongest proposition believed (to be true) in κ. Indeed, a proposition is believed in κ if and only if it is a superset of C_κ. Hence, the set of beliefs is *deductively closed* according to this representation.

There are two laws for the distribution of grades of disbelief. The *law of conjunction*: $\kappa(A \cap B) = \kappa(A) + \kappa(B \mid A)$, i.e., the grade of disbelief in A and the grade of disbelief in B given A add up to the grade of disbelief in A-and-B. And the *law of disjunction*: $\kappa(A \cup B) = \min\{\kappa(A), \kappa(B)\}$, i.e., the grade of disbelief in a disjunction is the minimum of the grades of the disjuncts. The latter is again only a consistency requirement, though a conditional one; if that law would not hold the inconsistency could arise that both $\kappa(A \mid A \cup B)$, $\kappa(B \mid A \cup B) > 0$, i.e., that both A and B are disbelieved given A-or-B.

According to the above definition, the law of disjunction indeed extends to disjunctions of arbitrary cardinality. I find this reasonable, since an inconsistency is to be avoided in any case, be it finitely or infinitely generated. Note that this entails that each countable set of ranks has a minimum and thus that the range of a ranking function is well-ordered. Hence, the range **N** is a natural choice.[4]

However, here we better avoid all complexities involved in infinity. Therefore I shall outright assume that we are dealing only with finitely many worlds and hence only with finitely many propositions. This entails that each world in W (or the set of its distinctive features) is finite in turn. Hence, as announced, they are small worlds. One may think that this is a strange start for an investigation of natural laws. However, an analysis of lawlikeness should work also under this finiteness assumption. After all, our world seems both to have laws and to be finite. Generalizing my observations below to the infinite case would require a separate paper.

There is no need here to develop ranking theory extensively. A general remark may be more helpful: ranking theory works in almost perfect parallel to probability theory. Take any probability theorem, replace probabilities by ranks, the sum of probabilities by the minimum of ranks, the product of probabilities by the sum of ranks, and the quotient of probabilities by the difference of ranks, and you are almost guaranteed to arrive at a ranking theorem. For instance, you thus get a ranking version of Bayes' theorem. Or you can develop the whole theory of Bayesian nets in ranking terms. And so on. The general reason is that one can roughly interpret ranks as the orders of magnitude of (infinitesimal) probabilities.

---

[2] $\bar{A}$ is the complement or the negation of A.

[3] I apologize for the double negation; after a while one gets used to it.

[4] In Spohn (1988)[here: ch. 1]. I still took the range to consist of arbitrary ordinal numbers. But the advantages of this generality did not make up for the complications.

The parallel extends to the laws of doxastic change, i.e., to rules of conditionalization. Thus, it is at least plausible that ranking theory provides a complete dynamics of doxastic states (as may be shown in detail; cf. Spohn 1988, sect. 5) [here: sect. 1.5].

It is still annoying, perhaps, that belief is not characterized in a positive way. But there is remedy: β is the *belief function* associated with κ (and thus a belief function) iff β is the function assigning integers to propositions such that $\beta(A) = \kappa(\bar{A}) - \kappa(A)$ for each $A \subseteq W$. Similarly, $\beta(B \mid A) = \kappa(\bar{B} \mid A) - \kappa(B \mid A)$. Recall that at least one of the terms $\kappa(\bar{A})$ and $\kappa(A)$ must be 0. Hence, $\beta(A) > 0$, < 0, or $= 0$ iff, respectively, A is believed, disbelieved, or neither; and A is the more strongly believed, the larger $\beta(A)$. Thus, belief functions may appear to be more natural. But their formal behavior is more awkward. Therefore I shall use both notions.

Above, I claimed that a full dynamics of belief is tantamount to an account of induction and confirmation. So, what is confirmation with respect to ranking functions? The same as elsewhere, namely *positive relevance: A confirms* or is a *reason for B* relative to κ iff $\beta(B \mid A) > \beta(B \mid \bar{A})$, i.e., iff $\kappa(\bar{B} \mid A) > \kappa(\bar{B} \mid \bar{A})$ or $\kappa(B \mid A) < \kappa(B \mid \bar{A})$ or both.[5]

There is an issue here whether the condition should require $\beta(B \mid A) > \beta(B)$ or only $\beta(B \mid A) > \beta(B \mid \bar{A})$, as stated. In the corresponding probabilistic case, the two conditions are equivalent if all three terms are defined, but the first condition is a bit more general, since it may be defined while the second is not. That is why the first is often preferred. In the ranking case, however, all three terms are always defined, and the second condition may be satisfied while the first is not. In that case the second condition on which my definition is based seems to be more adequate.[6]

A final point that will prove relevant later on: Ranking functions can be mixed, just as probability measures can. For instance, if κ_1 and κ_2 are two ranking functions for W and if κ^* is defined by

$$\kappa^*(A) = \min\{\kappa_1(A), \kappa_2(A) + n\} \text{ for some } n \in \mathbf{N} \text{ and all } A \subseteq W,$$

then κ^* is again a ranking function for W. Or more generally, if K is a set of ranking functions for W and ρ a ranking function for K, then κ^* defined by

$$\kappa^*(A) = \min\{\kappa(A) + \rho(\kappa) \mid \kappa \in K\} \text{ for all } A \subseteq W$$

is a ranking function for W. The function κ^* may be called the *mixture* of K by ρ.

---

[5] I believe that if epistemologists talk of justification and warrant, they should basically refer to this relation of A being a reason for B; cf. Spohn (2001b). That's, however, a remark about a different context.

[6] A relevant argument is provided by the so-called problem of old evidence. The problem is that after having accepted the evidence it can no longer be confirmatory. However, this is so only on the basis of the first condition. According to the second condition, learning about A can never change what is confirmed by A, and hence the problem does not arise. This point, or its probabilistic analogue, is made by Joyce (1999, sect. 6.4) with the help of Popper measures.

This is all the material we shall need. I hope that the power and beauty of ranking theory is apparent already from this brief introduction. I have not argued here that if one wants to state a full dynamics of plain belief or acceptance-as-true, one must buy into ranking theory. I did so in Spohn (1988, sect. 3) [here: sect. 1.3]. Even that argument may not be entirely conclusive. However, I guess the space of choices is small, and I would be very surprised if a simpler choice than ranking theory were to be available.

Be this as it may, let us finally turn to our proper topic, the epistemology of laws.

6.3 Laws

Let me start with a simple formal observation. Given some ranking function κ, to believe $A \wedge B$ means that $C_\kappa \subseteq A \cap B$, i.e., $\kappa(\neg A \vee \neg B) > 0$, i.e., $\min\{\kappa(\neg A), \kappa(\neg B)\} > 0$. This, however, can be implemented in many different ways. In particular, it leaves open how $\kappa(\neg A \vee \neg B)$ relates to $\kappa(\neg A)$ and $\kappa(\neg B)$ and thus whether or not $\kappa(\neg B \mid \neg A) = 0$. Hence, if you start with believing $A \wedge B$, but now learn that $\neg A$ obtains, you may, or may not, continue to believe B, depending on the value of $\kappa(\neg B \mid \neg A)$.

Basically the same point applies to believing a universal generalization. This, I propose, is the clue to understanding laws. Let us take $G = \bigwedge x(Px \rightarrow Qx)$ as our prototypical generalization ($\rightarrow$ always denotes material implication). I have already simplified things by assuming the worlds in W to be finite. This entails that the quantifier in G ranges over some finite domain D. For $a \in D$, let G_a be the instantiation of G by a, i.e., $G_a = Pa \rightarrow Qa$. Now to believe G in κ means that $C_\kappa \subseteq G$, i.e., $\kappa(\neg G) > 0$, i.e., $\min\{\kappa(\neg G_a) \mid a \in D\} > 0$. Thus, the generalization is believed as strongly as the weakest instantiation.[7]

Let us assume, moreover, that this is the only belief in κ, i.e., that $C_\kappa = G$; thus, no further beliefs interfere. This entails in particular that $\kappa(Pa \wedge Qa) = \kappa(\neg Pa \wedge Qa) = \kappa(\neg Pa \wedge \neg Qa) = 0 < \kappa(Pa \wedge \neg Qa)$ for each $a \in D$ and hence that $\kappa(\neg Qa \mid Pa) > 0$, i.e., that Pa is positively relevant for Qa. In other words, under this assumption the belief in the material implication $Pa \rightarrow Qa$ is equivalent to the positive relevance of Pa for Qa.

Again, the belief in G can be realized in many different ways. Let me focus for a while on two particular ways, which I call the "persistent" and the "shaky" attitude. If you learn about positive instances, G_a, G_b, etc. you do not change your beliefs according to κ, since you expected them to be positive, anyway.[8] The crucial

[7] Note, by the way, that this would also hold for an infinite domain of quantification. Hence, for ranking theory there is no problem of null confirmation for universal generalizations which beset Carnap's inductive logic.

[8] I am using here a technical notion of positive instance: a is a positive instance of G iff G_a, i.e. $Pa \rightarrow Qa$, is true. If $Pa \wedge Qa$, a positive instance in the intuitive sense, would be learnt, the beliefs would change, of course (at least given our assumptions that nothing except G is believed in κ).

difference emerges when we look how you respond to negative instances, $\neg G_a$, $\neg G_b$, etc. according to the various attitudes.

If you have the *persistent* attitude,[9] your belief in further instantiations is unaffected by negative instances, i.e., $\kappa(\neg G_b) = \kappa(\neg G_b \mid \neg G_a)$ $(b \neq a)$, and indeed $\kappa(\neg G_b) = \kappa(\neg G_b \mid \neg G_{a_1} \wedge \ldots \wedge \neg G_{a_n})$ for any $n \in \mathbf{N}$ $(b \neq a_1,\ldots,a_n)$. If, by contrast, you have the *shaky* attitude, your belief in further instantiations is destroyed by a negative instance, i.e., $\kappa(\neg G_b \mid \neg G_a) = 0$ and, a fortiori, $\kappa(\neg G_{\neq a} \mid \neg G_a) = 0$.[10]

The difference is, I find, characteristic of the distinction between lawlike and accidental generalizations. Let us look at two famous examples. First the coins:

(1) All German coins are round
(2) All of the coins in my pocket today are made of silver

It seems intuitively clear to me that we have the persistent attitude towards (1) and the shaky attitude towards (2). If we come across a cornered German coin, we wonder what might have happened to it, but our confidence that the next coin will be round again is not shattered. If, however, I find a copper coin in my pocket, my expectations concerning the further coins simply collapse; if (2) has proved wrong in one case, it may prove wrong in any case.

Or look at the metal cubes, which are often thought to be the toughest example:

(3) All solid uranium cubes are smaller than 1 mile^3.
(4) All solid gold cubes are smaller than 1 mile^3.

What I said about (1) and (2) applies here as well, I find. If we bump into a gold cube this large, we are surprised – and start thinking there might well be further ones. If we stumble upon an uranium cube of this size, we are surprised again. But we find our reasons for thinking that such a cube cannot exist unafflicted and will instead start investigating this extraordinary case (if it obtains for long enough).

As far as I see, this difference applies as well to the other examples prominent in the literature (cf., e.g., the overview in Lange 2000, pp. 11f.). However, my wording is certainly more determined than my thinking. According to my survey, intuitions are often undecided. In particular, the attitude seems to depend on how one came to believe in the regularity; there may be different settings for one and the same generalization. However, at the moment I am concerned with carving out what appears to me to be the basic difference. Therefore I am painting black and white. As we shall see, ranking theory will also allow for a more refined account.

In any case, what the examples suggest is this: We treat a universal generalization G as lawlike if we have the persistent attitude towards it, and we treat it as accidental if we have the shaky attitude towards it. Hence, the difference does not

---

[9] "Resilient" might be an appropriate term as well, but I do not want to speculate whether this would be a use of "resilient" similar or different to the one introduced by Skyrms; cf., e.g., Skyrms (1980).

[10] Here, $G_{\neq a}$ stands for $\bigwedge x(x{\neq}a \rightarrow G_x)$. Note that we have $\kappa(\neg G \mid \neg G_a) = 0$ according to both the persistent and the shaky attitude, simply because $\neg G_a$ logically implies $\neg G$.

lie in the propositional content, it lies only in our inductive attitude towards the generalization or, rather, its instantiations.[11]

Given how much we have learned from Popper about philosophy of science, this conclusion is really ironic, since it says in a way that it is the mark of laws that they are *not* falsifiable by negative instances; it is only the accidental generalizations that are so falsifiable. Of course, the idea that the belief in laws is not given up so easily is familiar at least since Kuhn's days (and even Popper insisted from the outset that falsifications of laws proceed by counter-laws rather than simply by counter-instances). But I cannot recall having seen the point being stripped down to its induction-theoretic bones.

What I have said so far may provoke a confusion that I should hurry up to clarify. The persistent attitude towards $G = \bigwedge x(Px \to Qx)$ is characterized, I said, by the *independence* of the instantiations; experience of one instance does not affect belief about the others. In this way, belief about an instance G_b, i.e., the positive relevance of Pb for Qb, is persistent. But didn't we learn that one mark of lawlikeness is *enumerative induction*, i.e., the confirmation of the law by positive instances? Surely, enumerative induction outright contradicts the independence I claim.

Herein lies a subtle confusion. Belief in a law is more than belief in a proposition, it is a certain doxastic attitude, and that attitude as such is characterized by the independence in question. If I would have just this attitude, just this belief in a law, my κ would exhibit this independence. Enumerative induction, by contrast, is not about what the belief in a law *is*, but about how we may acquire or confirm this belief. The two inductive attitudes involved may be easily confused, but the confusion cannot be identified as long as one thinks belief in a law is just belief in a proposition.

However, what could it mean at all to confirm a law if it does not mean to confirm a proposition? Indeed, my definition in Section 6.2 applies only to the latter, and to talk of the confirmation of laws, i.e., of a second-order inductive attitude towards a first-order inductive attitude, is at best metaphorical so far; enumerative induction or falsificationism do not seem to make sense within this setting. In Section 6.5 I shall make a proposal for translating and saving enumerative induction and the falsification of laws. But here and in the next section I am concerned only with the attitude in which the belief in a law itself consists.

Is my explanation of lawlikeness a deep one? No, it is just as plain as, for instance, that of the counterfactual theorist who says that lawlikeness *is* support of counterfactuals or that a law *is* a universally quantified subjunctive conditional. Analysis has to start somewhere, and it acquires depth only by showing how to explain other features of laws by the basic ones. That is a task that cannot be

---

[11] In arriving at this conclusion, I am obviously catching up with Ramsey (1929) who states very early and very clearly: "Many sentences express cognitive attitudes without being propositions; and the difference between saying yes or no to them is not the difference between saying yes or no to a proposition" (pp. 135f.). "... laws are not either" [namely propositions] (p. 150). Rather: "The general belief consists in (*a*) A general enunciation, (*b*) A habit of singular belief" (p. 136).

pursued here.[12] But I would like to insist that, as a starting point, the present analysis is to be preferred. There are good reasons for feeling uneasy about starting with subjunctives or a similarity relation between worlds. By contrast, ranking theory is a very plain theory with a very obvious interpretation.

The only doubt one may have about my starting point may concern its sufficiency as a basis of analysis. In particular one may feel that the crucial property of laws is one which justifies the inductive attitude I have described, say, some kind of material or causal necessity. Maybe. But I am skeptical and refer to my second decision in Section 6.1.

This does not mean that I have to sink into subjectivism, that I am bound to say that it is merely a matter of one's inductive taste what one takes to be a law. There may be objectivizations and rationalizations for our beliefs in laws. I do not intend to start speculating about this, but one very general rationalization is quite obvious. It is of vital importance to us to have persistent attitudes to a substantial extent. Something is almost always going wrong with our generalizations, and if we always had the shaky attitude, our inductions and expectations would break down dramatically and we could not go on living.

But of course, it is high time to admit that the distinction between the persistent and the shaky attitude is too coarse. It is not difficult, though, to gain a systematic overview within ranking theory. Let us see how many ways there are to believe the generalization G, i.e., for $\kappa(\neg G) > 0$. A natural and strongly simplifying assumption is

Symmetry: For all $a_1,\ldots,a_n,b_1,\ldots,b_n \in D$

$$\kappa(\neg G_{a_1} \wedge \ldots \wedge \neg G_{a_n}) = \kappa(\neg G_{b_1} \wedge \ldots \wedge \neg G_{b_n}).$$

In obvious analogy to inductive logic, symmetry says that the disbelief in violations of a generalization depends on their number, but not on the particular instances. For $n = 1$ symmetry entails that there is some $r > 0$ such that for all $a \in D$ $\kappa(\neg G_a) = \kappa(\neg G) = r$. More generally, symmetry entails, as is easy to see, that there is some function c from **N** to **N** such that for any $n + 1$ different $a_1,\ldots,a_n,b \in D$ the equality $\kappa(\neg G_b \mid \neg G_{a1} \wedge \ldots \wedge \neg G_{an}) = c(n)$ holds, where $c(0) = r$. Indeed, all ranks of all Boolean combinations of the G_a are uniquely determined by the function c.

Another plausible assumption familiar from inductive logic is

Non-negative instantial relevance: For all $a_1,\ldots,a_n,a_{n+1},b \in D$

$$\kappa(\neg G_b \mid \neg G_{a_1} \wedge \ldots \wedge \neg G_{a_n}) \geq \kappa(\neg G_b \mid \neg G_{a_1} \wedge \ldots \wedge \neg G_{b_n} \wedge \neg G_{a_{n+1}}).$$

This is tantamount to the function c being non-increasing.

[12] But see my account of causal explanation in terms of ranking functions in Spohn (1991) [here: ch. 9].

Given the two assumptions there remain not so many ways to believe G; any non-increasing function c with $c(0) = r$ stands for one such way. Hence, the persistent attitude characterized by $c(n) = r$ for all n stands for one extreme, whereas the shaky attitude for which $c(n) = 0$ for $n \geq 1$ stands for the other. So, one may think about whether any ways in between fit the examples better than the extreme ones. Still, the consideration shows that the two attitudes I have discussed at length are suited best for marking the spectrum of possible attitudes.

6.4 Other Things Being Equal, Normal, or Absent

It is commonplace by now that laws or their applications are often to be qualified by some kind of *ceteris paribus* condition. As long as a law is conceived of as a proposition, the nature of this qualification is hard to understand. It seems to make the proposition indeterminate or trivial. But when we conceive of belief in a law as more than belief in a proposition, at least some mysteries dissolve in quite a natural way. Indeed, the account of laws given above almost yearns to be amended by such qualifications.

We should start, though, with the observation, often made in the literature, that we are dealing here with a mixed bag of qualifications. "*Ceteris paribus* condition" seems to have established itself as the general term, although it is clear to everyone that it really refers only to one kind of qualification. "*Ceteris paribus*" = "other things being equal" is obviously a relational condition. But what does it relate to? We shall return to this question. Another frequent qualification is that a law holds only in the absence of disturbing influences.[13] Still another way of hedging is to say that a law holds only under normal conditions.[14] A fourth kind are ideal conditions that are assumed by idealized laws though they are known not to obtain strictly. And there are other kinds, perhaps.

Yet another thing unclear is what exactly the qualifications are to act on. Some say it is the laws themselves that are hedged by the various conditions, while Earman and Roberts (1999) insist that the conditions exclusively pertain to the applications of laws to particular situations. Hence, provisoes in the sense of Hempel (1988, p. 151) which are "essential, but generally unstated, presuppositions of theoretical inferences" and hence part of the applications do not cover the phenomenon in full breadth, either.

This shows that the topic is not so uniform. Indeed, the inhomegeneity is common theme in this collection. Still, let us squarely approach the topic from the

[13] Some call this a *ceteris absentibus* condition. My Latin expert informs me, though, that "*ceteris absentibus*" usually means only "other men (and not women or non-human things) being absent."

[14] My Latin expert also tells me that there is not really a good translation of "other things being normal" into Latin.

vantage point reached so far. This will illuminate at least normal conditions and the absence of disturbing factors.

We have arrived at the result that the belief in the generalization $G = \bigwedge x(Px \rightarrow Qx)$ as a law is represented by having $\kappa(\neg G_a) > 0$ for each a $\in D$ in a persistent way, i.e., unshattered by violations of the law. I have praised persistence as a virtue. But, to be honest, does it not appear just narrow-minded? Violations of a law are cause for worry, not for stubbornness. Sure, but the worry should concern the violation, not the future. Indeed, ranking functions provide ample space for such worries. There may yet be a ramified substructure of additional conditions. Let me explain.

Suppose $\kappa(Pa) = 0$ and $\kappa(\neg Qa \mid Pa) = r > 0$, that is, you do not exclude *Pa* and believe *Qa* given *Pa* according to κ. This allows for there being an exceptional condition *Ea* such that $\kappa(\neg Qa \mid Pa \wedge Ea) = 0$. This is due to the non-monotonicity of defeasible reasoning embodied in a ranking function. Of course, this entails via the ranking laws that $\kappa(Ea \mid Pa) \geq r$, i.e., that the exceptional condition *Ea* is at least as strongly disbelieved as the violation of the law itself.

This, I find, is quite an appropriate schematic description of what actually goes on. We encounter a violation of a law, we are surprised, we inquire more closely how this was possible, and we find that some unexpected condition is realized under which we did not assume the law to hold, anyway. In this way, hence, each ranking function representing the belief in the law *G* automatically carries an aura of *normal conditions* which is implicit at the level of belief, i.e., the function's core, and becomes explicit only if we look more deeply at the substructure below the core.

This substructure may indeed dispose to further changes of opinion. There may, e.g., be a further condition *E′a* such that the law *G* is reinstalled, i.e., $\kappa(\neg Qa \mid Pa \wedge Ea \wedge E'a) > 0$ for all $a \in D$. Defeasible reasoning may have arbitrarily many layers according to a ranking function.

Relative to a given κ embodying the belief in the law *G* we can even define the normal conditions hedging *G*. For, if *Ea* and *Fa* are exceptional conditions, $Ea \vee Fa$ is so as well. $\kappa(\neg Qa \mid Pa \wedge Ea) = \kappa(\neg Qa \mid Pa \wedge Fa) = 0$ is easily seen to imply $\kappa(\neg Qa \mid Pa \wedge (Ea \vee Fa)) = 0$. Hence, the disjunction E^* of all exceptional properties *E* for which $\kappa(\neg Qa \mid Pa \wedge Ea) = 0$ for all $a \in D$ (or for some $a \in D$, if symmetry is given) is the *weakest* exceptional property, and we may thus define the normal conditions N^* pertaining to *G* (relative to κ) as the complement or negation of E^*.

Note that N^* is not simply the disjunction *N* of all maximal properties *M* such that the law *G* holds given *M*, i.e., $\kappa(\neg Qa \mid Pa \wedge Ma) > 0$ for all $a \in D$. N^* is at least as strict as *N* and usually stricter. For instance, the condition $E \wedge E'$ under which the law *G* was assumed to be reinstalled two paragraphs above would be a specification of *N*, but not of N^*. The example also shows that normal conditions are more adequately explicated by N^*, because the condition $E \wedge E'$ should count as doubly exceptional and indeed counts as exceptional according to N^*, whereas it would count as normal according to *N*.

In any case, I find it entirely appropriate that normal conditions are thus explicated relative to a given doxastic state. Normalcy is something in the eye of the observer, in the first place, and therefore it is best described via its epistemic functioning. And ranking functions are particularly suited to grasp this.

However, this specifies only the statics of normal conditions. But we are rather interested in their dynamics, i.e., in the way in which our conception of them changes. After all, if we encounter a violation of a law, closer inspection of the case will often not confirm our previous understanding of exceptions, but will instead inform and revise it. This issue, however, belongs under the heading "confirmation of laws", which I address only in the next section.

So much for the ramifications of the belief in a single law G. The next issue to face, hence, is: How to believe in several laws at once, in particular if they pertain to the same property? Let us look at the simplest example: Often we seem to believe in the law $G = \bigwedge x(Px \rightarrow Qx)$ *and* in a further law $G' = \bigwedge x(P'x \rightarrow \neg Qx)$ predicting *non-Q* for circumstances P'.[15] How can we do this?

This is the problem of the *superposition of laws* or, if the laws are causal, of the *interaction of causes*.[16] In mechanics the problem finds an elegant solution: the total force acting on a body is just the vector sum of the individual forces, each of which is governed by a specific force law. But in general there is no general solution. Only so much can be said.

It is possible to believe both in G and G', though only if one also believes that $\neg\bigvee x(Px \wedge P'x)$. This is simply a matter of logic.

From the ranking perspective two remarks must be added. First, both laws can also be believed in the sense explained here, but only if the disbelief in each instance $Pa \wedge P'a$ is sufficiently strong. Second, and more importantly, even if a ranking function κ represents the belief in both G and G' as laws it still contains a prediction for the unexpected case that a instantiates both P and P'; $\beta(Qa \mid Pa \wedge P'a)$ must take some value. Hence, if two competing laws are believed in κ, they are automatically superposed in κ in some way (which may well be suspension of judgment, i.e., $\beta(Qa \mid Pa \wedge P'a) = 0$).

Even though this description is very unspecific (and is bound to be so), there is one point where it seems to be false. The description assumes that for each law it is exceptional in the above sense that the other law applies as well in a given case. But this is not how we normally look at the laws. We should be able to account for the superposition of G and G' even if $\kappa(Pa \wedge P'a) = 0$. This is why the present problem cannot be subsumed under the problem of normal conditions. But what else could be the account?

The only way seems to be to make the laws exclusive, i.e., to modify G into $\bigwedge x(Px \wedge \neg P'x \rightarrow Qx)$ and G' into $\bigwedge x(P'x \wedge \neg Px \rightarrow \neg Qx)$ and to modify κ correspondingly. The laws did not make any prediction for the case $\neg Pa \wedge \neg P'a$, anyway. What is left open, hence, is the case $Pa \wedge P'a$, for which one may, and has to, assume some degree of (dis-)belief in Qa. The resulting κ, according to which three

[15] The more familiar case will be that the laws do not predict that a quality Q is present or absent, but rather that a magnitude assumes different values in a given object. From a logical point of view this does not make much of a difference. Let us stick here to the simplest case.

[16] For the following discussion see in particular Cartwright (1983, chs. 2 and 3).

laws, the modified G and G' and the new one, are believed, may also be called a superposition of the laws G and G'.[17] This consideration shows that the belief in a law as such, as I have described it, is implicitly understood in abstraction from other things, i.e., other relevant laws, and this abstraction is made explicit in the superposition in the second sense; i.e., in the modifications of G and G'.[18]

So, in which way do these remarks bear on the hedgings of laws familiar from the literature? Let me briefly summarize.

The account of normal conditions I have given is exactly the one compellingly suggested by the literature on non-monotonic reasoning, default logic, or whatever the labels were, which has been richly produced since 1975. What I add is only the conviction that ranking theory, owing to its completeness concerning induction or belief revision, provides the optimal base for studying these phenomena.

The absence of disturbing influences or factors may stand for various things. It may simply mean the presence of normal conditions. Or it may mean that the case at hand is not governed by a further law which would require some guess or knowledge as to how the laws involved superimpose. To this extent, at least, this kind of hedge is covered by my remarks.

What about *ceteris paribus* clauses? As already mentioned, they require a standard of comparison which is usually left implicit. The default standard, I guess, is given by the normal conditions. In this case, other things being equal just means other things being normal. If, however, the standard of comparison is taken as variable, then the clause yields what Schurz (2002) calls comparative CP-laws, or it amounts to some such principle like "equal causes, equal effects" or "induction goes by suchnesses, not thisnesses" which might be explicated by symmetry principles like the one above. But I shall not pursue this issue.

Finally, I have not said anything about idealizations. This seems to be a somewhat different topic. But I should at least mention that it is accessible to the belief revision perspective as well, as has been shown by Rott (1991).

6.5 On the Confirmation of Laws

At several crucial points we missed an account of the confirmation of laws, and it was quite unclear how to give one, since the issue is not about the confirmation of propositions, which was already well handled by ranking functions. My paper would be badly incomplete without such an account.

But I have a proposal. Indeed, it will not be a surprise to anyone who is aware of the close similarity between probability and ranking theory, who has in particular noticed that a law according to my conception is analogous to a sequence of

---

[17] The superposition in the second sense could also be conceived of as the contraction of a superposition in the first sense by $\neg\bigvee x(Px \wedge P'x)$.

[18] An alternative way to remove the apparent conflict between G and G', which was envisaged by Cartwright (1983, pp. 57ff.), is to say that G and G' are not about the same Q. Rather, G is about Q-as-caused-by-P, and G' about Q-as-prevented-by-P'. In substance, though, the problem of superposition remains the same under this alternative.

independent, identically distributed random variables, and who knows the work of de Finetti (1937). In his famous theorems de Finetti showed that there is a one-one-correspondence between symmetric probability measures for an infinite sequence of random variables and mixtures of Bernoulli measures according to which the variables are independent and identically distributed, and that the mixture concentrates more and more on a single Bernoulli measure as evidence accumulates. He thus showed to the objectivist that subjective symmetric measures provide everything he wants, i.e., beliefs about statistical hypotheses that converge toward the true one with increasing evidence.

The issue between objectivism and subjectivism is not my concern. Ranking functions are thoroughly epistemological and have as such no objective interpretation.[19] Still, we can immediately extract an account of the confirmation of laws from de Finetti's theory. Since this will look a bit artificial and formalistic, I shall demonstrate this with the basic construction and not discuss variants and ramifications.[†2]

Let us start with n mutually exclusive and jointly exhaustive properties or predicates $Q_1,\ldots,Q_n$ (these are Carnap's Q-predicates). For each $i \le n$ we have the elementary law $G^i = \neg\bigvee x Q_i x = \bigwedge x \neg Q_i x$. For any proposition $A \subseteq W$ we may now count how often the law G^i is violated if A obtains; this is done by the function $v(A, i) = \text{card}\{a \in D \mid A \subseteq Q_i a\}$.[20] So, if we define the ranking function κ^i for W by $\kappa^i(A) = v(A, i)$, κ^i precisely represents the belief in the law G^i. Without any evidence, though, we do not believe in any law G^i. Our attitude towards the laws is rather represented by the ranking function ρ_0 for which $\rho_0(\kappa^i) = 0$ for each $i = 1,\ldots,n$. Hence, our doxastic attitude towards the propositions $A \subseteq W$ is represented by the mixture κ_0 of the κ^i with respect to ρ_0, as defined by

$$\kappa_0(A) = \min_{i \le n} \kappa^i(A) + \rho_0(\kappa^i) = \min_{i \le n} \nu(A, i).$$

Now, how does this attitude change by experience? Via conditionalization, as always. But let us describe this in detail. Let $\mathbf{r} = \langle r_1,\ldots,r_n\rangle$ stand for any sequence of n non-negative integers, and let $r = r_1 + \ldots + r_n$. Define next $E(\mathbf{r})$ to be the proposition (evidence) that among the first r objects precisely r_i instantiate Q_i ($i = 1,\ldots,n$); the order of instantiation is irrelevant. Clearly, $\kappa_0(E(\mathbf{r})) = \min r_i$. Let B range over propositions about the remaining objects and not the first r ones, and let $\kappa_{\mathbf{r}}$ be the ranking function that we have for those propositions after receiving evidence $E(\mathbf{r})$. Then we have:

$$\begin{aligned}\kappa_{\mathbf{r}}(B) &= \kappa_0(B \mid E(\mathbf{r})) = \kappa_0(B \cap E(\mathbf{r})) - \kappa_0(E(\mathbf{r}))\\ &= \min_{i \le n}(\nu(B,i) + r_i) - \min_{i \le n} r_i = \min_{i \le n}(\kappa^i(B) + (r_i - \min_{i \le n} r_i)).\end{aligned}$$

That is, if we define $\rho_{\mathbf{r}}$ by $\rho_{\mathbf{r}}(\kappa^i) = r_i - \min_{i \le n} r_i$, we have

[19] But see Spohn (1993a) [here: ch. 5].

[†2] The subsequent considerations are thoroughly elaborated in Chapter 7.

[20] I am still jumping between sentences and the corresponding propositions as seems convenient to me.

$$\kappa_{\mathbf{r}}(B) = \min_{i \leq n}(\kappa^i(B) + \rho_{\mathbf{r}}(\kappa^i)).$$

Hence, $\kappa_{\mathbf{r}}$ is the mixture of the κ^i with respect to $\rho_{\mathbf{r}}$. So, the evidence $E(\mathbf{r})$ makes us change our attitude towards the laws from ρ_0 to $\rho_{\mathbf{r}}$, and $\rho_{\mathbf{r}}$ represents the degrees to which the various laws have been confirmed or rather disconfirmed. If $\rho_{\mathbf{r}}(\kappa^i) > 0$, we might say that κ^i is falsified, but note that falsification is never conclusive in this construction.

This account is essentially a translation of de Finetti's results into the framework of ranking functions. I find the translation basically plausible, and it strongly suggests following its course. One should characterize the class of ranking functions which represent mixtures of laws, and one should inquire the extent to which the representation is unique (for instance, there is an obvious one-one-correspondence between the $\kappa_{\mathbf{r}}$ and the $\rho_{\mathbf{r}}$ in the above mixtures). One should look at de Finetti's representation results for the infinite as well as for the finite case (recall the finiteness assumption made in this paper). The ranking analogue to de Finetti's notion of partial exchangeability would be particularly interesting. And so forth.[21]

On the other hand, the translation still looks artificial and quite detached from actual practice. For instance, if $\min r_i$ is large, one would tend to say that all of the laws G^i are disconfirmed by $E(\mathbf{r})$ and to conclude that none of the laws holds. One might account for this point by defining some κ^0 representing the belief in lawlessness, by mixing it into κ_0, say with the weight $\rho_0(\kappa^0) = s$, and by finding then that as soon as $\min r_i > s$ we have $\rho_{\mathbf{r}}(\kappa^i) = 0$ only for $i = 0$. Moreover, one might wonder how precisely this story of mixtures carries over to the belief in a given law and its possible hedgings by various possible normal conditions, since one would like to be able to account for one hedging rather than another being confirmed by the evidence. And so on.

All this shows that there is a lot of work to do in order to extend the proposal and to apply it to more realistic cases. Still, the message should be clear already from the case I have explained in detail. The theory of mixtures provides a clear account of what it means to confirm and disconfirm not only propositions, but also inductive attitudes such as ranking functions representing belief in laws. Hence I was not speaking metaphorically when I talked about such confirmation earlier in the paper.

6.6 Some Comparative Remarks

The literature on *ceteris paribus* laws is rich and disharmonious, and so far I have only added to the polyphony. Since the idea of this ERKENNTNIS issue was to promote harmony (which does not require everybody to play the same melody), I should close with some comparative remarks.

[21] See, e.g., the rich results collected in the papers in Carnap and Jeffrey (1971) and Jeffrey (1980).

So far, Schurz (1995) and Silverberg (1996) were the only ones to decidedly use the resources of non-monotonic reasoning for our topic (cf. also Schurz 2002, sect. 5). I emphatically continue on this line of thought, but we certainly have an argument about the most suitable account of non-monotonic reasoning.

What is novel to me is that the topic may also be approached from the learning-theoretic perspective. Indeed, I feel that Glymour (2002) and the present paper sandwich, as it were, the paper by et al. (2002), which is the focal challenge of this collection. How the two sides stick together is not clear. However, Kelly (1999) has established a general connection between formal learning theory and ranking theory, and the relation should become closer when one compares Kelly (forthcoming) with the present Section 6.5. So, let me briefly sketch my part of the pincer movement towards Earman et al. (2002), which will lead me across some other positions.

Clearly, my position is very close to that of Lange (2000), who says, for instance, that "the root commitment that we undertake when believing in a law involves the belief that a given inference rule possesses certain objective properties, such as reliability" (p. 189), and who reminds us on that occasion of the long tradition of the conception of laws as inference rules.[22] From a purely logical point of view, it was always difficult to see the difference between the truth of $\bigwedge x(Px \rightarrow Qx)$ and the validity of the rule "for any a, infer Qa from Pa". However, I find that the aspect of persistence, which was so crucial for me, is more salient in the talk of inference rules. Thus, what appeared to be merely a metaphorical difference turns out to have a precise induction-theoretic basis. It should have been clear, in any case, that ranking functions *are* (possibly very complex) inference rules, indeed, as my analysis of normal conditions has shown, defeasible inference rules that are believed to be reliable, but not necessarily universally valid. Hence, my account may perhaps be used to underpin Lange's much more elaborated theory, and conversely his many applications to scientific practice may confer liveliness and plausibility on my account.

To put the point differently, one might say that the emphasis in my account of laws is on the single case. The mark of laws is not their universality, which breaks down with one counter-instance, but rather their operation in each single case, which is not impaired by exceptions. Here, I clearly join Cartwright (1989) and her repeated efforts to explain that we have to attend to capacities and their cooperation taking effect in the single case. Her objective capacities or powers thus correspond to my subjective reasons as embodied in a ranking function, a correspondence which is salient again in the comparison of Cartwright (2002, sect. 2) with my Sections 6.3 and 6.4. However, as I already said, I am content here with my subjective correlate and do not discuss its objectivization.

This is what separates me from Cartwright also according to the classification of Earman and Roberts (1999). They distinguish accounts that try to provide truth conditions for *ceteris paribus* laws from accounts that focus rather on their

[22] The insight that the issues concerning laws fundamentally rest on the theory of induction rather than the theory of counterfactuals is more salient in Lange (2000) than in Lange (2002). However, the theory of induction takes a probabilistic turn in Lange (2000, ch. 4), a move about which I have already expressed my reservations.

pragmatic, methodological, or epistemological role, and they place Cartwright in the first group, whereas my account clearly belongs to the second. Hence, I appear to be exempt from their criticism. However, though I agree with many of their descriptions, e.g., when they say that "a '*ceteris paribus* law' is an element of a 'work in progress'" (p. 466), I feel that pragmatics is treated by them, as by many others before them, as a kind of waste-basket category that consists of a morass of important phenomena defying clear theoretical description.

This feeling is reinforced by Earman et al. (2002), who motivate their pragmatic or non-cognitivist turn in Section 6.4 by their finding in Section 6.3 that there is no solution to the "real trouble with CP-laws" that we have "no acceptable account of their semantics" and "no acceptable account of how they can be tested" (p. 292). In a way, the main purpose of this paper was to answer this challenge. To be sure, I did not provide a semantics in the sense of specifying truth conditions. But I gave an "epistemic semantics" in the sense of describing the doxastic role of unqualified as well as hedged laws, and I gave an account of how things having this role can be confirmed and disconfirmed. Of course, I did so on a fairly rudimentary formal level not immediately applicable to actual practice. But often, I find, the gist of the matter stands out more clearly when it is treated from a logical point of view.

Chapter 7
Enumerative Induction and Lawlikeness[†1]

7.1 Introduction*

Enumerative induction says that a law is confirmed by its positive instances or may be inductively inferred from them (in the absence of negative instances). It is, for sure, the most venerable and primitive of all inductive rules. But it has a bad press. It is very crude; science does not seem to proceed with such simple rules. Goodman's new riddle of induction has shown that enumerative induction is inconsistent, if generally applied; and it seems impossible to say what the appropriate restrictions are. On the face of it, it is a rule of qualitative confirmation theory; but philosophers have despaired of constructing such a theory.

The rule has finally found a Bayesian home. It is true, though, that at least within inductive logic as developed by Carnap (1971/80) nothing can confirm a law because each law has probability 0 (if its domain of quantification is infinite). The natural idea was then to turn enumerative induction into the Principle of Positive Instantial Relevance according to which each positive instance confirms that the next instance is also positive. This seems reasonable, and accepted. So, why bother any longer?

Well, "primitive" is ambiguous. It may indeed mean "not workable". But it also means "basic". If we do not fully understand the basic things, how can we ever hope to come to terms with the more complicated things? So whoever is concerned with inductive, plausible, or uncertain reasoning should be concerned to understand such a primitive rule as enumerative induction. The goal of the paper is to enhance this understanding. The way to reach the goal is to bring enumerative induction home from quantitative to qualitative confirmation theory, and the reason why this is feasible is that in the meantime we have a fully general qualitative confirmation theory at our disposal, namely ranking theory. This needs a little explanation.

[†1] This paper was originally published in: *Philosophy of Science* 72 (2005) 164–187. It is reprinted here with kind permission of the Philosophy of Science Association and the University of Chicago Press.

*I am indebted to two anonymous referees whose rich remarks led to numerous improvements and clarifications of this paper.

Traditionally, confirmation theory is a field within philosophy of science. Its quantitative or probabilistic version, i.e. Bayesianism, has been a major option from the beginning. In the 1950s and 1960s we also saw forceful attempts to construct a qualitative confirmation theory. However, the project was abandoned in the 1970s, for reasons nicely summarized in Niiniluoto (1972). Thus, at least within philosophy of science Bayesianism had won the day. However, logicians and computer scientists were very active since around 1975 in producing alternatives, though rarely under the labels 'induction' or 'confirmation' (see, e.g., the many theories collected in Gabbay and Smets 1998–2000, in particular vols. 1 and 3) and hence scarcely noticed in epistemology and philosophy of science. The multiplicity of proposals developed there is quite confusing. Still, I believe that ranking theory as developed by me in Spohn (1983a, sect. 5.3; 1988, [here: ch. 1]), though under a different name, is the most suitable qualitative account of induction or confirmation.

This introduction is not the place for extensively arguing the case; the old reasons given in Spohn (1988) still apply. Let me state only the most important point. The central notion in this connection is the notion of *conditional belief*. In order to say whether some evidence would qualitatively confirm some hypothesis we have, to put it vaguely, to look at whether the hypothesis would be believed given the evidence and not given the evidence. If we want to give an account of induction, we have to give an account of belief change; so I have argued in Spohn (2000a). And belief change best works by conditionalization rules that essentially refer to conditional beliefs, just as probabilistic conditionalization rules refer to conditional probabilities. We do need an adequate notion of conditional belief.

Hence, we should look at the various attempts to explain it. Belief revision theory (cf., e.g., Gärdenfors 1988) makes a plausible proposal: B is believed given A in a certain belief state iff B is believed in the revision of that state by A. But as I have argued in Spohn (1988) [here: ch. 1], belief revision theory, as it is presented up to date, is defective and the proposal therefore inadequate. One might say that B is believed given A iff $P(B \mid A) = 1$, but this proposal is incomplete, because in standard probability theory this conditional probability is undefined if $P(A) = 0$. One might insist on the proposal by interpreting P as a Popper measure that fills this incompleteness by taking conditional probability as an undefined primitive. However, as shown in Spohn (1986), this idea is defective in just the way belief revision theory is. And so on. In the end, I claim, one must turn to ranking theory that offers the most adequate account of conditional belief.

Here, I simply want to proceed on the basis of this scarcely redeemed claim. The point of the introductory remarks was only to suggest that the most promising way to study enumerative induction is in terms of ranking functions. This is what I want to do here. Hence, the plan of the paper is this. In Section 7.2 I shall introduce the theory of ranking functions as far as we need it here. Section 7.3 will then apply ranking theory to enumerative induction which, as we shall see, may realize in a variety of schemes. This will turn out to be a brief and rather boring exercise; the insights come later. In Section 7.4, I shall propose a ranking theoretical explication of what a possible law or a nomological hypothesis is. In Section 7.5, we shall be able to show that there is a one-one-correspondence between schemes of enumerative induction as found in Section 7.3 and mixtures of nomological hypotheses as explained in Section 7.4. Thus, our ranking theoretic analysis will result in transferring de Finetti's deep account of the confirmation of statistical hypotheses to the deterministic or qualitative realm. Section 7.6 will conclude with some remarks on the defeasible or unrevisable apriority of lawfulness or the uniformity of nature.

7.2 Ranking Functions

Let us start with a set W of possible worlds, small worlds in the sense of Savage rather than maximally large worlds in the sense of Lewis. Each subset of W is a truth condition or *proposition*. Hence, the set of propositions forms a complete Boolean algebra. I shall outright assume propositions to be the objects of doxastic attitudes, thereby taking these attitudes to be intensional. We know well that this is problematic, that the so-called propositional attitudes are presumably hyperintensional. But we scarcely know what to do about the problem. Hence, my assumption is just to signal that I do not want to worry here about these kinds of problems.[1]

Moreover, I assume that there is a distinguished class of (logically independent) *atomic propositions*. The paradigmatic atomic proposition states that a certain object has a certain property. Finally, I shall assume that the complete algebra of propositions is generated by the atomic propositions. Thus, each possible world is tantamount to a maximally consistent and possibly infinite conjunction of atomic propositions. A proposition is called *molecular* iff it is a member of the Boolean algebra generated by the atomic propositions, i.e., iff it is generated from the atomic propositions by *finitely* many Boolean operations.[2]

This is all we need to introduce our basic notion:

Definition 1: κ is a *ranking function* (for W) iff κ is a function from W into the set of extended non-negative integers $\mathbf{N}^+ = \mathbf{N} \cup \{\infty\}$[3] such that $\kappa(w) = 0$ for some $w \in W$. For each proposition $A \subseteq W$ the *rank* $\kappa(A)$ of A is defined by $\kappa(A) = \min \{\kappa(w) \mid w \in A\}$ and $\kappa(\emptyset) = \infty$. For $A, B \subseteq W$ the (*conditional*) *rank* $\kappa(B \mid A)$ of B *given* A is defined by $\kappa(B \mid A) = \kappa(A \cap B) - \kappa(A)$.

Since singletons of worlds are propositions as well, the point and the set function are interdefinable. The point function is simpler, but auxiliary, the set function is the one to be interpreted as a doxastic state.

---

[1] The *locus classicus* concerning (hyper-)intensionality is Carnap (1947); cf. in particular sects. 11–15. He there proposed to solve the problem of hyperintensionality with his notion of intensional structure. Quine responded by directly taking sentences as objects of belief. And till today the issue has remained obscure and undecided.

[2] Cf. also Carnap (1971/80) who proceeds with a similar algebraic framework.

[3] This is a deviation from the definition I have given in earlier papers. It will be explained below.

Indeed, ranks are best interpreted as *degrees of disbelief*. $\kappa(A) = 0$ says that A is not disbelieved at all; $\kappa(A) = 1$ says that A is disbelieved (and hence $\bar{A}$ believed) to degree 1; etc. Note that $\kappa(A) = 0$ does not say that A is believed; this is rather expressed by $\kappa(\bar{A}) > 0$, i.e., that non-A is disbelieved (to some degree). The clause that $\kappa(w) = 0$ for some $w \in W$ is thus a *consistency* requirement. It guarantees that at least some proposition, and in particular W itself, is not disbelieved (and hence that some proposition, e.g. $\emptyset$, is not believed). This entails the

law of negation: for each $A \subseteq W$, either $\kappa(A) = 0$ or $\kappa(\bar{A}) = 0$ or both.

The set $C_\kappa = \{w \mid \kappa(w) = 0\}$ is called the *core* of κ (or of the doxastic state represented by κ). C_κ is the strongest proposition believed (to be true) in κ. In fact, a proposition is believed in κ if and only if it is a superset of C_κ. Hence, the set of beliefs is *deductively closed* according to this representation.[4]

These observations make clear the following essential point: On the one hand, the degrees of disbelief are the basic notion. On the other hand, these degrees also contain an all-or-nothing notion of disbelief (and thus belief): disbelief *is* disbelief to some positive degree. If we would confine ourselves to a static perspective, this all-or-nothing notion, which I sometimes called *plain* (dis-)belief and which is well studied in doxastic logic, would be good enough. However, in order to define an adequate notion of conditional belief and thus to account for the dynamics of the all-or-nothing notion, we have to introduce the degrees. I emphasize this point because it marks an important advantage of ranking over probability theory. The latter cannot offer an adequate notion of plain belief, and hence those raised in probabilistic thinking tend to find the notion disreputable. But, intuitively, we have the notion, and it is basic to large parts of epistemology. Ranking theory satisfies both needs here, the one for the all-or-nothing notion and the other for the degrees.

There are two laws for the distribution of degrees of disbelief: the

law of conjunction: $\kappa(A \cap B) = \kappa(A) + \kappa(B \mid A)$.

That is, the degree of disbelief in A and the degree of disbelief in B given A add up to the degree of disbelief in A-and-B; this appears highly intuitive. With Definition 1 we may say conversely that this is precisely how conditional degrees of disbelief are to be understood. And there is the

law of disjunction: $\kappa(A \cup B) = \min\{\kappa(A), \kappa(B)\}$.

That is, the degree of disbelief in a disjunction is the minimum of the degrees of the disjuncts. Given the definition of conditional ranks, this law is nothing but a conditional

[4] Consistency and deductive closure are standard in doxastic logic; they have been often attacked and equally often defended. The issue of logical omniscience is indeed highly problematic and closely related to the issue of hyperintensionality of propositional attitudes already mentioned. We have, however, decided the issue already by assuming propositions as objects of doxastic attitude; under this assumption consistency and deductive closure are quite trivial rationality requirements.

consistency requirement; if it would not hold the inconsistency could arise that both $\kappa(A \mid A \cup B)$, $\kappa(B \mid A \cup B) > 0$, i.e., that both A and B are disbelieved given A-or-B.

According to Definition 1, the law of disjunction indeed extends to disjunctions of arbitrary cardinality. I find this reasonable, since an inconsistency is to be avoided in any case, be it finitely or infinitely generated. Note that this entails that each countable set of ranks must have a minimum (not only an infimum) and that the range of a ranking function must therefore be well-ordered. Hence, the range $\mathbf{N}^+$ is a natural choice. This point will become important later on.[5]

I immediately add:

Definition 2: A ranking function is *regular* iff all consistent molecular propositions have finite ranks.

In the sequel we shall consider only regular ranking functions. In earlier papers I have assumed a stronger form of regularity by outright defining a ranking function to be a function from W into $\mathbf{N}$ so that only $\varnothing$ receives infinite rank. If all propositions are molecular, there is no difference. In this paper, however, we want to consider possibly infinite and thus non-molecular generalizations, and then this stronger form of regularity is not feasible. Whence the present weaker assumption.

There is no need here to develop ranking theory more extensively. A general remark may be more helpful: ranking theory works in almost perfect parallel to probability theory. Take any probabilistic theorem, replace probabilities by ranks, the sum of probabilities by the minimum of ranks, the product of probabilities by the sum of ranks, and the quotient of probabilities by the difference of ranks, and you are almost guaranteed to arrive at a ranking theorem. Additivity of probabilities thus translates into the law of disjunction for ranks. The probabilistic law of multiplication translates into the above law of conjunction. It is easy to prove the ranking analogue to the formula of total probability, the

$$\textit{formula of the total rank:}\ \kappa(A) = \min_{i \le n} [\kappa(A \mid B_i) + \kappa(B_i)],$$

which says for a partition $\{B_1, \ldots, B_n\}$ of W how to compute the rank of some proposition A from the rank of A given various hypotheses B_i and the ranks of the hypotheses B_i themselves. One may continue with a ranking version of Bayes' theorem.[6] One can even develop the whole theory of Bayesian nets in ranking terms.[7] And so on.

---

[5] It is obvious that one has various options at this point. For instance, in Spohn (1988) [here: ch. 1]. I still took the range to consist of arbitrary ordinal numbers, but the advantages of this generality did not make up for the complications. By contrast, Hild (t.a., sect. 3.2) does not extend the law of disjunction to the infinite case and is thus free to adopt non-negative reals as values.

It is also obvious that the issue about infinite disjunctions is closely related to the discussion of the Limit Assumption in Lewis (1973, sect. 1.4). Without this assumption, it may happen that "if A were the case, then B_i would be the case" is true for infinitely many B that are jointly unsatisfiable. Lewis finds reason to accept this situation. I prefer to accept the Limit Assumption instead.

[6] This point is strongly developed in Hild (forthcoming).

[7] This was my original motivation. The basis of this theory, namely the so-called graphoid axioms of conditional independence, are proved for ranks in Spohn (1983, sect. 5.3) and (1988, sect. 6) [here: sect. 1.6].

The general reason is that ranks may roughly be interpreted as orders of magnitude of (infinitesimal) probabilities. Consider a non-standard probability measure taking non-standard reals as values. The logarithm of a product of such probabilities is the sum of the logarithms of the factors, w.r.t. any base. And the order of magnitude (= the logarithm in round figures) of a sum of such probabilities is the minimum of the orders of magnitude of its terms, at least w.r.t. an infinitesimal base. This perspective explains the translatability. However, I should emphasize that the translation is only an excellent rule of thumb, but not perfectly reliable, as we shall see later on (cf. also Spohn 1994). The matter is not fully cleared up.

It is still annoying, perhaps, that belief is not characterized in a positive way. But there is remedy.

Definition 3: β is the *belief function* associated with κ (and thus a *belief function*) iff β is the function assigning integers to propositions such that $\beta(A) = \kappa(\bar{A}) - \kappa(A)$ for each $A \subseteq W$. Similarly, $\beta(B \mid A) = \kappa(\bar{B} \mid A) - \kappa(B \mid A)$.

Recall that at least one of the terms $\kappa(\bar{A})$ and $\kappa(A)$ must be 0. Hence, $\beta(A) > 0$, < 0, or $= 0$ iff, respectively, A is believed, disbelieved, or neither; and A is the more strongly believed, the larger $\beta(A)$. Thus, belief functions may appear to be more natural. But their formal behavior is more awkward. I shall use both notions.

Since this is an essay about confirmation theory, we must ask: what is confirmation with respect to ranking functions? The same as elsewhere, namely *positive relevance*.

Definition 4: A *confirms* or is a *reason for* B relative to κ iff A is positively relevant to B, i.e., iff $\beta(B \mid A) > \beta(B \mid \bar{A})$, i.e., iff $\kappa(\bar{B} \mid A) > \kappa(\bar{B} \mid \bar{A})$ or $\kappa(B \mid A) < \kappa(B \mid \bar{A})$ or both.[8]

There is an issue here whether the condition should require $\beta(B \mid A) > \beta(B)$ or only $\beta(B \mid A) > \beta(B \mid \bar{A})$, as stated. In the corresponding probabilistic case, the two conditions are equivalent if all three terms are defined, but the first condition is a bit more general, since it may be defined while the second is not. That is why the first is often preferred. In the ranking case, however, all three terms are always defined, and the second condition may be satisfied while the first is not. In that case the second condition on which my definition is based seems to be more adequate.[9]

Let me close my presentation of ranking theory with formally introducing a point that will receive great importance later on: Ranking functions can be mixed, just as probability measures can. For instance, if κ_1 and κ_2 are two ranking functions for W and if κ^* is defined by

$$\kappa^*(A) = \min\{\kappa_1(A), \kappa_2(A) + n\} \text{ for some } n \in \mathbf{N}^+ \text{ and all } A \subseteq W,$$

[8] I believe that if epistemologists talk of justification and warrant, they ought to refer basically to this relation of A being a reason for B; cf. Spohn (2001b). That's, however, a remark about a different context.

[9] A relevant argument is provided by the so-called problem of old evidence. The problem is that after having accepted the evidence it can no longer be confirmatory. However, this is so only on the basis of the first condition. According to the second condition, learning about A can never change what is confirmed by A, and hence the problem does not arise. This point, or its probabilistic analogue, is made by Joyce (1999, sect. 6.4) by using Popper measures, relative to which the second condition is defined even if $P(\bar{A}) = 0$. However, cf. my skeptical remark about Popper measures in Section 7.1.

then κ^* is again a ranking function for W. Or more generally:

Definition 5: Let K be a set of ranking functions for W and ρ a ranking function for K. Then κ^* defined by

$$\kappa^*(A) = \min\{\kappa(A) + \rho(\kappa) \mid \kappa \in K\} \text{ for } A \subseteq W$$

is (obviously) a ranking function for W and is called the *mixture* of K by ρ.

Note the similarity of this definition with the formula of the total rank; the various κ take here the role of the various hypotheses B_i in that formula.

7.3 Symmetry and Non-negative Instantial Relevance

Now we are well prepared turn to our proper topic, enumerative induction. Let us start with simplifying the propositional structure as far as our topic allows: by considering an infinite series of objects and just one property P. So, each object can either have or lack P, and there are just two universal generalizations: "all objects are P", and "all objects are not P". Concerning the objects this is all the generality we need; concerning the properties we proceed minimally. This will facilitate our business. It will be clear, though tedious to prove, that the results below generalize to any finite number of properties.[†2] So, the results are considerably stronger than they appear. However, I don't know how things stand with an infinity of properties that may be generated, e.g., by a real-valued magnitude.

This simplification allows us to represent each possible world by a sequence $z = (z_1, z_2, \ldots)$ of 1s and 0s, where $z_n = 1$ or 0 means, respectively, that the n-th object has or lacks P. $\{\boldsymbol{x}$ takes $z_{i_1}, \ldots, z_{i_n}\}$ is short for the proposition $\{\boldsymbol{x} \mid x_{i_j} = z_{i_j}$ for $j = 1, \ldots, n\}$.

The most basic assumption ranking functions will be supposed to satisfy is *symmetry*. This means that ranking functions should be able to distinguish different objects only with respect to the properties considered, in our case P and non-P. Let us make this precise in:

Definition 6: κ is *symmetric* iff for any sequences $\boldsymbol{y}$ and $\boldsymbol{z}$ and any permutation π of $\mathbf{N}$ $\kappa(\boldsymbol{x}$ takes $y_1, \ldots, y_n) = \kappa(\boldsymbol{x}$ takes $z_{\pi(1)}, \ldots, z_{\pi(n)})$ if $y_i = z_{\pi(i)}$ for $i = 1, \ldots, n$.

Regular symmetric ranking functions take a particularly simple form, as stated in the obvious:

Theorem 1: For each regular symmetric κ there is a *representative function* f from $\mathbf{N} \times \mathbf{N}$ into $\mathbf{N}$ such that $\kappa(\boldsymbol{x}$ takes $z_1, \ldots, z_{m+n}) = f(m,n)$ if $\sum_{i=1}^{m+n} z_i = m$, i.e., if exactly

[†2] The matter is in fact more complicated than I thought. The correct generalization will be found in Spohn (in preparation, ch. 12). The case of one property and its negation dealt with here is just very simple, but not incorrect or misleading.

m of the first $m+n$ objects have P and the others lack P. This function satisfies $f(0,0) = 0$ and the minimum property $f(m,n) = \min\,[f(m+1,n), f(m,n+1)]$ (for a proof apply the law of disjunction to the fact that $\{\boldsymbol{x}$ takes $z_1,\ldots,z_{m+n}\} = \{\boldsymbol{x}$ takes $z_1,\ldots,z_{m+n+1}$ and $z_{m+n+1} = 1\} \cup \{\boldsymbol{x}$ takes $z_1,\ldots,z_{m+n+1}$ and $z_{m+n+1} = 0\})$. Conversely, any function f from $\mathbf{N} \times \mathbf{N}$ into $\mathbf{N}$ with these two properties represents a regular symmetric ranking function.

This entails that f can be visualized as in infinite triangle of non-negative integers

$$\begin{array}{ccccccc} & & & f(0,0) & & & \\ & & f(1,0) & & f(0,1) & & \\ & f(2,0) & & f(1,1) & & f(0,2) & \\ \ldots & & & \ldots & & & \ldots \end{array}$$

If a *path* in such a triangle is any sequence which starts at any point $f(m,n)$ and in which each member is succeeded by its left or right neighbor immediately below, then the minimum property entails that each such path is non-decreasing and that whenever a path increases by going left any path going right at this point does not increase, and vice versa.

Symmetry has a long and venerable history. Indeed, van Fraassen (1989) even went so far to argue that lawlikeness is a confused idea we should dispense with and that symmetry takes the key role in scientific reasoning in its place. This paper will in fact confirm van Fraassen's view, with the minor divergence that lawlikeness need not be dispensed with, but will receive an appropriate account through the notion of symmetry. In any case, we shall pursue our investigation of enumerative induction only in terms of symmetric (and regular) ranking functions.

The first noteworthy observation in this pursuit is that given symmetry there is no difference between belief in the next instance and belief in the universal generalization about all further instances. Suppose that after some evidence concerning the first n objects you believe that the n+1st object will have P; that is, your κ is such that $\kappa(\boldsymbol{x}$ takes $z_{n+1} = 0) = s$ for some $s > 0$. Because of symmetry you then believe that any further object will have P, that is, $\kappa(\mathbf{x}$ takes $z_{n+k} = 0) = s > 0$ for any $k \geq 1$. And because of the infinite variant of the law of disjunction this entails that you believe in *all* further objects having P *with the same strength*; that is, your disbelief that *some* future object lacks P is as strong as your disbelief that a specific future object lacks P; i.e., $\kappa(\bigcup_{k\geq 1} \mathbf{x}\,\text{takes}\, z_{n+k} = 0) = s > 0$. If this sounds counter-intuitive,[10] we have to return arguing about the law of disjunction and the conditional consistency it reflects. However, don't be confused; your disbelief that *all* further objects lack P may still be much stronger or even infinite.

This means that as far as positive confirmation is concerned, i.e., confirmation that generates or strengthens belief instead of merely diminishing disbelief, there is no difference between the next or any other positive instance and the universal

[10] It is not unlikely, though, that your intuitions are probabilistically trained, and then it is difficult to tell apart the intuitions and the training.

generalization about all further instances. Hence, Carnap's problem of the null confirmation of universal generalizations disappears in the ranking theoretic context, and the recourse to instantial relevance which was only a substitute in the Bayesian framework is fully legitimate here.[11]

Instantial relevance can take a stronger and a weaker form. The *principle of positive instantial relevance* (*PIR*) says that, given any evidence concerning the first n objects, the n+1st object having or lacking P confirms, respectively, the n+2nd object having or lacking P. The weaker *principle of non-negative instantial relevance* (*NNIR*) requires only that the contrary is not confirmed. Hence, let us state:

Definition 7: A regular symmetric ranking function κ *satisfies PIR* iff $\beta(\boldsymbol{x}$ takes $z_{n+1} \mid \boldsymbol{x}$ takes $z_1, \ldots, z_n) < \beta(\boldsymbol{x}$ takes $z_{n+2} \mid \boldsymbol{x}$ takes $z_1, \ldots, z_{n+1})$ whenever $z_{n+1} = z_{n+2}$, i.e., iff for the relevant representative function f and all $m, n \in \mathbf{N}$ $f(m+2, n) - f(m+1,n+1) < f(m+1,n) - f(m,n+1) < f(m+1,n+1) - f(m,n+2)$. κ *satisfies NNIR* iff the weak inequalities hold instead.

PIR may look like the correct formalization of enumerative induction; alas, we have:

Theorem 2: There is no regular symmetric ranking function satisfying PIR.

Proof: Let us try to satisfy PIR by an appropriate representative function f. So we start with $f(0,0) = 0$ and, without loss of generality, $f(1,0) = 0$ and $f(0,1) = r \geq 0$. This entails $f(2,0) = 0$. Hence, if we set $f(1,1) = r$, we already violate PIR. So, we must choose $f(1,1) = s > r$ and $f(0,2) = r$. This in turn entails $f(3,0) = 0$ and $f(0,3) = r$. But we cannot complete, then, the fourth line of our triangle: we must set $f(2,1)$ or $f(1,2) = s$, and both choices violate PIR.

This failure should not come as a surprise. If we try to satisfy PIR with respect to the positive instances and increase the disbelief in a negative instance with increasing positive evidence, we cannot at the same time satisfy PIR with respect to the negative instances. For, many negative instances are then just as disbelieved as a single one, and hence the negative instances cannot be positively relevant to further negative instances. We cannot have it both ways.

Hence, we are forced to settle for the weaker NNIR. It is easily seen to be consistent. Within a probabilistic setting non-negative instantial relevance is in fact entailed by symmetry (cf. Humburg 1971). Thus it is worth noting that this is not the case here; it is obvious that there are symmetric ranking functions violating NNIR (because there are representative functions violating the additional condition of Definition 7).

Where do we stand? If we want to account for enumerative induction within the ranking theoretic setting, we have to accept the second best explication, i.e., NNIR. We should also keep in mind that, within this setting, positively confirming the next instance is tantamount to confirming the corresponding generalization.

---

[11] Within the probabilistic context, the strongest proposal to overcome Carnap's problem of the null confirmation of universal generalizations is the *K*-dimensional system of Hintikka and Niiniluoto (1976). It would be interesting to compare it with the ranking theoretic approach.

Thus, we may preliminarily conclude that each symmetric ranking function satisfying NNIR is a way to realize enumerative induction, there being indeed an infinity of such ways.

Still, the preliminary conclusion does not look right. There is a definite loss in the retreat from PIR to NNIR. Even partial instantial irrelevance does not really seem compatible with enumerative induction; it is strange that the confirming effect of a positive instance must fail at least sometimes. This is a negative illusion, though. In Section 7.5 all doubts dissolve. We shall find that NNIR is exactly right and that, contrary to appearance, positive relevance can be fully reestablished.

Our investigation has remained superficial so far. The topic gains depth only when we remind ourselves of the fact that enumerative induction was never taken to apply to all universal generalizations whatsoever, but rather only to laws or potential laws; at most with respect to laws it may claim to be a reasonable rule of inductive inference. Where is this crucial point reflected in our ranking theoretic explication? Well, it *is* reflected, but not at all in an obvious way. In order to uncover it, we have to think about what lawlikeness may mean in ranking theoretic terms.

7.4 Laws

In our simple setting we considered just two universal generalizations: $G_1 = (1,1,\ldots)$ and $G_0 = (0,0,\ldots)$. What could it mean to treat G_1, say, as a law and not as an accidental generalization? I think, quite unoriginally, that this shows in our inductive behavior. To believe in G_1 as a law is, first, to believe in G_1, as expressed by $\kappa(\overline{G}_1) > 0$. But, as we already know, the belief in G_1 can be realized in many different ways; this belief alone does not fix the inductive relations between the various instances. Which forms may they take? Well, if you learn about positive instances of G_1, you do not change your beliefs about the further instances according to κ, since you expected them to be positive, anyway. Crucial differences emerge only when we look at how you respond to negative instances according to the various attitudes. Let me focus for a while on two particular responses, which I call the 'persistent' and the 'shaky' attitude.

If you have the *persistent* attitude, your belief in further positive instances is unaffected by negative ones, i.e., $\kappa(\boldsymbol{x}$ takes $z_{n+1} = 0) = \kappa(\boldsymbol{x}$ takes $z_{n+1} = 0 \mid \boldsymbol{x}$ takes $z_1 = \ldots = z_n = 0)$. If, by contrast, you have the *shaky* attitude, your belief in further positive instances is destroyed by a negative instance, i.e., $\kappa(\boldsymbol{x}$ takes $z_2 = 0 \mid \boldsymbol{x}$ takes $z_1 = 0) = 0$, and, due to symmetry, also by several negative instances.

The difference is, I find, characteristic of the distinction between lawlike and accidental generalizations. Let us look at two famous examples. First the coins:

(1) All Euro coins are round
(2) All of the coins in my pocket today are made of silver

It seems intuitively clear to me that we have the persistent attitude towards (1) and the shaky one towards (2). If we come across a cornered Euro coin, we wonder what might have happened to it, but our confidence that the next coin will be round again is not shattered. If, however, I find a copper coin in my pocket, my expectations concerning the further coins simply collapse; if (2) has proved wrong in one case, it may prove wrong in any case.

Or look at the metal cubes, which are often thought to be the toughest example, because they display no perspicuous syntactic or semantic difference:

(3) All solid uranium cubes are smaller than 1 mile^3
(4) All solid gold cubes are smaller than 1 mile^3

What I said about (1) and (2) applies here as well, I find. If we bump into a gold cube this large, we are surprised – and start thinking there might well be further ones. If we stumble upon a uranium cube of this size, we are surprised again. But we find our reasons for thinking that such a cube cannot exist unafflicted and will instead start investigating this extraordinary case (if it obtains for long enough). As far as I see, the difference between the shaky and the persistent attitude applies as well to the other examples prominent in the literature.[12]

I am well aware that this sounds at best partially convincing. I am deliberately painting black and white here in order to elaborate the opposition between the persistent and the shaky attitude. Obviously, one would be prepared to say how one would respond in such cases only if they would be described in much more detail, especially concerning the evidence which led one to believe in the relevant generalizations in the first place. So, there is also a lot of grey.

There are at least two different kinds of grey. First, there is a broad range of attitudes between the two extremes I have described. Being shaky means to be *very* shaky; the belief in further positive instances may instead fade more slowly. And being persistent means to be *strictly* persistent; the belief in further positive instances may instead fade so lately that we never come to the point of testing it. Second, if confronted with such cases, we would in a sense widen our perspective. Take the uranium cube again. If we would really bump into such a large uranium cube, we would not simply mumble "impossible!" and stick to the belief that there will be no further exceptions. Rather, we would say that our original law was qualified by a *ceteris paribus* clause, anyway, and that a thorough investigation of the case will allow us to get clearer about normal and exceptional conditions. However, as fascinating as it is, the issue of *ceteris paribus* laws is certainly beyond the scope of this paper.[13]

There are now two ways to respond. One may either say there is too much grey not decomposable into black and white. Or one may say that there is an important

---

[12] Cf., e.g., the overview in Lange (2000, pp. 11f.). As Köhler (2004) pointed out to me, Bode's law of the logarithmic distribution of the planets in the solar system aptly illustrates my dichotomy. This law appeared to be accidental, and one counter-instance would have destroyed the confidence in it. Only recently it has acquired lawlike status via very sophisticated considerations, and the discovery of an anomaly would not impair this status.

[13] But I am convinced that ranking theory helps understanding this bewildering issue. At least I have argued so in Spohn (2002) [here: ch. 6].

insight in my black and white distinction which opens a fruitful way to analyse the shades of grey. I hope I have given at least some plausibility to proceeding on the second response.

If this is the right way to see the matter, treating a generalization strictly as a law is really to take the strictly persistent attitude towards it. This conclusion leads us to a further consequence, namely that the characteristic of lawlikeness is not something to be found in the propositional content of the generalization; it rather lies in our inductive attitude towards it or its instantiations. This consequence will be of crucial importance in the sequel.

The account given so far is obviously very close to the old idea that laws are not general statements, but rather inference rules or inference licenses. The idea goes back at least to Ramsey (1929) who stated it very clearly: "Many sentences express cognitive attitudes without being propositions; and the difference between saying yes or no to them is not the difference between saying yes or no to a proposition" (pp. 135f.). And "…laws are not either" [namely propositions] (p. 150). Rather: "The general belief consists in (a) A general enunciation, (b) A habit of singular belief" (p. 136). The idea has become quite popular among philosophers.

From a purely logical point of view, however, it is hard to see the difference between accepting the generalization as an axiom and accepting the corresponding inference rule for each instantiation. The only difference is that the rule is logically weaker; the rule is made admissible by the axiom, but the axiom cannot be inferred with the help of the rule. What else beside this unproductive logical point could be meant by the slogan "laws are inference rules" has been little explained.

Still, one might say that the inference-license perspective emphasizes the single case. This emphasis has now been stripped of its merely rhetorical character; it is reflected, I think, in my central notion of persistence and thus finds a precise induction-theoretic basis. In this perspective, the mark of laws is not their universality which breaks down with a single counter-instance, but rather their operation in each single case, which is not impaired by exceptions. Here, my account meets with Cartwright (1989) and her continuous efforts to explain that physical laws are deceptive and that we should rather attend to the single case and to the capacities (co-)operating in it. In Spohn (2002) [here: ch. 6] the point is argued a bit more extensively.

So much for some striking agreements. The most obvious disagreement is with Popper, of course. There is no doubt about how much philosophy of science owes to Popper. In view of this, my account is really ironic, since its conclusion is, in a way, that the mark of laws is their *not* being falsifiable by negative instances; only accidental generalizations are subject to such falsification. To be a bit more precise: Of course, any generalization is falsified by a single counter-instance. But falsified generalizations are to be rejected according to Popper. By contrast I have argued that the belief in the further instances is shattered by the falsifying instances only in the case of accidental generalizations, but not in the case of laws. No doubt, the idea that the belief in laws is not given up so easily is familiar at least since Kuhn (1962), and already Popper (1934, ch. IV, sect. 22) has insisted that the falsification of laws proceeds by more specialized counter-laws rather than by mere counter-instances. Here, however, the point is boiled down to its induction-theoretic essence.

7.5 Laws and Enumerative Induction

There is a striking and severe tension between Sections 7.3 and 7.4. We saw that, given symmetry, PIR is not feasible. Hence, we retreated to NNIR as a preliminary explication of enumerative induction. Then we noticed that enumerative induction applies only to laws. Finally, I have proposed an explication of laws according to which instances are *independent* of each other; this is what persistence amounts to. Thus we arrived at complete instantial *irrelevance* which is rather a caricature of NNIR and not in agreement with enumerative induction at all. Something must have gone badly wrong.

No, there is only a subtle confusion. Belief in a law is more than belief in a proposition. It is a certain doxastic attitude, and that attitude as such is characterized by the independence in question: if I would have just this attitude, just the belief in a strict law and no further belief, my κ would exhibit this persistence or independence. Enumerative induction, by contrast, is not about what the belief in a law *is*, but about how we may acquire or *confirm* this belief. The two inductive attitudes involved may be easily confused, but the confusion cannot be identified as long as one thinks that belief in a law is just belief in a proposition.

However, what could it mean to confirm a law if it does not mean to confirm a proposition? My definition of confirmation in Section 7.2 applies only to the latter. Hence, the talk of the confirmation of laws, i.e., of a second-order inductive attitude towards a first-order inductive attitude, is so far mere metaphorics. Can we do better?

Yes, we can. There is fortunately clear precedent in the literature. Given the close similarity between probability and ranking theory, one might notice that *a law* as I conceived it is nothing but *a sequence of independent, identically distributed random variables* translated into ranking terms. It thus becomes obvious that de Finetti (1937) addresses exactly our problem in the probabilistic context. In his celebrated theorems de Finetti showed that there is a one-one correspondence between symmetric probability measures for an infinite sequence of random variables and mixtures of Bernoulli measures according to which the variables are independent and identically distributed; and he showed that the mixture focusses more and more on a single Bernoulli measure as evidence accumulates. He thus showed to the objectivist that subjective symmetric measures provide everything he wants: beliefs about statistical hypotheses that converge toward the true one with increasing evidence.

De Finetti's issue between objectivism and subjectivism is not my concern. Ranking functions are thoroughly epistemological and have as such no objective interpretation.[14] Still, we can immediately translate de Finetti's theory into an account of the confirmation of laws as conceived here. The basic construction is, I find, illuminating, despite its formalistic appearance.

---

[14] But see Spohn (1993a) [here: ch. 5], where I tried to reduce the tension between my ranking theoretic and hence subjective explication of causation and the hardly deniable view that causation is an objective relation in the world.

Let us return to our simple one-property frame. We had two universal generalizations G_1 and G_0. But there are infinitely many persistent, lawlike attitudes. If we define for all $r,s \in \mathbf{N}^+$

$$\lambda_{-r}(\mathbf{x}\,\text{takes}\,z_1,\ldots,z_n) = r \cdot \sum_{i=1}^{n} z_i, \text{and}\, \lambda_s(\mathbf{x}\,\text{takes}\,z_1, \quad ,z_n) = s \cdot (n - \sum_{i=1}^{n} z_i),$$

then $\Lambda = \{\lambda_t \mid t \in \mathbf{Z}^+)$ includes all and only the persistent attitudes (where $\mathbf{Z}^+ = \mathbf{Z} \cup \{\infty, -\infty\}$). Λ contains precisely the 'Bernoullian' ranking functions which are symmetric and according to which each instance is independent from all others. For $t > 0$ λ_t believes in G_1 and disbelieves in each negative instance with rank t. For $t < 0$ it is just the other way around; such a λ_t believes in G_0 and disbelieves in each positive instance with rank t. In short, each λ_t counts the number of counter-instances within $\{\boldsymbol{x}$ takes $z_1, \ldots z_n\}$ to the generalization it believes in and multiplies it by t (or $-t$).

What then is the difference between, e.g., λ_1 and λ_2? There is none in content and none in persistence. The only difference lies in the disbelief in negative instances; λ_2 is firmer a law, one might say, than λ_1. Rather for technical reasons we have to include λ_∞ and $\lambda_{-\infty}$. λ_0, finally, does not represent a law at all. It rather represents *lawlessness*, indeed complete agnosticism; nothing (except the tautology) is believed in λ_0. Its special role will be discussed in the final section.

Now, believing in laws, confirming and falsifying laws, etc. are doxastic attitudes towards laws, which will here be modelled, of course, by a ranking function ρ over the set Λ of possible laws. If the possible laws are possible first-order attitudes, then ρ is a second-order attitude, which, however, induces a first-order attitude. What, according to ρ, is the rank of a proposition $A \subseteq W$, i.e. the degree of disbelief in A? It is the minimum of all the disbeliefs in A according to the possible laws in Λ weighed by the disbelief in the laws according to ρ; that is, the first-order attitude induced by ρ is just the mixture of Λ by ρ as defined in Definition 5.

Are we talking about a specific second-order attitude ρ? No, you may have any ρ you like. The following considerations are perfectly general in this respect. Let us call ρ *proper*, though, iff at most one of $\rho(\lambda_\infty)$ and $\rho(\lambda_{-\infty})$ is finite. Now we can start translating de Finetti's theorems.

First, we have:

Theorem 3: For each proper ρ over Λ, the mixture of Λ by ρ is a regular symmetric ranking function satisfying NNIR.

Proof: Regularity and symmetry are obvious since all λ_s are regular and symmetric. The proof of NNIR is essentially a tedious exercise. And since ρ is to be proper, the mixture is regular.

Second, we have: For each regular symmetric ranking function κ satisfying NNIR there is a proper ranking function ρ over Λ such that κ is the mixture of Λ by ρ. We may indeed strengthen the claim. Suppose we mix, e.g., λ_1 and λ_2 by some ρ with $\rho(\lambda_1) = \rho(\lambda_2) = 0$. Then λ_2 is obviously a redundant component of the mixture; it never determines the result of the mixture, i.e., the relevant minimum.

Because of such redundant components mixtures are never unique.[15] Uniqueness can be achieved only with minimal mixtures, as we might call them. However, we must be careful in catching the right kind of minimality. The point of the following definition will become fully clear only with Theorem 5 below.

Definition 8: λ_s is a *redundant component* of the mixture of Λ by ρ w.r.t. a proposition A iff there is no proposition B such that $\min\{\lambda_t(A \cap B) + \rho(\lambda_t) \mid t \in \mathbf{Z}^+\} < \min\{\lambda_t(A \cap B) + \rho(\lambda_t) \mid t \in \mathbf{Z}^+ - \{s\}\}$, i.e., iff λ_s does not determine the value of the mixture for any $A \cap B$. λ_s is a *strongly redundant component* of the mixture of Λ by ρ iff λ_s is a redundant component of the mixture w.r.t. to all $A_{m,n}$ $(m,n \geq 0)$, where $A_{m,n}$ is the proposition that (in some order) m of the first m+n objects have P and the other n objects lack P. Finally, the mixture of Λ by ρ is called *minimal* iff for all its strongly redundant components λ_s $\rho(\lambda_s) = \infty$.

Hence, in a minimal mixture all strongly redundant components get weight ∞ and cannot enter the mixture at all. The strengthened claim then is:

Theorem 4: For each regular symmetric ranking function κ satisfying NNIR there is a unique ρ over Λ such that κ is the minimal mixture of Λ by ρ.

Proof: Let κ be a regular symmetric function satisfying NNIR, let f be its representative function forming an infinite triangle of non-negative integers, and let $c = \sup f$ be the supremum of f, which may be finite or infinite. Let us focus on *simple* paths starting at the boundary of the triangle and making no turns. These paths take two forms. For each $m \geq 0$ there is the *right* path $f(m,0), f(m,1), f(m,2), \ldots$ starting at the left and going always right, and for each $n \geq 0$ there is the *left* path $f(0,n), f(1,n), f(2,n), \ldots$ starting at the right and going always left. We know that the simple paths are non-decreasing (like all others). NNIR entails, moreover, that the simple paths do not *accelerate*; whenever i, j, k are three consecutive members of such a path, then $k - j \leq j - i$.

Each simple path either goes to infinity or reaches a maximum and then remains constant. Let us define a_m to be the supremum of the m-th right path $f(m,0), f(m,1), \ldots$ and b_n to be the maximum of the n-th left path $f(0,n), f(1,n), \ldots$ $(m,n \geq 0)$. Again, both sequences $\boldsymbol{a} = (a_0, a_1, \ldots)$ and $\boldsymbol{b} = (b_0, b_1, \ldots)$ must be non-decreasing and, due to NNIR, also non-accelerating. Either $a_0 = 0$ or $b_0 = 0$ or both, and $c = \sup \boldsymbol{a} = \sup \boldsymbol{b}$.

With the help of the two sequences $\boldsymbol{a}$ and $\boldsymbol{b}$ we can construct now the relevant minimal mixture ρ. If $a_1 - a_0 := r$, we set $\rho(\lambda_{-r}) = a_0$; and if a_m is any point at which $\boldsymbol{a}$ decelerates, i.e., such that $a_m - a_{m-1} > a_{m+1} - a_m := r$, we set $\rho(\lambda_{-r}) = a_m - mr$. Similarly, if $b_1 - b_0 := s$, we set $\rho(\lambda_s) = b_0$; and if b_n is any point at which $\boldsymbol{b}$ decelerates, i.e., such that $b_n - b_{n-1} > b_{n+1} - b_n := s$, we set $\rho(\lambda_s) = b_n - ns$. If for any $t \in \mathbf{Z}^+$ $\rho(\lambda_t)$ is not thereby defined, we set $\rho(\lambda_t) = \infty$. This completes the construction of ρ.

[15] This is a noticeable difference to probabilistic mixtures where every ingredient with positive weight contributes to the mixture, however slightly.

Note, in particular, that this entails $\rho(\lambda_0) = c$. Hence, the lawless λ_0 is a relevant component of the mixture only if c is finite.

Since either $a_0 = 0$ or $b_0 = 0$, there is some $t \in \mathbf{Z}^+$ with $\rho(\lambda_t) = 0$. Since at least one of a_1 and b_1 is finite, either $\rho(\lambda_\infty) = \infty$ or $\rho(\lambda_{-\infty}) = \infty$ or both. Hence, ρ is a proper ranking function over Λ.

The mixture of Λ by ρ indeed generates the representative function f: For all $m,n \geq 0$ we have either $f(m,n) = a_m$ or $f(m,n) = b_n$, since either $f(m,n+1) = f(m,n)$ or $f(m+1,n) = f(m,n)$, and thus either the right or the left simple path through $f(m,n)$ does not increase after $f(m,n)$. Now suppose $f(m,n) = b_n \leq a_m$, and let us check whether our mixture yields the same result:

As above, let $A_{m,n}$ be the proposition that (in some order) m of the first $m+n$ objects have P and the other n objects lack P. Hence,

$$\begin{aligned} f(m,n) &= \kappa(A_{m,n}) = \min\{\lambda_t(A_{m,n}) + \rho(\lambda_t) \mid t \in \mathbf{Z}^+\} \\ &= \min_{r,s\geq 0}[mr + \rho(\lambda_{-r}),\ ns + \rho(\lambda_s)] \end{aligned}$$

How to calculate this minimum? Let b_{n^*} be the largest point before b_n where $\boldsymbol{b}$ decelerates and let $s^* = b_{n^*+1} - b_{n^*}$. Hence, $b_{n^*} = b_n - s^*(n - n^*)$. What about $ns^* + \rho(\lambda_{s^*})$? We have:

$$\begin{aligned} ns^* + \rho(\lambda_{s^*}) &= ns^* + b_{n^*} - n^* s^* \text{ (according to the definition of } \rho) \\ &= ns^* + b_n - s^*(n - n^*) - n^* s^* = b_n. \end{aligned}$$

Moreover, it is clear from the construction that $ns + \rho(\lambda_s) \geq ns^* + \rho(\lambda_{s^*})$ for all other $s \in \mathbf{N}^+$. The same reasoning shows that $mr + \rho(\lambda_{-r}) \geq a_m \geq b_n$ for all $r \in \mathbf{N}^+$. Thus, indeed, $f(m,n) = b_n$ according to the mixture.

Of course, if $f(m,n) = a_m \leq b_n$, the corresponding argument holds.

Some λ_t ($t \in \mathbf{Z}^+$) receiving finite rank by ρ may be a redundant component of the mixture of Λ by ρ w.r.t. $A_{0,0}$ (= the tautology); this always happens when two successive members a_m and a_{m+1} of $\boldsymbol{a}$ or b_n and b_{n+1} of $\boldsymbol{b}$ are points of deceleration. But none of them is strongly redundant, and the mixture is indeed minimal in the sense of Definition 8. This, however, will become clear only with the next theorem. It will also be obvious, then, that the ρ we have constructed is unique, i.e., provides the only minimal mixture generating the representative function f.

The final step in our translation of de Finetti is to inquire how the mixture is changed by evidence. This can be directly read off from the results above. Suppose that we collect the evidence $A_{m,n}$ that m of the first $m+n$ objects have and the other n objects lack P. If we start with the regular symmetric κ with representative function f, what is then the a posteriori ranking function $\kappa_{m,n}$ on the space of possibilities for the infinitely many remaining objects? Well, we learn by conditionalization; hence, for any proposition B within this space $\kappa_{m,n}(B) = \kappa(B \mid A_{m,n})$. The representative function $f_{m,n}$ of $\kappa_{m,n}$ is thus given by $f_{m,n}(p,q) = f(m+p, n+q) - f(m,n)$.

Now, suppose that κ is the minimal mixture of Λ by ρ. What is then the unique $\rho_{m,n}$ so that $\kappa_{m,n}$ is the minimal mixture of Λ by $\rho_{m,n}$? We know that *f* is the result of the mixture by ρ, i.e.,

$$f(m,n) = \min_{r,s\geq 0} [\lambda_{-r}(A_{m,n}) + \rho(\lambda_{-r}), \lambda_s(A_{m,n}) + \rho(\lambda_s)]$$
$$= \min_{r,s\geq 0} [\rho(\lambda_{-r}) + mr, \rho(\lambda_s) + ns].$$

Thus, we have for all $p,q \in \mathbf{N}$:

$$f_{m,n}(p,q) = f(m+p, n+q) - f(m,n)$$
$$= \min_{r,s\geq 0} [\rho(\lambda_s) + (n+q)s,\ \rho(\lambda_{-r}) + (m+p)r] - f(m,n)$$
$$= \min_{r,s\geq 0} [\rho(\lambda_s) + ns - f(m,n) + qs,\ \rho(\lambda_{-r}) + mr - f(m,n) + pr].$$

This already suggests how to define $\rho_{m,n}$. However, $\rho_{m,n}$ has to be a minimal mixture, and therefore we still need to eliminate some of the components originally having finite rank. For this purpose, let a_{m^*} be the largest member of $\boldsymbol{a}$ up to a_m where $\boldsymbol{a}$ decelerates and b_{n^*} the largest member of $\boldsymbol{b}$ up b_n where $\boldsymbol{b}$ decelerates (thus, possibly $a_{m^*} = a_m$ and $b_{n^*} = b_n$), and let $r^* = a_{m^*+1} - a_{m^*}$ and $s^* = b_{n^*+1} - b_{n^*}$. Now we can state:

Theorem 5: Define for $r,s \in \mathbf{N}^+$:

$$\rho_{m,n}(\lambda_{-r}) = \rho(\lambda_{-r}) + mr - f(m,n)\ for\ r \leq r^*\ and$$
$$\rho_{m,n}(\lambda_s) = \rho(\lambda_s) + ns - f(m,n)\ for\ s \leq s^*;$$

and if $r > r^*$ and $s > s^*$, then $\rho_{m,n}(\lambda_{-r}) = \rho_{m,n}(\lambda_s) = \infty$. Then $\kappa_{m,n}$ is the minimal mixture of Λ by $\rho_{m,n}$.

Proof: It is obvious from the construction for Theorem 4 that λ_{-r} and λ_s are strongly redundant components of $\rho_{m,n}$ for $r > r^*$ and $s > s^*$. Thus the minimality of the mixture of Λ by $\rho_{m,n}$ carries over to $\rho_{m,n}$. Therefore, the above calculations already prove that $f_{m,n}$ is generated by $\rho_{m,n}$.

The point of defining minimality as we did in Definition 8 now becomes clear. As mentioned, some components of the mixture of Λ by ρ may be initially redundant, i.e., w.r.t. to $A_{0,0}$. Still, they may become non-redundant after conditionalization by $A_{m,n}$. Hence, they have to be included already in the original mixture. Otherwise, we could not have obtained $\rho_{m,n}$ from ρ so easily as in Theorem 5.

The theorem has three important consequences. First, it helps to reestablish positive instantial relevance. Suppose, we find the $m+n+1$st object to have *P*; thus, our evidence increases from $A_{m,n}$ to $A_{m+1,n}$. How does the mixture change from $\rho_{m,n}$ to $\rho_{m+1,n}$? Insofar $\rho_{m+1,n}$ is finite we have for $r,s \geq 1$:

$$\rho_{m+1,n}(\lambda_{-r}) = \rho(\lambda_{-r}) + (m+1)r - f(m+1,n) \text{ and}$$
$$\rho_{m+1,n}(\lambda_s) = \rho(\lambda_s) + ns - f(m+1,n).$$

Hence, in any case $\rho_{m+1,n}(\lambda_{-r}) - \rho_{m+1,n}(\lambda_s) = r + \rho_{m,n}(\lambda_{-r}) - \rho_{m,n}(\lambda_s)$. That is, each λ_{-r} as opposed to any of the λ_s is more disbelieved in $\rho_{m+1,n}$ than in $\rho_{m,n}$ (by r ranks). In other words, the additional positive instance is positively relevant to the positive lawlike attitudes. So, on the level of the second-order attitudes we indeed have exceptionless positive instantial relevance, which is blurred, though, by the mixture and thus weakens to NNIR on the level of first-order attitudes. Theorem 2 has shown that this weakening is unavoidable, but now we see that it is only an artifact of the mixture.

This observation teaches us, secondly, that as more and more positive instances accumulate and $m - n$ diverges to infinity, $\rho_{m,n}(\lambda_{-r}) - \rho_{m,n}(\lambda_s)$ ($r,s \geq 1$) diverges to infinity as well, i.e., the disbelief in the negative lawlike attitudes heads for infinite firmness. This parallels de Finetti's observation in the probabilistic case.

So, all in all, we have seen that de Finetti's account of the confirmation of statistical hypotheses may be perfectly translated into ranking theoretic terms, thus deepening our understanding of enumerative induction and lawlikeness.

There is still a third lesson, which has in fact no probabilistic analogue. It thus goes a little step beyond de Finetti and deserves a brief concluding section of its own.

7.6 The Apriority of Lawfulness

This lesson concerns the special role of λ_0. We noticed already that λ_0 is total agnosticism expressing lawlessness instead of lawfulness. Now, we either have $\rho(\lambda_0) = \infty$, which entails $\rho_{m+n}(\lambda_0) = \infty$ for all $m,n \in \mathbf{N}$. Then ρ embodies the maximally firm belief that some law or other will obtain. This belief would indeed be invariable, not refutable even by very long sequences of apparent random behavior of the instances with respect to P. This does not appear reasonable.

The alternative is that we give $\rho(\lambda_0)$ some finite value; hence, $\rho_{m,n}(\lambda_0) = \rho(\lambda_0) - f(m,n)$. This entails that with each unexpected realization of an instance λ_0 gets less disbelieved. After too many disappointments we shall eventually have lost our belief in lawfulness and any belief about the behavior of new objects concerning P, the belief in lawlessness being the only remaining option. This may also sound implausible. However, $\rho(\lambda_0)$ may be very large so that the agnostic state is in fact never reached.

The more relevant observation, though, is that the whole story I have told about the single property P can be generalized to any finite number of properties $P_1, \ldots, P_m$ in a straightforward way. We can define Carnap's Q-predicates, i.e., the atoms of the Boolean algebra of properties generated by $P_1, \ldots, P_m$; for each Q-predicate Q_k we can consider the generalization "there is no Q_k" and the corresponding laws, i.e., persistent attitudes; and then all the theorems of Section 7.4 continue to hold. So, what we would really do if lawlessness with respect to P threatens is to try to correlate P with some other properties and to pursue the investigation within a larger space of properties.

Within such a larger space also more complex forms of laws become available going beyond persistent attitudes towards "there is no Q_k". As already mentioned, the ranking theoretic framework in particular allows of an analysis of ceteris paribus laws (cf. Spohn 2002, sect. 4). So, there are rich prospects of generalization. I don't know, though, whether and how the de Finettian story I have told concerning simple laws (about P or the Q_k) carries over to such more complex laws. And I don't know of any working account of conceptual change answering the threat of lawlessness within any given set of properties or conceptual framework. So, there is still a lot to do as well.

However, let me finally emphasize what my brief discussion of λ_0 means in more traditional terms. Kant tried to overcome Hume's objectivity skepticism generally with his transcendental logic and its synthetic principles a priori and Hume's inductive skepticism particularly with his a priori principle of causality. This principle ascertained rather only the rule- or law-guidedness of everything happening and was thus as well called the principle of uniformity of nature (cf., e.g., Salmon 1966, pp. 40ff.). As was often observed, this principle did not offer any constructive solution of the problem of induction, since it does not give any direction as to specific causal laws or specific inductive inferences. Still, it provided, if a priori true, an abstract guarantee that our inductive efforts are not futile in principle. *Is* it a priori true?

Nowadays, two notions of apriority are usually distinguished. A proposition is *unrevisably a priori* if it must be believed and cannot be given up under any evidential circumstances. This is certainly the notion which Kant used, though did not express it in this way, and which Quine attacked when attacking analyticity. By contrast, a proposition is *defeasibly a priori* if it is to be believed initially, prior to any experience (and may be given up later on). The prior probabilities discussed by Bayesians are a paradigm of defeasible apriority because they are, of course, expected to change.

Now, our initial ranking function is some regular symmetric κ satisfying NNIR. Via Theorem 4, κ uniquely corresponds to some ranking function ρ over Λ. The belief in lawfulness, then, is the same as the disbelief in lawlessness, i.e. $\rho(\lambda_0) > 0$. We saw that this is an extremely reasonable assumption. And we now see that it is tantamount to the defeasible apriority of lawfulness: we must start believing in the uniformity of nature.

The unrevisable apriority of lawfulness, however, is expressed by the stronger condition $\rho(\lambda_0) = \infty$. We also saw that this condition does not appear reasonable, at least if one relates it to the property P or, more generally, to any fixed set of properties. Still, it may be unrevisably a priori that there is *some* set of properties with respect to which nature is uniform. I am not prepared to decide whether or not the unrevisable apriority of lawfulness is defensible in this sense. But I think the issue is more clearly arguable on the basis provided here.

Chapter 8
Chance and Necessity: From Humean Supervenience to Humean Projection†

8.1 Introduction*

Probability abounds in the natural and social sciences. Yet, science strives for objectivity. Scientists are not pleased when told that probability is just opinion and there is no more sense to it. They are prone to believe in objective probabilities or chances. This is an essay about how to understand them.

Indeed, it is my first serious attempt in English[1] to come to terms with the notion of chance or objective probability. I cannot help feeling that this is a presumptuous enterprise. Many great minds have penetrated the topic. Each feasible position has been ably defended. No philosophically relevant theorem remains to be discovered. What else should there be to say? Yet, the issue is not settled. Even though all pieces are on the table, no one missing, how to compose the jigsaw puzzle is still not entirely clear. Philosophical uneasiness continues. Everybody has to try anew to put the puzzle together. So, here is my attempt to do so.

Let me lay my cards on the table right away. An event, or a state of affairs, is *chancy* iff it is partially determined by its past, to some specific degree; some might call this an Aristotelian conception of chance. Chance laws, then, generalize over such singular partial determinations. Likewise, a state of affairs is *necessary* (in the sense not of metaphysical, but of *natural* necessity) iff it is fully determined

† The original publication of this paper is in: E. Eells, J Fetzer (eds.), *The Place of Probability in Science*, Chicago: Open Court, to appear. It is reprinted here with kind permission of the Open Court Publishing Company.

* I am most grateful to Ludwig Fahrbach and Jacob Rosenthal for thorough-going discussions of earlier drafts of this paper; it gained immensely thereby.

[1] I have written a minor note, Spohn (1987), which foreshadows the general line of thought, and a German paper Spohn (1999b), of which the present paper is a substantial elaboration.

(i.e., sufficiently caused) by its past.[2] Deterministic laws generalize over such singular full determinations, so that we may reversely say that a state of affairs is necessary iff it is entailed by the laws and its past. This parallel will become important later on.

There *is* determination. There are deterministic laws, or so we believed at least for ages. And there are chance laws and hence chancy events, as modern physics tells us. Objective probabilities may thus be conceived as single-case propensities of a radical kind: propensities of the entire world as it has developed up to now to realize not only this or that current state of affairs, but in effect this or that entire future evolution.[3] The localization of propensities is a secondary, though, of course, important issue. The primary and really vexed issue is how at all to understand partial and full determination.

Given that there is partial determination, subjectivism or eliminativism concerning objective probabilities, a position associated with Bruno de Finetti and his positivistic predilections, is out of place. (Still, the most basic truths lie in his insights, and this essay will end up as hardly more than a projectivistic reinterpretation of de Finetti's views.)

Reductionism concerning objective probabilities seems ill-guided, too, whether in the analytical form trying to define chances in non-probabilistic terms as, e.g., (hypothetical) frequentism does or in the weaker ontological form as displayed in the doctrine of Humean Supervenience championed by David Lewis. Indeed, the failure of Humean Supervenience is nowhere clearer, I find, than in the case of chances.

Hence, realism without reductionism is perhaps what we should be heading for. I am indeed attracted by the picture as sketched, e.g., by Black (1998, pp. 371f.) who argues against Lewis that the world is more than "a vast mosaic of local matters of particular fact" (Lewis 1986a, p. ix), more, as it were, than a pattern of inert colors; it is also a pattern of pushes, hard deterministic as well as soft chancy ones. Maybe we should accept realism about primitive laws, dispositions, capacities, propensities, etc. (or their categorical bases), as has been vigorously defended by Armstrong (1983, in particular ch. 9 and 1997, ch. 15) and in quite a different way by Cartwright (1989).

Yet I share the widespread epistemological concerns about Australian realism that are as old as Hume's criticism of necessary connexion or determination. What we need to get explained, at least, is the theoretical web within which chances get

[2] There presumably are deep connections between metaphysical and natural necessity. Still, the two kinds of necessity must at first be kept apart. Metaphysical necessity is tied up with identity and existence, natural necessity is not, prima facie. Here, I shall deal only with the latter without worrying about its connection to the former.

[3] Similar phrasings may be found in Popper (1990, pp. 18f.) and Miller (1995, p. 138).

their role to play.[4] However, the explanations given by propensity theorists are generally not in good shape.[5] And so I appear to be torn by my various dissatisfactions, finding no place to rest.

No other than David Hume has suggested a position possibly comforting everyone, with his doctrine that causation is an idea of reflexion and that necessary connexion is nothing but determination or customary transition *in thought*. The doctrine has received its most extraordinary shape in Kant's transcendental idealism. Nowadays, it is rather called projectivism and defended by Simon Blackburn under the label 'quasi-realism' and summarized thus:

> Suppose we honor the first great projectivist by calling 'Humean Projection' the mechanism whereby what starts life as a non-descriptive psychological state ends up expressed, thought about, and considered in propositional form. Then there is not only the interest of knowing how far Humean Projection gets us. There is also a problem generated even if the mechanism gets us everywhere we could want. If truth, knowledge, and the rest are a proper upshot of Humean Projection, where is it legitimate to invoke that mechanism? Perhaps everywhere, drawing us to idealism, or nowhere, or just somewhere, such as the theory of value or modality. (1993, p. 5)

This 'mechanism', I shall argue, is operative at least in the case of chance and natural necessity. It is thus no accident that I am referring twice to Hume within one page. The move from Humean Supervenience to Humean Projection will be *our* move in this paper. (Indeed, I find that the latter is much better anchored in Hume's writings than the former.)

The crux of projectivism, though, is that it may sound attractive as a general strategy, while remaining poor in constructive detail. Thus it is not likely to satisfy the probability community. Indeed, if one looks at recent surveys such as Gillies (2000), projectivism does not figure there under its own or any other name. This is the point where I hope to add a bit to the present discussion.[6]

As the reader may have guessed, this paper is largely an argument with David Lewis' philosophy of probability. This has a personal motive. I well recall how enthralled I was by Lewis (1980a) – and how bewildered by the continuation in Lewis (1986a, Introduction and Postscripts to 1980a, and 1994b). I just had to come

[4] For instance, Fetzer (2002) shares realism about propensities, but responds to such concerns by embedding propensities into an embracive account of explanation and abductive inference. While I am in sympathy to his general approach, I do not want to explicitly enter the topic of explanation. Of course, that topic is tightly interwoven with our present one, but it has its own intricacies, in particular, when it comes to saying what 'the best explanation' might be. As far as I can see, we shall be able to side-step these intricacies here without loss.

[5] My reference book is Rosenthal (2004) that offers forceful criticisms of prominent variants of the propensity interpretation.

[6] Logue (1995) apparently pursues the same goal. However, he insists on having only one notion of probability, a personalistic one, and he does not present an explicit projectivistic construal of objective probability. The only further probability book where the idea is taken up is Rosenthal (2004, pp. 199ff.). In fact, the challenge of understanding objective probability as it is built up in this book in a most pressing way provoked me to elaborate my (1999b) into the present paper.

to grips with his work. There is also a substantial reason. Lewis' account is peculiarly ambiguous. He starts inquiring the epistemology of chance and ends up investigating its ontological grounds. Thus, I find it most instructive to follow his line of thought and to search for the point of departure for a more adequate account.

There is a third reason. The parallel between deterministic and chance laws is obvious; it would be awkward to account for them in a wholly disparate manner. Lewis expressly pursues this parallel; after apparent success in the deterministic case, his strategy just had to carry over to chance laws, as elaborated in his (1994b). Therefore, Lewis is the natural point of contact on this score, too, and however I diverge from Lewis' account of chance, the divergence must work for deterministic laws as well. In fact, I see how it will. Contrary to appearances, natural necessity or full determination or lawlikeness is still less understood than partial determination; even the appropriate analytical means were missing. The theory of ranking functions will bring progress here. This remark, though, will be briefly outlined, and can be more easily grasped, I hope, after treating the actually more familiar probabilistic case.[7]

The paper will proceed in the following way: We shall start in Section 8.2 with recapitulating the central role the Principal Principle has, according to Lewis, for understanding chance. Lewis gives substance to this principle by claiming admissibility, as he calls it, for historical and chance information; this will be discussed and simplified in Section 8.3. The admissibility of chance information drives him into a contradiction, though, with the doctrine of Humean Supervenience. Lewis proposes to reform the Principal Principle, but I shall argue in Section 8.4 that it is rather Humean Supervenience that has to go. This provokes a closer look at that doctrine, and we shall see in Section 8.5 that it is inherently questionable. So, this will be the point where a projectivistic reconstruction of the notion of partial determination is likely to deliver a more coherent account. The reconstruction will be carried out in Section 8.6, via the observation that the Principal Principle may be taken, in a precise way, as a special case of the Reflection principle propagated by van Fraassen (1984); this is no deep formal insight, but of some conceptual interest. Section 8.7 will sum up the projectivistic doctrine and argue that it can meet familiar objections and serve the purposes for which Lewis had invoked Humean Supervenience. As explained, the whole line of reasoning must somehow carry over from chance to natural necessity or from partial to full determination. The appendix will indicate how this might go.

---

[7] Concerning deterministic laws, Ward (2002) also claims to give a projectivistic account which he extends to chance in Ward (2005). However, while I agree with his critical diagnosis, our constructive approaches widely diverge, as will become clear at the end of this paper.

8.2 Chance-Credence Principles

Let us approach our topic, objective probability, via the Principal Principle, which seems to its baptizer "*to capture all we know about chance*" (Lewis 1980a, p. 266, my emphasis) – a proper starting point, if this claim were true. There is in fact not only one principle relating chance and credence; subsequent literature has discerned a whole family of principles, which we do well to survey. So, let us start in a purely descriptive mood; we shall become involved into debate soon enough.

The basic idea relating chance and credence is very old and familiar; it is simply that if I know nothing about some proposition A but its chance, then my credence in A should equal this chance. This is the *Minimal Principle* (as Vranas 2004 calls it):

(MP) $C(A \mid P(A) = x) = x.$

Here, C stands for subjective probability or credence (the association with Carnap's 'confirmation' is certainly appropriate), and P for objective probability or chance (or propensity, if you like). The subject having the credence remains unspecified, since (MP) is, as it were, a generic imperative; (MP), like the subsequent principles, is a rationality postulate telling us how any credence function should reasonably behave.

(MP) is the starting point of the sophisticated considerations in Lewis (1980a). (MP) is also called "Miller's Principle", because Miller (1966) had launched a surprising early attack on it. However, (MP) is not an invention of the recent philosophical debate. It is known for long also under the label "direct inference".[8] In fact, it is implicit in each application of Bayes' theorem to statistical hypotheses; there the 'inverse' posterior probabilities of the hypotheses given some evidence are calculated on the basis of their prior (subjective) probabilities and the 'direct' probabilities or likelihoods of the evidence under the hypotheses; and these 'direct' probabilities hide an implicit use of (MP). The merits of the recent discussion pushed by Lewis (1980a) and others are rather to scrutinize variants of (MP).

Before proceeding to them there are, however, various things to clarify. Philosophy first, I propose. If Lewis is right that principles like (MP) "capture all we know about chance", then the philosophical interest of these principles is evident. Lewis does not really argue for this claim. In fact, he does not make it, it only seems true to him. Indeed, he cannot strictly believe it by himself. When, as we shall see later on, he claims chances to Humeanly supervene on particular facts, then he clearly transcends the Principal Principle. And I shall end up agreeing with Arntzenius and Hall (2003) that there must be more we know about chance.

The point should rather be seen as a challenge. For, what is true is Lewis' assertion "that the Principal Principle is indeed informative, being rich in consequences that are central to our ordinary ways of thinking about chance" (1980a, p. 288), as

[8] Often, direct inference is more narrowly understood as the more contested 'straight rule' that recommends credence to equal observed relative frequency.

is amply proved in his paper. For instance, it follows that chance conforms to the mathematical axioms of probability. The challenge then is what else there might be to say about chance. In default of an explicit definition of chance we seek for an implicit characterization, and it seems that we have already gone most of our way with the extremely neat Minimal Principle (which, as we shall see, is hardly strengthened by the other principles still to come). The more we are captivated by the Principal Principle, the harder the challenge.

The harder, though, also the philosophical puzzle posed by chances. It is strange that chances that are supposed to somehow reflect objective features of the external world should be basically related to our beliefs in a normative way. Our implicit characterizations of other theoretical magnitudes do not look that way. And the more weight is given to this relation, the more puzzling it is. How should we understand the peculiar normative power of that objective magnitude to direct our beliefs? If, indeed, the Principal Principle is all we know about chance, that power turns into sheer miracle. Why should we be guided by something the only known function of which is to guide us? Preachers may tell such things, but not philosophers.[9] One of Lewis' motives for the doctrine of Humean Supervenience is, we shall see, to solve this puzzle; indeed, he claims it to be the only solution. We need not take a stance right now, but we must always be aware in the subsequent discussion of the basic merits and problems of the Principal Principle. We are dealing here with high philosophical stakes.

At the moment, though, we must be a bit more precise about (MP). First, we must be clear about the domains of the functions involved. The chance measure P takes propositions, I said. We should not start, though, philosophizing about propositions. Let us simply assume that propositions are subsets of a given universal set W of possible worlds.

Is any kind of proposition in the domain of P? This is an open question. It is debatable which propositions may be chancy or partially or fully determined and which not. There may be entirely undetermined propositions and there may be propositions for which the issue makes no sense. Let us leave the matter undecided and grant, in a liberal mood, that at least all *matters of particular fact*, and hence all propositions algebraically generated by these facts, have some degree of determinateness, i.e., chance. Lewis (1994b, pp. 474f.) has an elaborate view on what particular facts are; here we may be content with an intuitive understanding.

In any case, a proposition saying that some factual proposition has some chance is not a particular fact in turn. This does not exclude that such a chance proposition is algebraically generated by particular facts, but neither does it entail it; it is crucial for this paper not to presuppose from the start that chance propositions are factual in the same way as particular facts. So, let us more specifically assume that each $w \in W$ is a complete possible course of particular facts. Whether we should be more liberal concerning the domain of chance will be an issue later on.

Credence is not only about particular facts, but also about possible chances; this is explicit in (MP) itself. Thus, if $\mathcal{P}$ denotes the set of possible chance

[9] The puzzle is vividly elaborated by Rosenthal (2004, sect. 6.3).

measures for W, then $W \times \mathcal{P}$ is the possibility space over which the credence C spreads.

Moreover, I shall be silent about the precise algebraic structure of the set of propositions[10] and just assume that each $P \in \mathcal{P}$ is defined on some algebra over W and C on some algebra over $W \times \mathcal{P}$. Accordingly, I shall be silent about the measures we are considering being finitely or σ-additive. This sloppiness will have costs, but rigorous formalization would have costs, too. I am just following the practice usually found in the literature I am mostly referring to.

For instance, one consequence of sloppiness is that (MP) does not make strict sense, since the condition will usually have probability 0. Lewis says that we should move then to non-standard analysis and hyperfinite probability theory where the condition in (MP) may be assumed to have infinitesimal probability. More easily said than done. Within standard probability theory one may circumvent the problem by stating (MP) in the more general form:

(MPI) $\quad C(A \mid P(A) \in I) \in I$ for any open interval I.[11]

This issue will return, and all the principles I am going to discuss should be restated accordingly.

There is another reason why (MP) will not do as it stands. C may not be any credence function. If C is already well informed about A, for instance by being based on the observation of A or of some effects of A, (MP) is obviously inadequate. As Lewis (1980a, pp. 267f.) explains, this concern is excluded for sure only if C is an initial or a priori credence function, as conceived as the target of further rationality constraints also by Carnap in his inductive logic. To indicate this, I shall denote an a priori credence by C_0 (0 being a fictitious first time).

Finally, in order to present Lewis' ideas we must note that chance evolves in time; this is particularly clear when chance is conceived as partial determination. Even full determination evolves in time, unless determinism holds and everything is fully determined at all times. Moreover, chance is world dependent; how chance evolves in time may vary from world to world. In order to make these dependences explicit we must replace P by P_{wt}, the chance *in w at t*. Thus we arrive at a slightly more explicit version of the Minimal Principle:

(MP*) $\quad C_0(A \mid P_{wt}(A) = x) = x$.[12]

---

[10] I shall even prefer sentential over set theoretical representations of propositions.

[11] This is Constraint 2 of Skyrms (1980, pp. 163–165), applied to degrees of belief and propensities.

[12] Even at the risk of appearing pedantic, let me at least once note what the correct set-theoretic representation of (MP*) is. There, credence is not about facts and chance, but rather about facts and evolutions of chance, i.e., about $W \times \mathcal{P}^T$, where T is the set of points of time. (Only at the end of the next section shall we be able to return to our initial simpler conception of credence.) (MP*) then says that $C_0(A \mid \{\pi \in \mathcal{P}^T \mid \pi(t)(A) = x\}) = x$, where the condition consists of all those evolutions of chance according to which the chance of A at t is x.

Having said all this, let us return to our descriptive path through the family of chance-credence principles (cf. also the useful overview in Vranas 2004). A first minor step proceeds to a conditional version of (MP) introduced by van Fraassen (1980b, pp. 106f.), the *Conditional Principle*:

(CP) $C_0(A \mid B \,\&\, P_{wt}(A \mid B) = x) = x$,

saying that, if you know nothing about A but its chance conditional on B, your conditional credence in A given B should equal this chance. (CP) is certainly as evident as (MP). We shall soon see that (CP) is hardly stronger than (MP).

David Lewis has taken a different, apparently bigger step. After retreating to the a priori credence C_0 in (MP) that is guaranteed to contain no information overriding the conditional chance information, Lewis poses the natural question which information may be added without disturbing the chance-credence relation stated in (MP). He calls such additional information admissible, and thus arrives at what he calls the *Principal Principle*:

(PP) $C_0(A \mid E \,\&\, P_{wt}(A) = x) = x$ for each admissible proposition E.

But what precisely is admissible information? The answer is surprisingly uncertain; the literature (cf. e.g., Strevens 1995; Vranas 2004) strangely vacillates between defining admissibility and making claims about it. I think it is best to start with a clear definition, which is obvious, often intimated (e.g., by Vranas 2004, footnote 5), but rarely endorsed in the literature (e.g., by Rosenthal 2004, p. 174):

(DefAd) E is *admissible w.r.t. A given D* iff $C_0(A \mid E \cap D) = C_0(A \mid D)$. Specifically, E is *admissible w.r.t. A in w at t* iff E is admissible w.r.t. A given $P_{wt}(A) = x$.

The first general part says that E is admissible w.r.t. A given D iff E does not tell anything about A going beyond D according to the a priori credence C_0. Admissibility is nothing but *conditional independence*. The second part gives the specification intended and used in (PP).

Obviously, the definiens states at least a necessary condition of admissibility; any admissible E not satisfying this condition would directly falsify (PP). I propose to consider the necessary condition to be sufficient as well. This strategy trivializes (PP); with (DefAd), (PP) reduces to nothing more than (MP), and the issue of admissibility is simply deferred. Still, I find the detour via (DefAd) helpful. It clearly separates the meaning of admissibility from the substantial issue which propositions E should be taken to satisfy (DefAd). This issue is our task in the next section.

One may still wonder why one should take the necessary condition for admissibility to be also sufficient. We may have stronger intuitions concerning admissibility. We may, for instance, think that two pieces of information admissible individually should also be jointly admissible, a feature not deducible from (DefAd). Or we may think that any E admissible w.r.t. A in w at t should be so for general reasons pertaining to w and t and not to idiosyncratic reasons pertaining to A.

And so on. However, the theoretical tactics is always to start with the weakest notion, which is (DefAd) in the case at hand. The substantial claims about admissibility will then also take a weak form, but we are always free to strengthen them. The point is that we would not have the reverse freedom when starting with a stronger notion right away.[13] A further worry may be that (DefAd) lets a priori credence C_0 decide about admissibility. However, we should read (DefAd) the other way around; whatever the substantial claims about admissibility, they are constraints on C_0 via (DefAd).

8.3 The Admissibility of Historic and Chance Information

Lewis (1980a) makes two substantial claims about admissibility. The first is that each piece of historic information is admissible. If I know the chance $P_{wt}(A)$ that A has in w at t, this knowledge cannot be improved by any information about what happened in w up to t; $P_{wt}(A)$ summarizes, as it were, all there is to know in w up to t. Let us denote the history of the world w up to time t by H_{wt}. $H_{wt} = \{v \in W \mid H_{vt} = H_{wt}\}$ is a proposition. Moreover, let us say that the proposition E is only about history up to t, or *t-historical*, for short, iff for each world w either $H_{wt} \cap E = \emptyset$ or $H_{wt} \subseteq E$. Then Lewis' claim is:

(AdH) If E is t-historical, then E is admissible w.r.t. A in w at t.

Note that reference to A is empty; only the relation of E to t is relevant to (AdH).

This claim is almost universally accepted. Lewis (1980a, p. 274) himself raises a doubt about (AdH). Could there not be a crystal ball that foretells me for sure whether or not A happens, even if A is chancy? I shall explain later why I am not worried by this alleged possibility. Here, I just join (AdH). Thus, the Principal Principle starts unfolding some strength. Let me add three remarks that deepen the understanding of the point.

First, Lewis (1980a) presents the case as if the admissibility of historical information were specifying (PP) and thus rationally constraining credence. So it does, but the core of the matter is thereby obscured in my view. (PP) and (AdH) immediately entail what might be called the *Determination Principle*:

(DP) $P_{wt}(H_{wt}) = 1.$

---

[13] Hall (2004, p. 103) arrives at another definition of admissibility. But it is clear that his move from his (3.12) to his (3.13) offers another sufficient condition the necessity of which is not argued for. In any case, his definiens entails mine, but not vice versa.

It is unfortunate that my paper was essentially finished before I could get aware of this paper of Ned Hall, which covers much of the same ground as mine, though with different twists and conclusions. Thus, my comparative remarks will be confined to some footnotes.

(This follows by replacing *A* as well as *E* in (PP) by H_{wt}.) (DP) simply says that history is no longer chancy. This consequence is, of course, intended. However, it is not about credence, but only about chance. In fact, it is an analytic truth about (partial) determination: what is past is fully determined.[14] Hence, it is more illuminating to realize that only this analytic truth needs to be added to (CP) to entail (AdH).[15] The matter will further simplify later in this section.

The second point is one I have not seen emphasized in the Principal Principle literature, though it deems important to me: In the prolonged efforts of understanding objective probabilities, whether as frequencies or propensities, the so-called *reference class problem* stood out as central and embarrassing.[16] The probability of a particular event seemed to depend on the reference class within which it was considered. Thus, that event could be assigned an objective probability only if one could distinguish the objectively correct reference class, apparently a dubious matter. The general recommendation was to rely on the narrowest reference class (or on the broadest reference class equivalent to the narrowest one); see also Hempel's so-called criterion of maximal specificity. This may be taken as the narrowest *available* reference class; but availability imports epistemic relativity. Or one may engage into the difficult business of Salmon (1984, pp. 60ff.) of distinguishing *objectively* homogeneous reference classes. If Lewis is to explain us objective probabilities, he must respond to this problem.

He does implicitly, his response is (AdH). For the objective probabilities at *t* the whole history, H_{wt}, *is* the objectively narrowest reference class; there could not be any more specific one. Indeed, it is hardly a class; it has only one member, actually unrepeatable and only counterfactually repeatable. Hence, this is at best a trivial and purely conceptual solution of the reference class problem; it does not even touch the real and deep methodological problem to specify sound and manageable reference classes. However, this is a problem the philosopher must leave to the scientist; the philosopher can only say what the ultimate standard is with which to compare all actually considered reference classes.

The third remark is related. If all the history up to *t* is admissible w.r.t. some proposition *A* in *w* at *t*, this means that up to *t* there is absolutely nothing more to know about *A* than its chance. You can learn absolutely everything up to *t* and you will be none the wiser concerning *A*. If you do not even know the chance of *A*, you are even more in the dark; the chance of *A* at *t* is the best you can know about *A* at *t*. This is the core intuition about partial determination: if *A* is in some way partially determined at *t*, there is nothing before *t* that would determine *A* in any other way. And knowledge before *t* can at best equal determination at *t*.

[14] Of course, I am presupposing classical time throughout. I do not venture speculating about the consequences of relativistic time for our topic.

[15] *Proof*: According to (CP) we have $C_0(A \mid H_{wt} \,\&\, P_{wt}(A \mid H_{wt}) = x) = x$. According to (DP) $P_{wt}(A \mid H_{wt}) = x$ expresses the same proposition as $P_{wt}(A) = x$. Hence, we also have $C_0(A \mid H_{wt} \,\&\, P_{wt}(A) = x) = x$. This just says that H_{wt} is admissible w.r.t. *A* in *w* at *t*. Since any *t*–historical *E* is the disjoint union of some H_{wt}, *E* is admissible, too, w.r.t. *A* in *w* at *t*.

[16] For a recent reinforcement of the problem see Hájek (2007).

This point is reflected in many accounts of probability, for instance in the old idea that genuine random processes cannot be outfoxed by a gambling system. The same thought is found in von Mises' (1919) definition of a collective as a sequence for which no place selection results in a subsequence with a deviating limit of relative frequencies and in the subsequent explications of this approach with recursion and complexity theoretic means (cf. Church 1940; Chaitin 1966). Salmon (1984, pp. 60ff.) realizes the same basic idea in terms of his objectively homogeneous reference classes, though in a different way. It is important to see all this connected with (AdH).

Let us turn to the second kind of admissible information acknowledged by Lewis (1980a): information about the chances themselves, not only about the actual ones, but also about the ones as they would have been at various times. If I know the actual chance of *A*, how could information how other chances would have been tell me more about *A*? It cannot, as Lewis (1980a, pp. 274ff.) argues.

To state this more precisely: Let T_w be the complete theory of chance holding at the world *w*, i.e., according to Lewis, the conjunction of all conditionals true at *w* having the form: "if the history of *w* up to t' had been $H_{vt'}$, then $P_{vt'}$ had been the chances in *w* at t'." Lewis assumes T_w to be a proposition over *W*; this is a controversial assumption to be discussed later. Is it at all a proposition over $W \times \mathcal{P}$ (or $W \times \mathcal{P}^T$ – cf. footnote 12)? Prima facie not, since the counterfactual conditional is not among the Boolean operations. Still, we may take T_w to be in the domain of C_0; the issue will be cleared up next page. Furthermore define *E* to be a *chance proposition* iff for each world *w* either $T_w \cap E = \emptyset$ or $T_w \subseteq E$. Then Lewis' second admissibility postulate is:

(AdP) If *E* is a chance proposition, then *E* is admissible w.r.t. *A* in *w* at *t*.

Note again that the reference to *A* and even to *t* is empty; all that matters is that *E* is a chance proposition.

Lewis assumed that separate admissibility of historic and chance information entails their joint admissibility. This does not follow, however, on the basis of (DefAd). Hence, we should better read (AdH) or (AdP) as saying that its kind of information is admissible *given* the other kind of information; this entails its unconditional admissibility.[17]

At this point, we can easily see that the Conditional Principle (CP) is hardly stronger than the Minimal Principle (MP). If $P_{wt}(A \cap B) = y$ is admissible w.r.t. *B* in *w* at *t* and $P_{wt}(B) = z$ is admissible w.r.t. $A \cap B$ in *w* at *t*, then (PP) yields $C_0(A \cap B \mid P_{wt}(A \cap B) = y \,\&\, P_{wt}(B) = z) = y$ and $C_0(B \mid P_{wt}(A \cap B) = y \,\&\, P_{wt}(B) = z) = z$ and both together yield (CP). In other words: We have to add to (MP) only the admissibility of a tiny bit of chance information in order to get (CP).

(PP) + (AdH) + (AdP) may finally be combined to what Lewis (1980a) called the Principal Principle reformulated and was later called the *Old Principle*, since it is not yet the end of the story.

[17] This follows from the graphoid axioms for conditional probabilistic independence; cf., e.g., Spohn (1978, pp. 102f.).

(OP) $C_0(A \mid H_{wt} \& T_w) = P_{wt}(A)$.

This follows from (PP) because H_{wt} & T_w is admissible according to (AdH) and (AdP), T_w contains "if H_{wt}, then $P_{wt}(A)$ is the chance of A in w at t", and thus H_{wt} and T_w entail what $P_{wt}(A)$ is. Conversely, (OP) entails (PP) + (AdH) + (AdP). So, (OP) is a very elegant summary of the foregoing discussion.

The story can be further simplified, though. Let us look at T_w again. It is not quite clear why it has to take the specific complicated form, perhaps because T_w is to allow that some histories leave some events not even partially determined. However, we wanted to ignore such complications and assumed that all matters of particular fact are partially determined (or almost fully determined via chance 1). Hence, T_w claims for each possible history H_{vt} a full chance measure P_{vt} for W. Then, however, we may condense the whole theory T_w into one big chance measure P_w such that the time-dependent chance $P_{w,vt}$ derives from P_w through conditioning by H_{vt}.[18] We thus simply replace the conditionals with probabilistic consequents by conditional probabilities.[19] That it is possible to so condense T_w is indeed a consistency requirement for T_w, which becomes explicit also in Lewis (1980a, pp. 280f.) in his discussion of the kinematics of chance.[20] P_w thus is the *time-independent chance law* or *scheme of partial determination* as it holds in w for all propositions over W, and T_w simply says that P_w is as it is.[21] Hence, we have arrived at the following reduction of Lewis' terminology:

(RED) $T_w = \{P_w\}$ (or rather $= W \times \{P_w\} \subseteq W \times \mathcal{P}$), and $P_{w,vt}(A) = P_w(A \mid H_{vt})$ for all A, v, and t.

This shows at the same time that T_w is indeed a proposition over $W \times \mathcal{P}$ (indeed over $\mathcal{P}$ alone) and is thus in the domain of C_0 as we have originally conceived it.

(RED) makes clear that all the considerations about time-dependent chance are perhaps intuitively helpful and perhaps required for more general chance theories, but merely a conceptual detour within our frame. (RED) also explains why the above Determination Principle is analytic; (DP) follows from the definition (RED). And (RED) reinforces the redundancy of (AdH); given (RED) and (AdP) (OP) is just an application of the Conditional Principle (CP). However, we just saw that (CP) is entailed by (MP*) and (a small part of) (AdP). So, the latter two are the only

[18] Why is P now double-indexed? Because we have to say for the world w not only what the chances are in w at t (= P_{wt}), but also what the chances would have been if H_{vt} had been its history up to t (= $P_{w,vt}$).

[19] This is different from the identification of probabilities of conditionals with conditional probabilities, of which Lewis (1976) has warned us.

[20] The satisfiability of the consistency requirement is obvious in the case of discrete time with a first point of time. In the other cases one has to allude to convergence theorems for descending martingales; cf., e.g., Bauer (1968, p. 281).

[21] Hall (2004, p. 96) undertakes the same reduction. P_w is what he calls ur-chance.

basic assumptions we need. (RED) finally helps us to express the Old Principle still more simply:

(OP*) $C_0(\cdot \mid T_w) = P_w$.

Indeed, (OP*) looks like the Minimal Principle itself; the only difference is that (MP*) refers to the chance of a single proposition, whereas (OP*) refers to a whole chance measure. It is only from the restricted perspective of (MP*) that (OP*) appears to additionally assume the admissibility of chance information. Initially, I suppose, intuition would have been indifferent between (MP*) and (OP*).

8.4 The Admissibility of Chance Information and Humean Supervenience

So far, so good. We might be happy with (OP*) and start discussing its philosophical significance. Alas, the story takes a most unexpected turn, for which it is important that we have discerned (AdP) as an additional assumption in (OP). (OP) thus becomes the starting point of considerable confusion. The source of the trouble is that Lewis not only takes chance-credence principles like (MP*) to provide the most basic understanding of chance, but also maintains the ontological doctrine of so-called Humean Supervenience – because this is an attractive metaphysical doctrine, and because such chance-credence principles seem to require it. The trouble is real, and therefore we shall have to scrutinize both grounds of Humean Supervenience. But let us first have a formal look at what the trouble is.

With respect to chance, Humean Supervenience consists in the claim:

(HS) T_w supervenes on the totality of particular facts in w.

With our reduction (RED) of T_w and our understanding of the worlds in W as mere totalities of particular facts, we might as well express this claim thus:

(HS*) P_w supervenes on w.

It is not quite clear for which worlds w (HS) is to hold. Certainly for the actual world we live in. One may think that (HS) applies to all worlds and is thus a necessary truth. Lewis (1994b) sees it only as a contingent truth; (HS) is to hold only for worlds *like ours*, certainly a more modest and a more mysterious view. We do not have to take a stance here.

Since we are a bit sloppy concerning the algebra of propositions, we may say that (HS) amounts to the claim that T_w is identical to a proposition over W.[22] (HS) thus says there are not two possibility spaces, one for possible facts (forming the domain of chances) and one for possible chances (jointly forming the domain of

[22] If this algebra were a complete one, this translation of (HS) would indeed be correct.

credences). The latter rather reduces to the former; there is only the space of possible facts. Chance propositions are in effect factual propositions – and thus in the domain not only of credences, but of chance measures themselves.

Now, however, we are caught in paradox. Imagine that our world w, after having started with H_{wt}, continues with some possible future F_{wt}. F_{wt} should have at least a tiny chance of coming about; so $P_{wt}(F_{wt}) > 0$ and, according to (OP), $C_0(F_{wt} \mid H_{wt}$ & $T_w) > 0$. On the other hand, F_{wt} may be an undermining future in the sense that $H_{wt} \cap F_{wt}$ (= $\{w\}$) is not in the supervenience base of T_w, i.e., according to H_{wt} and F_{wt} w would be governed by some chance law different from T_w. Then F_{wt} is impossible given H_{wt} & T_w, i.e., $C_0(F_{wt} \mid H_{wt}$ & $T_w) = 0$. To put the case very briefly with reference to (OP*): Consider the factual proposition $\overline{T}_w$ that T_w is false. Clearly, $P_w(\overline{T}_w) = C_0(\overline{T}_w \mid T_w) = 0$, However, if w is genuinely chancy, we should have $P_w(\overline{T}_w) > 0$. Somewhere, we have made a mistake.

It seems clear where. Given (HS), not all chance information can be admissible, since information about the future may well be inadmissible and since chance information *is* information about the future according to (HS).[23] Indeed, we should conclude that most chance information is inadmissible, though it is hard to be more precise because it is not so clear how supervenience exactly works, in which complex of particular facts T_w exactly consists.

However, as Lewis (1994b) argues, most chance information is at least nearly admissible, and (PP) and (OP) work approximately well even under the assumption of (AdP); the mistakes we incur are below noticeability. Still, the question is: if (OP) is only approximately valid, what is the standard it approximates? Following Thau (1994) Lewis (1994b) proposes that this standard is provided by the *New Principle*:

(NP) $$C_0(A \mid H_{wt} \& T_w) = P_{wt}(A \mid T_w),$$

or in our reduced form:

(NP*) $$C_0(\cdot \mid T_w) = P_w(\cdot \mid T_w).$$

This appears to solve our problem. The derivation of the paradox of undermining futures is blocked when we use (NP) instead of (OP), and the approximate validity of (OP) is explained by the fact that the difference between $P_{wt}(A)$ and $P_{wt}(A \mid T_w)$ is mostly below noticeability.

Is this an ad hoc solution? No. As Hall (1994, p. 511) and Strevens (1995, p. 557) observe and Hall (2004, pp. 104f.) insists, (NP) is a consequence of (CP) and

[23] We might, of course, strengthen (HS) to the effect that chances at t supervene on no more than factual history up to t; then chance information is only about the past, and the paradox cannot arise. Lewis (1980a, pp. 291f.) already mentions this option before clearly seeing the paradox. In (1986a, p. 131) he even expresses a preference for it after recognizing the paradox. In (1994b, sect. 6) he finally rejects it, rightly in my view.

(DP) which are uncontested.[24] Moreover, the admissibility of chance information that drove (OP) into paradox is guaranteed for (NP); T_w, and hence any weaker chance information, is trivially admissible w.r.t. A given T_w & $P_{wt}(A \mid T_w) = x$. Hence, (NP) appears to be the right way to reconcile Humean Supervenience with (PP), the admissibility of historic information and the general *in*admissibility of chance information.

However, Lewis (1994b) is not entirely satisfied. He says:

> A feature of Reality deserves the name of chance to the extent that it occupies the definitive role of chance; and occupying the role means obeying the old Principle, applied as if information about present chances, and the complete theory of chance, were perfectly admissible. Because of undermining, nothing perfectly occupies the role, so nothing perfectly deserves the name. But near enough is good enough. (p. 489)

And thus Lewis acquiesces in chances obeying (OP) not quite perfectly.

This remark provokes the final twist of the story. As Arntzenius and Hall (2003) point out, (NP) entails that there *is* a magnitude occupying the definitive role of chance perfectly, i.e. satisfying (OP) strictly. Suppose that the world w determines the chance theory T_w; according to (HS) it does so in some particular manner. And suppose that T_w allows for undermining futures so that (OP) does not apply. Now, define $P_w^* = P_w(\cdot \mid T_w)$. P_w and $T_w^* = \{P_w^*\}$. So, T_w and T_w^* obviously are incompatible chance theories – in one sense. However, change also the supervenience bases for chances; say for each w that it is not P_w, but rather P_w^* that is determined by (the facts of) w. So, in another sense, T_w is a factual proposition over W according to the initial way of determination, and T_w^* is so, too, according to the modified way of determination. And in this sense, they are not incompatible. On the contrary, T_w entails T_w^*, since whenever $T_v = T_w$ according to the initial way of determination, $T_v^* = T_w^*$ according to the modified one (though not necessarily vice versa). Moreover, T_w^* cannot be threatened by undermining futures. And, this is the upshot, if P_w, T_w satisfy (NP*), i.e., if $C_0(\cdot \mid T_w) = P_w(\cdot \mid T_w)$, then P_w^*, T_w^* satisfy (OP*), i.e., $C_0(\cdot \mid T_w^*) = P_w^*$.[25]

Hence, if the old principle is definitive of the chance role, as Lewis says, then P_w^*, rather than P_w, should be the chance law governing the world w. If we tend to say P_w is determined by the particular facts, we should say it is rather P_w^* that is determined by those facts. Thus, we face a new paradox, at least if we think that true chance theories must allow for undermining futures. And even if we deny this and rather attempt to choose P_w right away so that $P_w^* = P_w$, then Arntzenius and Hall (2003) complete their argument by showing that chances then behave in an unacceptable way.

Schaffer (2003) tries to escape by claiming vagueness. Chance may be given by P_w or by P_w^*, and disambiguation is of little importance, since the difference is

---

[24] *Proof*: Take (CP), specialize B to H_{wt} & T_w and apply (DP) for omitting H_{wt} from the condition of P_{wt}. Then you get $C_0(A \mid H_{wt} \;\&\; T_w \;\&\; P_{wt}(A \mid T_w) = x) = x$, which is nothing but (NP), since H_{wt} & T_w entail $P_{wt}(A \mid T_w) = x$.

[25] Cf. Arntzenius and Hall (2003, pp. 176f.) for a fuller explanation of the point.

small, anyway. However, it is not chance that is vague, I think, only our thinking about it is not clear enough. My conclusion is that we are in deep trouble and have not found any stable position concerning the admissibility of chance information and the possibility of undermining futures. What got us there? It was, of course, the assumption of Humean Supervenience unquestioned so far. It is high time to attend to it more closely.

(HS) assumed that the chance proposition T_w over $\mathcal{P}$ is supervenient upon, or, with sloppy algebra, identical with some factual proposition over W. As a consequence, we had to consider chance propositions as being in the domain of P_w and P_{wt}, at least under a liberal, though not exceptional conception of this domain, and hence we had to consider such chances as $P_w(T_w)$ or whether or not $P_{wt}(A) = P_{wt}(A \mid T_w)$. By contrast, if we give up (HS), we are free to reject such expressions and in particular the New Principle as meaningless. If we do so, the admissibility of chance information is rescued from paradox and perfectly acceptable. Indeed, at so many places philosophy ran into trouble in the past decades with iterating (the same kind of) modality. We should have been warned.

I raised the point in my (1999b, p. 170). But it is underrated in the literature. The worst Lewis (1994b) and Hall (1994) say about (NP) is that it is messy and user-hostile. Arntzenius and Hall (2003, p. 175) only say that the non-reductionist rejecting (HS) is free to assume $P_{wt}(T_w) = 1$ and to thus eliminate the discrepancy between (OP) and (NP). Hall (2004, p. 99) insists that this stipulation is harmless. Formally, this is correct, but the non-reductionist need not even take this step. And he should not; the harm done consists in blurring the issue. It creates the impression that the issue between the reductionist and the non-reductionist would be whether $P_{wt}(T_w)$ is equal to or smaller than 1; it creates the delusion of there at all being a meaningful issue. It simply makes no sense to say that there is some chance that our world is governed by this scheme of partial determination rather than that or that this atom has (at t) a propensity of .4 of having (at t) a propensity of .2 of decaying (within the next hour).

Hoefer (1997, p. 328) expressly agrees by saying:

> The laws are what they are because of the pattern of events in history, and not what they are "by law". This is just a restatement of the core idea of Humean analyses of law. For just the same reason, the chances are not what they are "by chance", and the quantity $P_{tw}(T_w)$ should be regarded by a Humean as an amusing bit of nonsense.

However, his argument is a different one. He doubts that all particular facts (and their Boolean combinations) are in the domain of the chance function. Hence, even if the chance of chancy facts supervenes on particular facts, the supervenience base will usually not be in the domain of the chance function. NP would only be guaranteed to make sense for the Human supervenientist, if all particular facts were chancy. By contrast, I am granting the latter and arguing that NP still does not make sense.

Vranas (2004, p. 373) tries to save the "arguably dubious" assumption that chance propositions are in the domain of P_{wt}. He notices the potentially vicious circularity in such expressions as $P_{wt}(T_w)$, which is indeed a point of worry for the *non*-reductionist, but not for the reductionist, and he proposes to make sense of such expressions within reflexive situation theory (cf. Barwise and Etchemendy 1987) and thus ultimately within set theory without the foundation axiom. But why at all should the non-reductionist try to overcome his worry and take recourse to such remote means? For the non-reductionist particular facts and Boolean combinations thereof are chancy and what lies outside this domain is not. It is up to the reductionist to give an argument for conceiving the domain more broadly, and the argument must not presuppose (HS), as one would if one praises the apparent progress from (OP) to (NP).

8.5 Humean Supervenience

So, let us squarely face Humean Supervenience itself. I propose first to look at how Lewis thinks it is feasible. Once we shall have seen the doubtfulness of Lewis' construction I can proceed with an alternative account and then with a brief discussion of Lewis' reasons for taking (HS) to be without good alternative.

The thesis of Humean Supervenience says, according to Lewis (1994b, p. 474)

> [T]hat in a world like ours, the fundamental relations are exactly the spatiotemporal relations … and … that in a world like ours, the fundamental properties are local qualities. … Therefore it says that all else supervenes on the spatiotemporal arrangement of local qualities throughout all of history, past and present and future.

Because this holds only for worlds *like ours* (HS) is contingent. Should alien qualities in Lewis' sense play a role – irreducible chance would be such an alien quality – the case may be different.

The bite of this claim emerges when we consider all the things that are extremely thorny for philosophers: laws, counterfactuals, causation – and objective probabilities. All this must be determined by the totality of particular facts, according to (HS). How? The crucial link is constituted by what Lewis calls the best-system analysis of law, which he takes over from F. P. Ramsey:

> Take all deductive systems whose theorems are true. Some are simpler, better systematized than others. Some are stronger, more informative, than others. These virtues compete: an uninformative system can be very simple, an unsystematized compendium of miscellaneous information can be very informative. The best system is one that strikes as good a balance as truth will allow between simplicity and strength. How good a balance that is will depend on how kind nature is. A regularity is a law iff it is a theorem of the best system. (Lewis 1994b, p. 478)

So far this applies only to deterministic laws. But Lewis suggests to expand the best-system analysis to cover chance laws as well, and he makes clear that the inclusion of chance laws in the best system is primarily governed by relative frequency and symmetry. Some say that Lewis' position thereby basically reduces to frequentism, others say that it essentially transcends frequentism. We need not decide. We may well accept the best-system analysis for the time being. It is plausible, as far as it goes; it is, to echo Lewis, simple, but uninformative.

There are two critical points, though. The first is that the team of the best-system analysis and the Principal Principle introduces not only an ontological, but also an epistemological double standard. We have already seen the ontological double standard. The best-system analysis somehow establishes T_w as the chance theory true of *w*, whereas (OP) rather requires T^*_w to be determined by *w*. In addition, we now face an epistemological double standard. On the one hand, our beliefs aim at the best system guided by standards of simplicity and strength and their balance. On the other hand, one should think that all these standards are encoded in the a priori credence function C_0 that we seek to constrain by (OP) and other rationality postulates. I do not see an incoherence here, but neither do I see how the two standards go together or what results from the circular procedure of letting C_0 decide about the best system and feeding in the decision into condition (OP) on C_0. These are unresolved frictions, to say the least.[26]

The second critical point is, of course, whether the best-system analysis can at all bolster up Humean Supervenience about laws and chance. Prima facie, it cannot. On the contrary, according to this analysis deterministic and probabilistic laws supervene not only on the totality of particular facts, but also on the measures for simplicity, for strength, and for the goodness of balance; and these measures are something *we* add (at least as far as simplicity and balance is concerned; strength has at least an objective partial order).

Surely, Lewis cannot be on good terms with this apparent consequence of the best-system analysis. He shies away from any idealistic tendency like the devil from the holy water, also in order to maintain Humean Supervenience. However, he sees a way out: Perhaps nature is kind to us, and "if nature is kind, the best system will be *robustly* best – so far ahead of its rivals that it will come out first under any standards of simplicity and strength and balance" (Lewis 1994b, p. 479 – his italics). If so, laws and chances do not depend on our inductive standards.

Yet, can there be a system that is robustly best under any standards? I guess even the kindest world is susceptible to transmogrification under gruesome standards. We may refer to factual human standards, but even there we find a lot of madness. Presumably, Lewis intends to quantify only over all reasonable inductive standards, and perhaps nature then has a better chance to be kind. Look, though, how wide the disagreement about reasonableness is, e.g., from the optimistic middle Carnap who had hoped for *the* inductive logic to the pessimistic subjectivists who plead for coherence and nothing more. It is quite obscure what a kind world is and how many of them there are.

In any case, Humean Supervenience turns out doubly constrained. It is ontologically restricted to worlds like ours devoid of alien matters, and it is epistemologically restricted to kind worlds free of indeterminateness concerning the best system.

[26] Sturgeon (1998) argues that the restrictions put on (HS) are indeed incoherent, however they are specified. Hall (2004, pp. 108ff.) also critically discusses how the inference of chances from facts is supposed to go.

The two restrictions appear to be independent, and together they turn Humean Supervenience into an uncomfortable doctrine. I think that the problems in the last section about undermining futures constitute a telling objection. On the whole, the doctrine seems in need of getting straightened out.

As to the second critical point concerning objectivity and independence of our standards Lewis had envisaged another solution:

> I used to think rigidification came to the rescue: in talking about what the laws would be if we changed our thinking, we use not our hypothetical new standards of simplicity and strength and balance, but rather our actual and present standards. (Lewis 1994b, p. 479)

Yes precisely. Rigidification is one salient strategy of objectification.[27] Alas, Lewis continues:

> But now I think that is a cosmetic remedy only. It doesn't make the problem go away, it only makes it harder to state. (Lewis 1994b, p. 479)

I did not understand this remark, so I requested him for clarification. Since I did not find the point explained elsewhere in his writings, let me quote extensively from his personal communication of February 13, 1996:

> Let me answer not your question but a generalization of it. The problem is that a certain analysis says that *X* (in this case, lawhood) depends on *Y* (in this case, our standards of simplicity, etc.) and yet we would ordinarily think this wasn't so. If *Y* were different, *X* would be just the same – or so we offhand think.
>
> A proposed answer is that '*X*' is a rigidified designator of the actual value of something that depends on *Y*, and of course it's not true that the *actual* value would be different if *Y* were different. That's supposed to explain our opinion that there's no dependence.
>
> Well, *if* that's so – I'd think that it well might be so under at least some legitimate disambiguation – let '†*X*' be a *derigidification* of the rigidified term '*X*'. Maybe there's some nice ordinary-language reading of the derigidifying modifier; or maybe not, but in any case we can introduce it into our language by a suitable semantic explanation (as is done, for instance, in Stalnaker's paper 'Assertion', *Syntax and Semantics* 9).[28] Then it might turn out that our original opinion that *X* doesn't depend on *Y* survives in modified form: as the opinion that even †*X* doesn't depend on *Y*. If so, the alleged rigidification of *X* ends up making no difference. I think that's what does happen in the case of lawhood and our standards of simplicity etc. And that's why the hypothesis of rigidification, even if true, doesn't make the problem of counter-intuitive dependence go away. It makes it harder to state, because to state it you must first introduce the notion of derigidification.

He did not further explain, however, why the intuition that lawhood is independent of our standards should be maintained under derigidification. Projectivism, which I am going to recommend, does not share this intuition. The projectivist rigidifies the result of his projection and thus legitimately claims objectivity for this result. But he is content with so much objectivity. He would immediately grant that

---

[27] Loewer (1996, pp. 114f.) also discusses the point and recommends rigidification.

[28] This refers to Stalnaker (1978).

derigidification brings the process of projection back into focus and thus displays the dependence on the cognitive subject. However, there is no need to decide the dispute about intuitions. The point rather is that Lewis' idea which was not good enough for himself helps projectivism to some arguably sufficient notion of objectivity while allowing to admit, in another sense, the dependence of lawhood on our inductive standards.[29]

8.6 Projection Turns the Principal Principle into a Special Case of the Reflection Principle

The last remark puts the cart before the horse. We still do not know what the projectivistic understanding of chances is actually supposed to be. In order to explain it, let us follow the Lewisian track of the best-system analysis, but let us avoid, contra Lewis, to give it an ontological turn, let us rather keep it within its epistemological home. This will lead us onto well-trodden paths, but I said right at the beginning that there are no new discoveries to be made.[30]

The best system is, first of all, based on complete experience, on complete knowledge of particular observable facts. If these should be only finitely many, then all statistical methodology tells us that they do not allow for guaranteed conclusions with respect to objective probabilities; to force a decision, for whatever reasons, is simply unjustified. This conclusion certainly remains true when we include the broader inductive considerations relevant to best systems. If the set of particular facts should be infinite, the situation is not really different. If a die is actually cast infinitely many times, the propensities of the throws will change, simply because the die will physically change, and then the limit of relative frequency does not help us to a definite conclusion concerning the propensities. This is our epistemic situation vis à vis a small die, and I do not see why it should be different with respect

[29] The point is indeed one of deep and general importance. It applies, I believe, to objecthood in general, certainly a most fundamental matter. *We* cut up the world into pieces, *we* constitute objects by saying which properties or kinds of properties are essential or constitutive for them. (This allows for the case that we fix only a space of possible essential properties of an object and leave it to the actual world to fix the actual essential properties.) Still, the objects thus constituted are objects independent of us, their objectivity is in no way impaired by our constituting them, in particular because we constitute objects in such a way that our constituting is *not* essential for them. The point extends to properties. In two-dimensional semantics each predicate expresses a (derigidified) concept and denotes a (rigidified) property, and while most concepts are, as I say, a priori relational, only few properties are necessarily relational – two notions of relationality that are particularly relevant vis à vis color predicates; cf. my (1997b, pp. 367ff.) [here: pp. 297ff.]. This footnote indicates the direction into which this paper would need most to be further thought through.

[30] This section elaborates the core of the predecessor paper Spohn (1999b).

to large worlds. In the strictest sense, nothing is repeatable. In saying this I flatly deny Humean Supervenience, of course.

Hence, it is actually unfeasible to precisely detect chances, even given complete knowledge of particular facts. The detectibility is rather merely counterfactual. Suppose we could run our world over and over again, indeed infinitely many times, suppose that all repetitions were governed by the same objective chance mechanism, and suppose we could learn all particular facts within not only one, but all repetitions. Then we would finally have established the chance law P_w of w, at least with probabilistic certainty. The last proviso is essential. If we live in a chancy world, we know a priori that there is a chance for misleading evidence, and we know a priori that even counterfactually ideal evidence cannot close the gap; the difference between probability 0 and impossibility is ineliminable.

If we want to describe this ideal detectibility of chances more formally, we obviously have to consider $W_0 \times W^\infty$, i.e., not only the original space $W = W_0$ of worlds of particular facts, but besides the space W^∞ of infinitely many possible counterfactual runs of the actual world; each $w^\infty \in W^\infty$ thus is an infinite sequence of possible worlds, each being a complete course of particular facts. (The term "W_0" is introduced only in order to distinguish that copy of W from its infinitely many counterfactual repetitions.) And we have to extend our probabilistic notions to $W_0 \times W^\infty$. If the actual world w is governed by the chance law $P_w \in \mathcal{P}$ defined for propositions over W_0, then these infinite sequences are governed by the product (or Bernoulli) measure $P_w^\infty \in \mathcal{P}^\infty$ which is the infinite product of P_w with itself and which is defined for propositions over W^∞. According to P_w^∞ the individual runs are governed by the same chance law P_w, and they are stochastically independent from one another; thus are our counterfactual suppositions for the ideal detectibility of P_w. Finally, we have to assume an a priori credence C_0^∞ also defined for propositions over $W_0 \times W^\infty$. C_0^∞ is not concerned with chances; it only captures our a priori expectations about all the particular facts in $W_0 \times W^\infty$. Of course, it extends the factual part of C_0; i.e., for each proposition $A \subseteq W_0$ we have $C_0^\infty(A) = C_0(A)$. I shall soon say a bit more about C_0^∞.

What we just said about the counterfactual detectibility of chances then condenses into what I would like to call the *Knowability Principle*:

(KP) $\quad C_0^\infty (A \mid w^\infty) = P_w(A)$ P_w^∞-almost surely for all P_w and all $A \subseteq W_0$.

The left-hand side is indeed a random variable with w^∞ as random argument. That the equation holds P_w^∞-almost surely is to say that the set of w^∞ for which the equation holds has P_w^∞-probability 1. The expression "$C_0^\infty (A \mid w^\infty)$" is once more sloppy mathematics; it is short for the limit of the conditional credence of A when the condition infinitely grows into w^∞.

Instead of an ontologically conceived Humean Supervenience of chances on the actual particular facts, we thus have (KP) asserting the counterfactual knowability on the basis of counterfactual particular facts. We shall soon see how (KP) reduces to still more basic rationality constraints on C_0^∞. "Knowability" is perhaps too strong a word; strictly speaking, we can never know the chances, we can only be almost sure of them. However, (KP) captures all what (counterfactual) particular facts can tell us about chances; even counterfactually there is no more to know; (KP) is our best approximation to knowability.

I introduced (KP) only as our epistemological substitute for the misguided ontological Humean Supervenience of chances. In fact, (KP) follows from standard principles. So far, we have not yet explicitly considered relative frequencies. This is easily done, though. Let $rf(A)(w^\infty)$ stand for the limit (if it exists) of the relative frequency of the realization of the proposition A in the infinite random sequence w^∞. Then two further principles hold, namely the (strong) Law of Large Numbers:

(LLN) $rf(A)(w^\infty) = P_w(A)$ P_w^∞ -almost surely for all P_w and all $A \subseteq W_0$,

and the so-called Reichenbach Axiom (recommended by Hilary Putnam to Carnap in 1953; cf. Carnap 1980, p. 120):

(RA) $C_0^\infty(A \mid w^\infty) = rf(A)(w^\infty)$ for all w^∞ and all $A \subseteq W_0$,

which says that our beliefs should increasingly and in the limit perfectly align with the observed relative frequencies, whatever they are. (KP), (LLN), and (RA) form a triangle connecting credence, chance, and relative frequency. Among the three, (LLN) and (RA) are the more basic ones. (LLN) is not a rationality postulate, but a mathematical theorem. Moreover, given (LLN), (KP) obviously follows from (RA), but not vice versa, because the equality of (KP) holds only almost surely.

Indeed, I find that de Finetti's representation theorem fits perfectly to my counterfactual set-up, thus providing further insight into the Reichenbach Axiom. This is why I have emphasized at the beginning of this paper that I do hardly more than rearrange de Finetti's philosophy of probability. The a priori credence C_0^∞ should be a symmetric measure over the product space, i.e., the event that n given propositions realize in the first n repetitions has the same credence as the event that these propositions realize in any other n repetitions. This seems even more compelling in our counterfactual set-up, where all repetitions are equal by fiat, than in any factual set-up. De Finetti's representation theorem tells that all and only symmetric measures are mixtures of product or Bernoulli measures, indeed unique mixtures. Hence, symmetry entails the principle of non-negative instantial relevance (cf. Humburg 1971, p. 228). Moreover, given symmetry, (RA) is equivalent to the assumption that the support or carrier of the mixture is the space of all product measures. This in turn makes clear that, given symmetry, (RA) entails the principle of positive instantial relevance (cf. Humburg 1971, p. 233). This may suffice as a brief reminder of the familiar epistemological home of the Reichenbach Axiom and thus of the epistemological grounds of the Knowability Principle.

My next point is that (KP) entitles us to project the credence C_0^∞ for $W_0 \times W^\infty$, i.e., for the actual world and its infinitely many counterfactual repetitions onto the credence C_0 for $W_0 \times \mathcal{P}$, i.e., for the actual world and its chance measure. The

Projection Rule tells for each proposition $A \subseteq W_0$ and each set $\mathcal{Q} \subseteq \mathcal{P}$ of chance measures for W that:

(PROJ) $$C_0(A \times \mathcal{Q}) = C_0^\infty(A \times \{w^\infty \mid C_0^\infty(. \mid w^\infty) \in \mathcal{Q}\}).$$

The Projection Rule thus says that a priori our credence that the true chance measure is in $\mathcal{Q}$ (and that some factual proposition A holds) is the same as our credence (that A holds and) that the counterfactual infinite evidence $w^\cdot$ moves us into some state in $\mathcal{Q}$.

Why is (PROJ) legitimate? (KP) says that for each possible P_w^∞ the set of w^∞ making C_0^∞ diverge from P_w is a P_w^∞-null set. Due to its symmetry, however, C_0^∞ is a mixture of all the P_w^∞. Hence, the set of w^∞ making C_0^∞ diverge from all measures in $\mathcal{Q}$ is also a C_0^∞-null set, because its C_0^∞-probability is a mixture of all the P_w^∞-null sets involved. Note, again, that (PROJ) is not an ontological thesis reducing chance to counterfactual infinite sequences of factual worlds. The ontological slack between truth and evidence is ineliminable. However, the ontological slack has not the slightest epistemological weight and cannot surface in the epistemological rule (PROJ); it is a genuine 'don't care'.

The upshot of these considerations is that the Minimal Principle is an immediate consequence of the Projection Rule. Take $\mathcal{Q} = \{P_w \mid P_w(A) = x\}$. Then (PROJ) specializes to

$$C_0(A \mid P_w(A) = x) = C_0^\infty(A \mid \{w^\infty \mid C_0^\infty(A \mid w^\infty) = x\}) = x.$$

And this is nothing but (MP*), which we have seen is all we need together with (RED) (and (AdP)) to duplicate Lewis' account. Thus, the replacement of the ontological doctrine of Humean supervenience by the epistemological Knowability Principle (which backed up the Projection Rule) at the same time replaces the conflict with (OP*) by a confirmation of (OP*).[31]

I find it illuminating to cast the point into a somewhat different form. For this purpose, we have to introduce the final player of my scenario, van Fraassen's so-called Reflection Principle. It is entirely about subjective probability. There we have static rationality postulates like Coherence or the axioms of mathematical probability, Regularity, Symmetry, etc., and we have dynamic rationality postulates

[31] Hall (2004, pp. 108f.) envisages the same kind of argument, also with reference to de Finetti's representation theorem, though without actually endorsing it. He ascribes the argument to a position he calls 'primitivist hypothetical frequentism', which, however, is not mine. As he describes it, this kind of frequentist equates chance with limiting hypothetical relative frequency and considers it to be a brute metaphysical fact that this equation is correct. By contrast, I emphasized the almost unnoticeable epistemological-ontological gap, and I do not see the necessity to close it per fiat.

the best known of which is, of course, the Rule of Conditionalization. About the most basic of these dynamic postulates is the *Reflection Principle*[32]:

$$\text{(RP)} \qquad C_t(A \mid C_{t'}(A) = x) = x.$$

Here, C_t is the subject's credence or subjective probability at time t, and it is understood that t' is later than t. In other words, C_t specifies the prior and $C_{t'}$ the posterior probabilities of the subject. The Reflection Principle thus says: Given the condition that my future probability for some proposition is x, my present probability for it is also x. In short: I trust now what I assume to be my future belief.

It is clear why (RP) is called an auto-epistemic principle; it assumes that my future beliefs are the objects of my present beliefs (even only as a supposition). If one accepts the richer auto-epistemic framework, then (RP) proves to be a most general dynamic doxastic law entailing conditionalization and its generalizations; it is even amenable to a Dutch Book justification (cf. Gaifman 1988; Hild 1998b). It is also obvious that (RP) is a rationality postulate of restricted validity. For instance, I should not now trust my future beliefs I will have when drunken, and when now reading the newspaper I should believe (within limits) what I have read even given that tomorrow I will have forgotten what I have read. Hence, I should reasonably trust only those of my future beliefs that I have acquired in a reasonable fashion and that I entertain from a *superior* point of view, which is certainly provided by experience (and maybe in other ways as well).

The similarity between the Minimal and the Reflection Principle strikes the eye, though they are about different subject matters. However, the similarity is easily turned into entailment. Take (RP), replace C_t by the 'first' a priori credence C_0^∞ and $C_{t'}$ by the 'last' credence $C_0^\infty(\cdot \mid w^\infty)$ counterfactually completely informed. (RP) thus specializes to

$$(\text{RP}^\infty) \qquad C_0^\infty(A \mid C_0^\infty(A \mid w^\infty) = x) = x.$$

Note that (RP^∞) is in fact a theorem, not merely a rationality postulate. As above, (PROJ) finally turns (RP^∞) into (MP*).[33] To summarize, in counterfactual 'future' we are completely informed about the counterfactual manifestations of the propensities in w of particular facts, thus completely informed we can infer the chances in w, and hence (MP*) turns out as a special case of (RP).

[32] The Reflection Principle is explicitly stated in van Fraassen (1984); there its deep philosophical relevance was fully recognized. He returns to it at length in van Fraassen (1995). Other references are Goldstein (1983) and Spohn (1978, pp. 162f.) where I stated an equivalent principle (called the Iteration Principle by Hild 1998a, p. 329) within an auto-epistemic or reflexive decision-theoretic setting and under the restrictions usually accepted nowadays. Penetrating discussions may in particular be found in Hild (1998a, b).

[33] Skyrms (1980, appendix 2) already observed that there is a common form to such principles that is open to various interpretations. Following Gaifman (1988), the common form might be called 'expert principle', since it describes trust in some kind of expert. For this unified view see in particular Hall (2004) and Hájek (2005). However, it is only our Projection Rule which establishes an entailment between the expert principles considered here, i.e., (RP) and (MP).

8.7 Humean Projection

What is the significance of these mathematically trivial transformations? If projectivism is the doctrine that some objective traits of the world can only be understood as objectified projections of human attitudes, how does the previous section support projectivism concerning chances? To resume, the story is as follows: We postulate chances, and we know that they are different from our subjective probabilities. Yet, we also know the rational shape of our credences, we know how we change and improve them, we know according to (KP) that we cannot say anything better than that the chances are what our credences would be after that infinite counterfactual information, not by necessity, but with probability 1, and we know according to (PROJ) that we may identify our credences about chances with our credences about that counterfactual information and what we learn from it. We are aware of the ontological gap between chance and credence, but our epistemological bridge over it leaves nothing to be desired. In this sense I take chance to be a projection from credence.

Jeffrey (1965, sect. 12.7) discusses the general idea that objective probabilities are *objectified* subjective ones, and in (2004, p. 19) he says, referring to Hume, that "chances are simply projections of *robust* features of judgmental probabilities from our minds out into the world" (his emphasis). Maybe he had the same picture in mind as the one developed here. However, objectification as he describes it in Jeffrey (1965, sect. 12.7) is admittedly not very objective; it just means conditioning subjective probabilities w.r.t. the true member of some partition of the possibility space considered (or the limit of these conditionings w.r.t. a sequence of the true members of ever finer partitions). Of course, the result depends on the initial subjective probabilities as well as on the chosen partition. Jeffrey argues that this latitude has some advantages, but it seems clear that the general idea needs refinement.

Lewis (1980a, pp. 278f.) is pleased that his account may be understood as offering such a refinement. According to (OP), it is the history-chance partition, as he calls it, which is *the* correct objectifying partition, and according to (OP*) it is more simply the chance partition consisting of all T_w themselves. Skyrms (1980, sect. IA4) makes, in effect, the same proposal, though he opts for more pragmatic flexibility than Lewis and rather hides the chance nature of his conditioning partition. It is a matter of taste whether one should call this a confirmation or a trivialization of Jeffrey's general idea. In any case, Jeffrey (2004, p. 20) reminds us that "on the Humean view it is *ordinary* conditions, making no use of the word 'chance,' that appear" in the condition of (MP) or in the conditioning partition (my emphasis). Jeffrey insists on the point because otherwise his objectification idea has no prospect of offering an analysis of chance, a prospect Lewis (1980a, pp. 288ff.) explicitly denies.

So, how does Jeffrey's general idea fare with Humean Projection as construed here? According to (PROJ) it is indeed the partition consisting of all T_w which is invoked in objectification; it is, however, to be conceived as the partition into all $\{w' \mid C_0^\infty(\cdot \mid w'^\infty) = P_w\}$. Hence, we have obeyed Jeffrey's reminder; we have used ordinary conditions making no use of the word "chance". Still, I am not sure whether

Jeffrey would be satisfied. His examples always use partitions of the original possibility space W of particular facts, whereas I move to a partition of the possibility space W^∞ of infinite counterfactual repetitions of W. Only there particular facts can get as close to objective chances as they can get; and if this is so, then Jeffrey's objectification within the space W can at best reach pragmatically weakened forms. The detour via W^∞ appears unavoidable to me.

My continuous massive invocation of counterfactuality may have raised, however, suspicions from the outset. Skyrms (1980, p. 31) has already warned that "attempts to construe propensities as modalized relative frequencies *only make things worse* in this regard" (his emphasis), the regard being the use of the law of large numbers as an analysis of propensity. Skyrms is right. We have seen that chances do not ontologically reduce even to propositions over the counterfactual space W^∞; the slack *is* ineliminable. However, W^∞ serves here only epistemological, not ontological purposes.

For the same reason I am not worried by Lewis (1994b, p. 477), when he says "I think that's a blind alley", thereby referring to "thinking of frequencies not in our actual world, but rather in counterfactual situations" (in order to deal with his puzzling case of unobtainium). Within his set-up he is indeed right. There, relative frequencies in counterfactual situations can inform us about the actual world only, if we have ascertained beforehand that the counterfactual situation is governed by the same chance law as the actual world. Thus, we would have to solve, according to Lewis, the supervenience issue for the counterfactual situation in order to solve it for the actual world; and this merely defers the issue. However, this is not our problem. We do not have the telescope view onto counterfactual situations, to use Kripke's terms; it was rather part of our counterfactual stipulation that all repetitions of W be governed by the same chance law P_w; there is no need to *ascertain* the chance law of the repetitions. I do not see why this counterfactual stipulation should be illegitimate. We always think about counterfactual situations and what we would believe given this or that situation, and in order to get Humean Projection running in our way, we only consider extreme cases of this kind. Specializing, or extending, (RP) to (RP$^\infty$), in order to derive (MP), is not a misuse of the Reflection Principle; it is an extreme, though legitimate use.

Well, it may be legitimate; still it hardly helps. Given the extreme counterfactual evidence we may be as certain about chances as we can. Our actual evidence, however, is infinitely poorer. Indeed doubly so; we can inquire only a tiny part of our actual world and never the counterfactual repetitions. The counterfactual construction may, and should, I think, satisfy philosophers, but it is of no use for scientists and statisticians who cannot do better than gathering actual evidence and drawing conclusions from this insufficient basis. This, however, is something to acknowledge, not to deplore. The philosophical account provides the ideal standard, and it then is a methodological issue how best to approximate the ideal within our factual limits. Statisticians have developed most sophisticated test methods, of which randomization is an important part. But there are also more general preconceptions.

In principle, the scheme of partial determination governing our world may be any chance measure whatsoever. In principle, the whole world has the propensity to move into this or that state, and propensities may vary from here to there and from now to then. In our counterfactual scenario we could discover any wild distribution of chances, but in the actual world we want to understand the 'mechanics' of partial determination. The ground rule is: equal causes, equal effects; or rather, equal conditions, equal propensities – which gets bite only by restricting "equal". The relevant conditions should be few, not many. If we are lucky, we have kept constant all relevant conditions during a row of some thousands throws of some die, and then we may take the actual row as approximating the counterfactual sequence. The relevant conditions should be local, or contiguous, to use Hume's term. Nonlocality is one of the mysteries created by quantum mechanics. Crystal balls are miraculous for the same reason. I find it incoherent to say that a given type of events is only partially determined, but can be unfailingly foreseen with a certain crystal ball. Rather, I would then take these events as fully determined – but would not understand how determination, i.e., the crystal ball works in these cases. If we are lucky, we shall be able to construe the chance law governing our world as a Markov process. If we develop different ideas about space and time, we have to adapt our preconceptions of the 'mechanics' of determination. And so forth.

If we do not succeed with our preconceptions, it is unclear how we would respond. In the extreme case, the idea of partial (or full) determination would dissolve. Thus it seems obvious to me that there is more to the notion of chance than just the Principal Principle. There are also all these preconceptions connecting chance with space and time, simplicity, orderliness, and whatnot. It is such things mentioned by Arntzenius and Hall (2003, pp. 177f.) when they arrive at the same conclusion. These preconceptions are modifiable, but only within limits; beyond the notion of chance will crumble.

Do such considerations reintroduce the epistemological double standard of which I have accused Lewis in Section 8.5? No. With regard to the ideal counterfactual evidence we can simply stick to de Finetti's story of the symmetric a priori credence satisfying the Reichenbach axiom and thus converging almost surely to the true chance measure, whatever it may be. Here, we do not need help from the additional considerations just mentioned. We have to rely on them when and because we try to make sense of our very restricted evidence. Thus, the second epistemological story that I have just indicated does not interfere with, but rather complements, the account I have extensively presented.

Since we have sacrificed Humean Supervenience, we also have avoided the ontological double standard and the resulting conflict between (OP) and (NP). We can, and do, simply stick to (OP*) and reject (NP*) as nonsense.

However, if we sacrifice (HS), we cannot do so without considering Lewis' two main reasons for it. The one consists in his ontological preferences. Without doubt, if (HS) were true, the resulting ontological picture would be most elegant and satisfying. Those rejecting Humean supervenience have different preferences and acknowledge irreducible dispositions, capacities, causes, necessities, or propensities. The projectivist, in particular, has a special story to tell about these matters that explains them ultimately with our subjective condition without diminishing their objectivity. I do not think that this ontological dispute can be resolved with general arguments. It is a matter of details, and there we have at least seen that Lewis had difficulties to maintain his prima facie elegance.

His second reason, though, is more pertinent and more urgent. It is best put in Lewis (1986a, pp. xvf):

> I could admit that … the chances … do not supervene on the arrangement of qualities. … Why not? I am not moved just by loyalty to my previous opinions. That answer works no better than the others. Here again the unHumean candidate for the job turns out to be unfit for its work. The distinctive thing about chances is their place in the 'Principal Principle,' which compellingly demands that we conform our credences about outcomes to our credences about chances. … I haven't the faintest notion how it might be rational to conform my credences about outcomes to my credences about some mysterious unHumean magnitude. Don't try to take the mystery away by saying that this unHumean magnitude is none other than *chance*! I say that I haven't the faintest notion how an unHumean magnitude can possibly do what it must do to deserve the name – namely, fit into the principle about rationality of credences – so don't just stipulate that it bears that name. Don't say: here is chance, now is it Humean or not? Ask: … Is there any way that an unHumean magnitude could [fill the chance-role]? … the answer is 'no'…

He repeats the point in Lewis (1994b, pp. 484f.) with more confidence, having been shown a way out of the paradox of undermining futures generated by (HS).

His own response to this challenge is (HS). It is no mystery how particular facts constrain credence; and if chance supervenes on particular facts, it is in principle no mystery how chance constrains credence. And thus he sets out to remove paradox by modifying (OP). Right at the beginning of Section 8.2 I indicated that this is the basic puzzle affecting the Principal Principle. The quotation indeed suggests that Lewis thinks that (HS) is the *only* solution of the puzzle (even though his challenge is directed foremost to the position of David Armstrong). How may the projectivist respond?

For the projectivist the puzzle has a straightforward solution.[34] This is clear from his general strategy. For him, chances are not alien features cognitive access to which is bound to be mysterious; they are of our own breeding. We need not speak figuratively, though; we have prepared a precise answer. Lewis is right; there is no mystery how particular facts constrain credence. However, van Fraassen is also right; there is in principle no mystery how future credence can constrain present credence. And we have seen that according to the projectivistic reconstrual the Principal Principle is nothing but an extreme application of the Reflection Principle. This was the whole point of my construction in the previous section. To be sure, in

[34] As mentioned in footnote 31, Hall (2004, p. 109) also envisages the solution defended here (with some doubts concerning its general feasibility). However, he envisages it only as a possibility in order to prove the point he is up to in his paper, viz., that the reductionist claiming (HS) need not have an advantage over the non-reductionist vis à vis this issue. For him (cf. p. 107), a no less acceptable response seems to be to declare the Principal Principle analytic and to reject any further justificatory demands. As I have explained in section 8.2, this will not do. We have a real challenge here which requires some substantial response.

that application a priori credence is constrained by an extremely counterfactual 'future' credence. However, it is mostly counterfactual future credence to which (RP) applies, and we should certainly not bother about being more or less extreme. In this way, the projectivist is able to remove the puzzling air from the Principal Principle. Chance, being almost surely identical to projected credence objectified, must constrain a priori credence precisely in the way summarized in (OP*).

8.8 Appendix on Ranking Functions and Deterministic Laws: The Same All Over Again

The whole of this paper immediately and perfectly carries over to full determination or natural necessity and deterministic laws. Lewis tells the same story, this story meets the same criticism, and I have a precise projectivistic substitute story. Indeed, all this is more or less a matter of routine; I do not have to write a twin paper. Let me just indicate the basic points.

A very common, and also Lewis', assumption is that laws are regularities which in turn are mere generalizations expressed by universally quantified sentences. However, not all regularities are laws; we have to be selective. Lewis offers his best-system analysis of laws in order to discriminate them from mere regularities. He thinks laws Humeanly supervene on particular facts, and he constrains the supervenience of laws in the same way as that of chance. The only point missing is that the Principal Principle and the ensuing discussion have no explicit deterministic counterpart.

The problems remain. (HS) is again ontologically as well as epistemologically constrained. Carroll (1994, ch. 3) and Ward (2002, sect. 3) attempt to specify examples of two worlds in which the same facts, but different laws obtain. Black (1998, p. 376) suggests "that laws can … undermine themselves, in that the laws of the universe might allow that the laws of the universe could have been otherwise." Hence, it looks like we are running into the same kind of problems with deterministic laws as we have extensively discussed here with respect to chance. Lewis' reasons for sticking to (HS) are also the same. Without (HS) we could not understand the idea of necessitation. Hence, the dialectic situation is as before. What, though, could be the constructive alternative? This is indeed much less clear than in the probabilistic case where subjective and objective probabilities and their delicate relation are perhaps not fully understood, but familiar for a very long time.

I think the basic mistake lies already in the common assumption that laws are (a special kind of) regularities. In this respect, laws are much more deceptive than chances. One immediately sees that chances are modalities; they take propositions as arguments and somehow assign numbers to them. By contrast, laws appear to be mere propositions, and modality is prima facie not involved. Any subsequent mounting of modality is then bound to create mysteries. The alternative, though, is not to start with a primitive necessitation operator, as Armstrong does in his analysis of lawhood. This is no less mysterious. Also, it will not do to conceive of deterministic laws as a limiting case of chance laws, not only for the reason that a chance of 1 is not quite necessitation. This is not the place, however, to go through all the various accounts of lawhood. Let me just say that I believe that the alternative must be somehow to tell the same kind of story as we did in the probabilistic case. But how?

The answer is: with the help of ranking functions (first presented in Spohn 1983a and 1988, where I called them ordinal conditional functions). As in probability, we must start with the subjective side, with the representation of belief. This is what a ranking function does. A *ranking function* κ for a given possibility space W is a function from W into the set of nonnegative integers such that $\kappa(w) = 0$ for some $w \in W$. The ranking is extended to propositions $A \subseteq W$ by defining $\kappa(A) = \min \{\kappa(w) \mid w \in A\}$. And conditional ranks are defined by $\kappa(B \mid A) = \kappa(B \cap A) - \kappa(A)$.

Ranks are degrees of disbelief. $\kappa(A) = 0$ says that A is not disbelieved at all; $\kappa(A) = n > 0$ says that A is disbelieved to degree n. Hence, $\kappa(\bar{A}) > 0$ expresses that $\bar{A}$ is disbelieved (to some degree) and hence that A is believed (to the same degree). Thus, ranking functions, unlike probability measures, represent belief (acceptance, holding to be true). This is their most distinctive feature due to which they can be related to deterministic as opposed to probabilistic laws. Unlike doxastic logic, or even AGM belief revision theory, ranking functions can also account for a full dynamics of belief; this means at the same time that they embody a full inductive logic. Basically, this dynamics consists in conditionalization, just as in probability theory.[35] The reason why this works perfectly is that conditional ranks as defined above behave almost exactly like conditional probabilities. Indeed, the parallel extends much farther. Practically all virtues of Bayesian epistemology can be carried over to ranking functions. (For a fuller explanation of these claims see Spohn 1988, [here: ch. 1].)

One thing we can now do, for instance, is to state the Reflection Principle in ranking terms:

$$(\text{RP}\kappa)\ \kappa_t(A \mid \kappa_{t'}(A) = n) = n,$$

which says that given you disbelieve A tomorrow to degree n you so disbelieve it already today. (RPκ) is indeed a strengthening of Binkley's principle.[36] All the remarks about the probabilistic version (RP) apply here as well.

I have emphasized that ranking functions must be interpreted as representing doxastic states. They represent what a subject takes to be true or false, but they are not true or false themselves. However, to some extent they can be objectified so that it makes sense to apply truth and falsity to them, just as to propositions. How this objectification works is a somewhat tricky story elaborated in Spohn (1993a) [here:

---

[35] The idea that belief is just probability 1 is not only intuitively unsatisfactory, but also theoretically defective, because conditionalization does not work for extreme probabilities and beliefs could then only be accumulated and never revised. (Popper measures solve this problem just as half-way as does AGM belief revision; see Spohn 1986). This is the essential reason why it does not work to correspondingly conceive deterministic laws as limiting cases of chance laws.

[36] It says that if I believe now that I shall believe tomorrow that p, I should already now believe that p. Binkley (1968) introduced it in relation to the surprise examination paradox.

ch. 5]. According to this story, most ranking functions cannot be objectified. This appears to be different with chances. Any credence measure for W could, it seems, also serve as a chance measure for W. But maybe not. We have seen above that there is more to chance and that one might, for instance, suggest that only probability measures representing a Markov process can be chance measures.

Anyway, what I have proposed in Spohn (1993a) [here: ch. 5] is that causal laws or, in the present terms, schemes of full determination are just such objectifiable ranking functions, a view I have philosophically more thoroughly explained in Spohn (2002) [here: ch. 6]. The crucial point is that the inductive behavior is thus directly built in into laws and not subsequently imposed on something propositional. Moreover, for laws so conceived we can tell de Finetti's complete story as shown in detail in Spohn (2005a) [here: ch. 7]. If the ranking function κ is such a scheme of full determination for W, we can again form the infinite product space $W^{\cdot}$ and the product ranking function $\kappa^{\cdot}$ independently repeating κ infinitely many times. Any symmetric ranking function over $W^{\cdot}$ is then a unique mixture of such product ranking functions, which will converge to the true law (= product ranking function) with increasing evidence. Hereby, the role relative frequencies have in the probabilistic case is taken over by the number of exceptions in the deterministic case.

In sum, we have here all the ingredients for telling exactly the parallel story about necessitation or full determination as we have told about partial determination. Deterministic laws are, in the way explained, projections of ranking functions, i.e., of subjective states representing beliefs and their dynamics.

Part IV
Coherence

Chapter 9
A Reason for Explanation: Explanations Provide Stable Reasons†

9.1 Introduction*

Why ask 'Why?'? Whence our drive for explanation? This is a bewildering question because it is hard to see what an answer might look like. I well remember having learnt in undergraduate courses that explanation is the supreme goal of science. So who would dare ask for more? Some fortunately did.[1] One prominent answer is that (scientific) explanation yields (scientific) understanding; and surely, we want to understand things. It is this answer which this paper is about.

When I first heard of this answer from Karel Lambert as being seriously discussed, it struck me as utterly tautological; and when arguing against it in Lambert (1988, 1991) he seemed to argue for a contradiction. However, there is one, and only one, way of rendering this answer sensible and sensibly doubtable: namely by giving independent characterizations of "(scientific) explanation" and of "(scientific) understanding" and checking how they relate. This is what Lambert (1988, 1991) did, thus recovering the full worth of the answer. But it is not what is usually done. Quite often the correctness of the answer has been presupposed, and ideas about what understanding might consist in have been built into the characterization of explanation.[2] But then the answer helps only to explicate, not to justify explanation.

---

† This paper was originally published in: W. Spohn, B. Skyrms, B. C. van Fraassen (eds.), *Existence and Explanation. Essays Presented in Honor of Karel Lambert*, Dordrecht: Kluwer, 1991, pp. 165–196.

* I am very grateful to Dan Hunter for carefully checking my English.

[1] Sketchy remarks about the utility of explanations may be found quite often. Much less often the question is explicitly addressed, e.g. by Salmon (1978), where he propounds his own answer to the question, and Salmon (1984, pp. 124ff., 259ff.), where he discusses also other answers.

[2] This is the declared strategy of, e.g., Friedman (1974) and Kitcher (1981).

Lambert (1991) concluded that the fact that an answer to a why-question is an explanation is neither sufficient nor necessary for it to yield understanding. I want to advance an argument in favor of the contrary conclusion. It needs a double stage-setting (Sections 9.2 and 9.4) and has two steps (Sections 9.3 and 9.5). Section 9.2 mainly presents a general theory of non-probabilistic induction. This is the basis for Section 9.3: a partial account of deterministic causation (which copies the probabilistic account I have given in my 1983b and 1990a, [here: ch. 2]) and a straightforward extension thereof to a partial account of causal explanation. Section 9.4 works up to some coherentistic principles in terms of the given theory of induction which involve what I shall call ultimately stable reasons. The notion of an ultimately stable reason cannot pretend to catch much of the rich notion of understanding; but, as Section 9.5 explains, it fits well the characterizations of understanding which have been given in this context and may thus serve as a substitute.[3] Section 9.5 finally proves a formal equivalence of causal explanations and ultimately stable reasons under some restrictions which require several comments. Since the epistemological relevance of ultimately stable reasons unfolds in a coherentistic picture of truth, this equivalence construes the search for explanation as the search for truth.

9.2 Induction and Causation

David Hume was the first to argue for an essential connection between induction and causation, so forcefully in fact that it has not ceased since to be in the focus of philosophical discussion. Indeed, for Hume induction and causation were virtually the same.

Although Hume himself struggled with the characterization of belief – believing, he said, is having ideas accompanied by a peculiar feeling of vivacity and firmness[4] – he had an elaborate theory of belief formation. Impressions as the most lively and forceful of all perceptions are the paradigms and the basis of belief; all other empirical beliefs are gained from them by inductive extension. How? Hume held that induction proceeds just by inferring causes from effects and vice versa, i.e. via causal inferences which sufficiently, though not completely, transfer the impressions' vivacity and firmness so characteristic of belief.[5] The realm of empirical belief therefore consists of nothing but causal inferences from impressions (and their recollections).

[3] Indeed, this paper originated from an observation of this fit.

[4] Cf., e.g., Hume (1739, pp. 94ff.), and (1777, pp. 47ff.). The struggle is most conspicuous in the appendix of (1739, pp. 623ff.).

[5] In (1777, p. 26) Hume writes: "All reasonings concerning matter of fact seem to be founded on the relation of *Cause and Effect*. By means of this relation alone we can go beyond the evidence of our memory and senses." In (1739, p. 107) he says equally clearly: "… we find by experience, that belief arises only from causation, and that we draw no inference from one object to another, except they be connected by this relation". However, the relevance of further principles of association, namely resemblance and contiguity, is not really clear. In (1739, pp. 107ff.) he argues that these are only assisting, but not basic principles; in (1777, pp.50ff.) he does not discriminate in this way.

Induction thus seems to reduce to causation. But one may as well view the matter the other way around. Hume defines causation, taken as what he calls a natural relation, as precedence, contiguity, and association, i.e. transfer of liveliness and firmness, the marks of belief.[6] Thus, if *A* precedes *B* and is contiguous to it, *A* is a cause of *B* if and only if *B* may be inductively inferred from *A*. This shows that induction and causation are in effect interdefinable for Hume.[7]

The imperfections of Hume's account are well known. It is certainly wrong to say that *A* is a cause of *B* if and only if *B* may be inductively inferred from *A*, even if *A* and *B* satisfy the other conditions; symptoms of later events are clear counterexamples. And if one gives up this equivalence, it is doubtful that causal inferences exhaust inductive inferences. However, I believe that such imperfections do not defeat Hume's fundamental insight into the essential connection between induction and causation; the task is to get it straight.

Since the ways of induction seem multifarious, it is implausible that induction should be definable in causal terms. Thus one part of this task, the one discussed in the rest of this section, must be to provide a general account of induction independent of causation. The other part, taken up in the next section, is then to say how causation relates to this account.

Concerning the first part, the first point to note is that induction and belief revision are one and the same topic: The input of an inductive scheme consists of all the information directly received, and it tells what to believe according to the input. The input of a scheme of belief revision consists of an old epistemic state and a new piece of information, and it yields a new epistemic state. Thus, a scheme of belief revision may be immediately inferred from an inductive scheme, and the latter follows from the former plus an initial epistemic state to start from. This congruence may not always have been clear because induction and revision have met different interests. Belief revisionists explicitly searched only for rationality constraints on belief, whereas the longer-standing discussion of induction tended to search for *the* correct inductive scheme, thereby presupposing, or perhaps only hoping, that there is just one such scheme possibly even independent of the initial epistemic state. History taught us, I think, that this presupposition, or hope, is misguided.[8] Therefore, the two fields have merged by now, and general accounts of induction may best be found by looking at accounts of belief revision.

---

[6]This is very explicit in Hume (1739, pp. 170–172).

[7]This observation raises questions: Does Hume take one of the two notions as primary? Or is there a circularity in Hume's account? Which role has Hume's definition of causation taken as what he calls a philosophical relation, which refers to regularity instead of association? Does it offer a way out of the possible circularity? Cf., e.g., Mackie (1974, ch. 1), and Beauchamp and Rosenberg (1981, ch. 1) for thorough discussions of these questions.

[8]This is the lesson, for instance, of Goodman's new riddle of induction and Carnap's acknowledged failure to distinguish even a small class of inductive methods. It is challenged, however, by the puzzling alternative set up in the final section of Lewis (1980a).

Within the representation of epistemic states as (subjective) probability measures, belief revision is a rich and lively topic.[9] However, I instead want to turn to a much less familiar representation of epistemic states. One essential weakness of the probabilistic representation is that it can hardly account for plain belief which simply holds propositions to be true or false or neither; this is the moral of the well-known lottery paradox.[10] Therefore I dismiss probability because I want to focus on plain belief – for several reasons: First, it is of intrinsic interest to examine the structure of inductive schemes for plain belief. Secondly, if induction and causation are indeed essentially connected, then, presumably, subjective probabilities are related to probabilistic causation, whereas (sufficient and/or necessary) deterministic causation relates to plain belief; and it is the latter kind of causation I am concerned with. Thirdly, the probabilistic counterparts of some of the assertions in the final section hold only under more restrictive conditions. Fourthly, and perhaps most importantly, subjective probabilities cannot be true or false; truth attaches only to plain belief; thus an important part of my argument will only work for plain belief.

Strangely enough, induction and revision with respect to plain belief is a much more experimental and less established field. Shackle's functions of potential surprise, Rescher's plausiblity indexing, and Cohen's inductive probability[11] are pioneering contributions, and revision of plain belief has been most thoroughly studied by Gärdenfors and his coauthors.[12] In (1988) [here: ch. 2] and (1990b) I have proposed a slight variant of these epistemic representations which has the advantage that it allows of generally and iteratedly applicable revision rules and thus in effect of a full account of induction for plain belief. Its basic concept is easily introduced.

Throughout, Ω is to be a set of *possible worlds* (as philosophers say without necessarily being so serious about it as is, e.g., David Lewis) or a sample space (as probability theorists prefer to say), i.e. just an exhaustive set of mutually exclusive possibilities; elements of Ω will be denoted by α, υ, ω, etc. As usual, *propositions* are represented by subsets of Ω, denoted by *A, B, C, D, E*, etc. The basic concept is then given by:

Definition 1: κ is a *natural conditional function*[13] (a *NCF*) iff κ is a function from Ω into the set of natural numbers[14] such that $\kappa^{-1}(0) \neq \emptyset$. A NCF κ is extended to propositions by defining $\kappa(A) = \min \{\kappa(\omega) \mid \omega \in A\}$ for each $A \neq \emptyset$ and $\kappa(\emptyset) = \infty$.[15]

[9] Cf., e.g., Hunter (1991).

[10] The further conclusion that plain belief is an illusion is unwarranted; it is drawn only in default of a more appropriate representation of epistemic states.

[11] See Shackle (1961/69), Rescher (1976), and Cohen (1977).

[12] He has summarized his work in Gärdenfors (1988).

[13] The only point of this technical label is that it be not confused with other notions. Perhaps the more suggestive term 'disbelief function' would be better, as Shenoy (1991) has proposed.

[14] My account in (1988) [here: ch. 2] is slightly more general insofar as the range there consists of ordinal numbers.

[15] Setting $\kappa(\emptyset) = \infty$ is a reasonable convention. But ∞ should not be allowed as value of possible worlds and consistent propositions because no good rules of belief revision can be devised for it.

A NCF κ is to be interpreted as a *grading of disbelief*. If $\kappa(\omega) = 0$, then ω is not disbelieved, i.e. ω might be the actual world according to κ. Because not every world can be denied to be the actual one, Definition 1 requires that $\kappa(\omega) = 0$ for some $\omega \in \Omega$. If $\kappa(\omega) = n > 0$, then ω is disbelieved with degree *n*. A proposition is then assigned the minimal degree of disbelief of its members.[16] Thus, if $\kappa(A) = n > 0$, then *A* is disbelieved with degree *n*. And if $\kappa(A) = 0$, then *A* is not disbelieved, i.e. *A* might be true according to κ. $\kappa(A) = 0$ does not mean that *A* is believed according to κ. Belief in *A* is rather expressed by disbelief in $\bar{A}$,[17] i.e. by $\kappa(\bar{A}) > 0$, i.e. $\kappa^{-1}(0) \subseteq A$. Thus, $\kappa^{-1}(0)$ determines what is plainly believed according to κ.

Two simple properties of NCFs should be noted: *the law of negation* that for each proposition *A* either $\kappa(A) = 0$ or $\kappa(\bar{A}) = 0$ or both, and *the law of disjunction* that for all propositions *A* and *B*, $\kappa(A \cup B) = \min(\kappa(A), \kappa(B))$.

According to a NCF κ, propositions are believed in various degrees. It is useful to explicitly introduce the function expressing these degrees, because it is more vivid than the above disbelief talk[18]:

Definition 2: β is *the belief function associated with* the NCF κ iff β is a function from the power set of Ω into the set of integers extended by $+\infty$ and $-\infty$ such that $\beta(A) = \kappa(\bar{A}) - \kappa(A)$.[19] β is a *belief function* iff it is associated with some NCF.

Thus, $\beta(\bar{A}) = -\beta(A)$, and *A* is believed true or false or neither according to β (or κ) depending on whether $\beta(A) > 0$ or < 0 or $= 0$.[20]

So far, the various degrees of belief did not really play a theoretical role. But they are crucial for an account of belief revision, the central notion of which is this:

Definition 3: Let κ be a NCF and *A* a non-empty proposition. Then *the A-part of* κ is the function $\kappa(\cdot \mid A)$ defined on *A* by $\kappa(\omega \mid A) = \kappa(\omega) - \kappa(A)$ for each $\omega \in A$. Again, this function is extended to all propositions by setting $\kappa(B \mid A) = \min\{\kappa(\omega \mid A) \mid \omega \in A \cap B\} = \kappa(A \cap B) - \kappa(A)$ for each $B \subseteq \Omega$. Finally, if β is the belief function associated with κ, we define, as in Definition 2, $\beta(B \mid A) = \kappa(\bar{B} \mid A) - \kappa(B \mid A)$.

Definition 3 immediately implies *the law of conjunction* that $\kappa(A \cap B) = \kappa(A) + \kappa(B \mid A)$ for all propositions *A* and *B* with $A \neq \emptyset$, and *the law of disjunctive conditions* that $\kappa(C \mid A \cup B)$ is between $\kappa(C \mid A)$ and $\kappa(C \mid B)$.[21]

---

[16] The various problems which cast serious doubt on the idea that belief takes propositions as objects are pressing, but must here be disregarded.

[17] $\bar{A}$ of course denotes the complement of *A* relative to Ω.

[18] I thereby follow a proposal of Shenoy (1991).

[19] This definition which is much simpler than my original one has been pointed out to me by Bernard Walliser. Note that because of the law of negation at least one of the terms of the definiens is 0.

[20] The reason why the more vivid belief functions are introduced only as a derivative concept is that their formal behavior is less perspicuous.

[21] This holds because $\kappa(C \mid A \cup B) = \kappa((C \cap (A \cup B)) - \kappa(A \cup B) = \min[\kappa(A \cap C), \kappa(B \cap C)] - \min[\kappa(A), \kappa(B)]$ and because $\min[y_1, y_2] - \min[x_1, x_2]$ is always between $y_1 - x_1$ and $y_2 - x_2$.

The A-part $\kappa(\cdot \mid A)$ of κ can be viewed as a NCF with respect to the restricted possibility space A and thus as a grading of disbelief *conditional on A*. Accordingly, $\beta(\cdot \mid A)$ expresses degrees of belief conditional on A.

It is obvious that a NCF κ is uniquely determined by its A-part $\kappa(\cdot \mid A)$, its $\bar{A}$-part $\kappa(\cdot \mid \bar{A})$, and the degree $\beta(A)$ of belief in A. This suggests a simple model of belief revision for NCFs. If a piece of information consists only in the proposition A, then it is plausible to assume that only the old degree $\beta(A)$ of belief in A gets changed to some new degree $\beta'(A) = n$, whereas the A-part and the $\bar{A}$-part of the old NCF κ are left unchanged; n, $\kappa(\cdot \mid A)$, and $\kappa(\cdot \mid \bar{A})$ then determine a new NCF κ', the revision of the old κ by that information.[22] There are also more complicated models in which the information need not concern a single proposition. These suggestions indicate that NCFs indeed allow for a theory of revision and induction for plain belief. But there is no need to further develop the theory of NCFs.[23] The sequel requires mainly an intuitive grasp of the notions introduced in Definitions 1–3.

A first useful application of these notions is the concept of a reason. Being a reason is always relative to an epistemic background, and given such a background a reason strengthens the belief in, or, in other words, is positively relevant to, what it is a reason for. This intuition can be immediately translated into formal terms:

Definition 4: Let β be the belief function associated with the NCF κ, and A, B, and C three propositions. Then A is *a reason for B relative to* β (or κ) iff $\beta(B \mid A) > \beta(B \mid \bar{A})$. And A is *a reason for B conditional on C relative to* β (or κ) iff $\beta(B \mid A \cap C) > \beta(B \mid \bar{A} \cap C)$.

Note that, according to this definition, the relation of being a reason is *symmetric, but not transitive*, in analogy to probabilistic positive relevance, but in sharp contrast to the narrower relation of being a deductive reason (which is just set inclusion between contingent propositions[24]). Note also that, according to this definition, being a reason does not presuppose that the reason is actually given, i.e. believed; on the contrary, whether A is a reason for B relative to β is independent of the degree $\beta(A)$ of belief in A.

Since the value 0 has the special role of a dividing line between belief and disbelief, different kinds of reasons can be distinguished:

[22] In my (1988, p. 117) [here, p. 30] I have defined this process as A, n-conditionalization.

[23] For further details see my (1988) or my (1990b). There it is made clear why, given certain assumptions, revision schemes for plain belief have to take the form of NCFs; it is shown that NCFs behave very much like probability measures with respect to conditionalization and (conditional) independence; and the justification for more general forms of conditionalizations of NCFs closely parallels that for Jeffrey's generalized probabilistic conditionalization given by Teller (1976).

[24] A proposition A is contingent iff $\varnothing \neq A \neq \Omega$.

Definition 5:

$$A \text{ is a } \begin{Bmatrix} \textit{additional} \\ \textit{sufficient} \\ \textit{necessary} \\ \textit{weak} \end{Bmatrix} \textit{ reason for } B \textit{ relative to } \beta \text{ (or } \kappa\text{) iff } \begin{Bmatrix} \beta(B \mid A) > \beta(B \mid \bar{A}) > 0 \\ \beta(B \mid A) > 0 \geq \beta(B \mid \bar{A}) \\ \beta(B \mid A) \geq 0 > \beta(B \mid \bar{A}) \\ 0 > \beta(B \mid A) > \beta(B \mid \bar{A}) \end{Bmatrix}.$$

Conditional reasons of the various kinds are defined similarly. If A is a reason for B, it belongs at least to one of these four kinds; and there is just one way of belonging to several of these kinds, namely by being a necessary and sufficient reason. Though the emphasis will be on sufficient and on necessary reasons, the two other kinds, which do not show up in plain belief and are therefore usually neglected, well deserve to be allowed for by Definition 5.

9.3 Causation and Explanation

Ultimately, this section will arrive at a (partial) explication of causal explanation. But this will be only a small step beyond saying how causation is related to the general account of induction for plain belief just formally introduced. So, let me turn to the latter task.

A is a cause of B, as a first approximation, iff A and B both obtain, A precedes B, and under the obtaining circumstances A raises the epistemic or metaphysical rank of B. Most people can agree on this vague characterization, the disagreement is only about how to precisely understand it. It's thus a good start; four points call for comment.

(1) 'A and B obtain': The precise nature of the causal relata A and B is a serious problem beyond the scope of this paper. I just take them to be propositions; since I did not say much about what propositions are except that they are subsets of Ω, this can hardly be wrong. No one doubts that the causal relata have to obtain, to be facts, or to be actual. This entails that causation is world-relative, i.e. that the explicandum rather is 'A is a cause of B in ω'. In the given framework, the condition that A and B obtain in ω is simply expressed by the clause that $\omega \in A \cap B$.

(2) 'A precedes B': Some think that backwards causation should not be excluded by definition, and some more think that at least instantaneous causation should be allowed. I am not sure. But since this is not my present concern, I will just stick to the temporal precedence of the cause.

But so far, there is no time in possible worlds; they need a bit more structure: Let I be a non-empty set of factors or variables; each variable $i \in I$ is associated with a set Ω_i containing at least two members; Ω_i is the set of values i may take. The set Ω of possible worlds is then represented as the Cartesian product of all the Ω_i

($i \in I$). Thus, each $\omega \in \Omega$ is a course of events, a function assigning to each variable $i \in I$ the value $\omega(i)$ which i takes in the possible world ω.

I shall call I a *frame* and say that Ω and its elements are *generated* by the frame I and that a NCF on Ω and its associated belief function are *for* the frame I. Already here it is clear, and to be emphasized, that the explication of causation given below will be frame-relative. This is unavoidable, if the explication is to be expressed in formally well-defined terms. Though this frame-relativity seems to me to be natural, one may find it awkward that A is a cause of B within one frame, but not within another. From the present position this relativity can only be overcome by moving into a fictitious universal frame I^* which is not further extensible. Since we shall have occasion in the next section to indulge in that fiction, we may at present be content with this relativity.

Time may now simply be represented by a weak order., i.e. a transitive and connected relation, on the frame I (since metric properties of time are irrelevant); $<$ denotes the corresponding irreflexive order on I; and for $j \in I$, $I_{<j}$ is to be the set $\{k \in I \mid k < j\}$. I shall neglect the complications of continuous time and assume that time is discrete.

Time should be associated not only with variables, but, if possible, also with propositions. Therefore we define a proposition A to be a *J-measurable* or, in short, a *J-proposition* for a set $J \subseteq I$ of variables iff for all $\upsilon, \omega \in \Omega$ agreeing on J, i.e. with $\upsilon(i) = \omega(i)$ for all $i \in J$, $\upsilon \in A$ iff $\omega \in A$. Maximally specific J-propositions will be called J-states; thus, A is a *J-state* iff A is J-proposition and any two $\upsilon, \omega \in A$ agree on J. In particular, ${}^{\omega}J$ is to denote the J-state $\{\upsilon \in \Omega \mid \upsilon$ agrees with ω on $J\}$.

There are many contingent propositions which are about a single variable, and the temporal order of variables is easily carried over to them. Indeed, I see no loss at all in restricting causes and effects to be such, so to speak, logically simple propositions which are about one variable. The condition that A precedes B will thus be expressed by requiring that there are $i,j \in I$ with $i < j$ such that A is a contingent i-proposition and B a contingent j-proposition.

(3) '*A* raises the epistemic or metaphysical rank of *B*': The clumsiness and obscurity of this phrase is due to its intended generality. That A raises the rank of B simply means that the rank of B given A is higher than given $\bar{A}$; thus the phrase makes sense only if conditional ranks are defined. With respect to probabilistic causation, these ranks are probabilities, of course; and they are epistemic or metaphysical ranks according to whether probabilities are interpreted subjectively as degrees of belief or objectively as chances. With respect to deterministic causation, the phrase covers all kinds of approaches – regularity theories, counterfactual theories, analyses in terms of necessary and/or sufficient conditions (however these are understood in turn), etc. – which differ on the interpretation of metaphysical and epistemic ranks. What I shall propose is easily anticipated: I shall take ranks to be epistemic ranks as given by belief functions in the sense of Definition 2. Thus, this is the point where I, following Hume, trace the essential connection between induction and causation.

Why should one follow Hume and conceive causation as an idea of reflexion, as he calls it?[25] Why construe the apparently realistic notion of causation as essentially epistemically relativized? Why try to say not what causation *is*, but rather what the causal conception of a subject in a given epistemic state is? After all, Hume himself was not so unambiguous; his definition of causation as a philosophical relation is pure regularity theory void of any epistemic elements, and when stealing the realist's thunder in (1739), pp. 167–169, his emphasis is on that definition. I cannot do justice here to this profound problem, which even provoked Kant's so-called Copernican revolution; let me just mention my two main motives for taking Hume's side.

One reason is quite concrete. The literature is full of examples presenting problems to various explications of causation, and an explication of causation relative to belief functions is, I believe, more successful in coping with these problems than rival accounts. I will expand a bit on this claim after the formal explication.

The other reason is that there is not only a strong realistic intuition of causation, but also an urgent need for explaining the most prominent epistemological role of the notion of causation. If causation is epistemically relativized, this explanation ensues naturally. But without such a relativization, I do not know of a good explanation. If causation is conceived as a kind of physical ingredient of the world (say, energy transfer), the explanation would have to go like this: "There are a lot of people around, and I can't fail to notice them; therefore, people play an important role in my world picture. Similarly, there is a lot of causation around, and I can't fail to notice it; this explains the prominent epistemological role of the notion of causation." But that parallel sounds wrong to me; it underrates the peculiar epistemological importance of causation, which is different from that of people and other ubiquitous things. And if causation is conceived as a kind of structural component of the world (say, a deductive relation between laws of nature and singular facts, or a relation of counterfactuality, or a certain relation between universals), the explanation must be given in terms which cannot be accepted without further elucidation. Such terms may be lawlikeness for a regularity theory, similarity between possible worlds for the counterfactual account of Lewis (1973a),[26] a theoretical relation of causal necessitation between universals for Tooley (1987, sect. 8.3.2), etc.; and I am not convinced that there are unproblematic ways of objectively understanding these terms.

Of course, the realistic intuition of causation should not be forgotten because of the epistemological concern. Hume did not forget it, as his two definitions of causation as a philosophical and a natural relation show, in which regularity is the objective

[25] This is Hume's most influential conclusion of the crucial sect. XIV of Book I of his (1739). "The idea of necessity arises from some impression. … It must … be derived from some internal impression, or impression of reflexion", he writes on p. 165, and it is clear that necessity here includes causal necessity.

[26] If taken subjectively, Lewis' similarity relations are similar, but not equivalent to my NCFs; see my (1988, p. 127).

counterpart to subjective association. Any more adequate implementation of Hume's general strategy has to make the same kind of move. In particular, it is incumbent on me to say under which conditions there is a kind of objective counterpart to NCFs or belief functions.[27] However, here I will be content with the primary, epistemically relativized explication.[28] These remarks may also make the above-mentioned frame-relativity more plausible.

(4) 'Under the obtaining circumstances': This phrase is also beset with difficulties. In particular, it seems that the relevant circumstances of *A*'s causing *B* are all the other causes[29] of *B* which are not caused by *A*; and this renders the initial characterization of causation patently circular.[30] However, the circularity dissolves, if only *A*'s being a *direct* cause of *B* is considered; there are then no intermediate causes, i.e. no causes of *B* which are caused by *A*, and thus the relevant circumstances may be conceived as consisting of all other causes of *B*. Moreover, it seems to do no harm when all the irrevelant circumstances are added, i.e. all the other facts preceding, but not causing *B*; and thereby the obtaining circumstances of *A*'s directly causing *B* may be conceived as consisting of all the facts preceding *B* and different from *A*. This is what I propose:

Definition 6: Let $\omega \in \Omega$, $i, j \in I$, *A* be an *i*-proposition, and *B* a *j*-proposition. Then *A* is *a direct cause of B in* ω *relative to* the belief function β (or the associated NCF κ) iff $\omega \in A \cap B$, $i < j$[31], and $\beta(B \mid A \cap {}^{\omega}(I_{<j} - \{i\}) > \beta(B \mid \bar{A} \cap {}^{\omega}(I_{<j} - \{i\})$, i.e. *A* is a reason for *B* conditional on ${}^{\omega}(I_{<j} - \{i\})$ relative to β. More specifically, *A* is *an additional, sufficient, necessary, or weak direct cause of B in* ω according to whether *A* is an additional, sufficient, necessary, or weak reason for *B* conditional on ${}^{\omega}(I_{<j} - \{i\})$.

In my (1980, pp. 79ff.) and (1983b, pp. 384ff.) I have more fully argued that ${}^{\omega}(I_{<j} - \{i\})$, i.e. the state in ω of all the variables preceding the effect and differing

---

[27] My (1993a) [here: ch. 5] is an attempt to meet this obligation and thus to do justice to the realistic intuition within the present framework.

[28] In the third paragraph of this section I claimed to have a neutral usage of 'proposition'. This may seem objectionable because I assume propositions to be objects of belief *and* of causation and thus to play a double role which is arguably unsatisfiable. This is indeed a problem. But the problem arises within a realistic conception of causation and is thus part of and additional burden to the objectivization problem just put aside.

[29] It should be clear that a fact may have several causes. Thus I follow the common understanding which construes 'cause' as 'partial cause' and not as 'total cause'.

[30] This is precisely the idea and the conclusion of Cartwright (1979) concerning probabilistic causation.

[31] Even for direct causation it would not be reasonable to generally require that *i* immediately precedes *j*, because the frame *I* may be any wild collection of variables; in particular, *I* may contain many variables which are temporally, but not causally, between *i* and *j*. This requirement at best characterizes nice frames. It should be noticed that the assumption of discrete time is important for Definition 6; given continuous time, direct causes presumably do not exist or are simultaneous with their effects.

from the cause is indeed the correct proposition to conditionalize on; that is, I have argued that whenever we base our judgment about the direct causal relation between *A* and *B* on fewer facts, it could be just the neglected facts which would change the judgment. This is confirmed by the more detailed investigation into the relevant circumstances of causal relations in sect. 4 of my (1990a) [here: sect. 2.4].

I believe that causation in general should be defined as the transitive closure of direct causation, as seems quite natural and as many have assumed. A fuller defense of this view, however, is a long story, parts of which I have told in my (1990a). For the present purpose, it suffices to consider only direct causation.

To make Definition 6 a bit more vivid, it may be helpful to briefly explain how it deals with three standard difficulties. The first is the problem of irrelevant law specialization introduced by Salmon (1970, pp. 177ff.) which says that, according to the original Hempel-Oppenheim account, John's regularly taking birth control pills explains his not becoming pregnant. Regularity theories of causation are of course threatened by this problem, too. But there is no problem for Definition 6. Given that John is a man (before the given time of his non-pregnancy), his regularly taking contraceptives (before that time) is just irrelevant to, and not a reason for, his non-pregnancy at that time, at least relative to our educated belief functions.

The second problem is the distinction between causes and symptoms which is a graver obstacle to regularity accounts of causation and explanation. Take, e.g., C. D. Broad's Manchester hooters and London workers discussed at length by Mackie (1974, pp. 81ff.). Whenever the factory hooters in Manchester and London sound, which is the case every working day at 6 p.m., then the workers in Manchester and London leave their work shortly afterwards. But only the London and not the Manchester hooters have an impact on the London workers. Again, this case presents no problem for Definition 6. Unconditionally, the proposition that the Manchester hooters sound (at a particular day) is relevant to the proposition that the London workers leave; but given that the London hooters sound (or do not sound), the former is just irrelevant to the latter. Again this is true relative to our normal belief functions. Of course, one may have a different belief function yielding also a conditional relevance; but then the sounding of the Manchester hooters is not treated as a mere symptom of the London workers' leaving.

The general scheme should be clear by now. NCFs and belief functions help us to notions of (conditional) relevance and irrelevance which are much more sensitive than the relevance notions provided by other approaches to deterministic causation and which behave almost the same as probabilistic relevance notions.[32] Thus, they enable us to copy the methods of dealing with these problems which have been so successfully developed for probabilistic causation.

[32] Cf. my (1988, sect. 6) [here: sect. 1.6].

A third problem further illustrating this scheme distinguishes Definition 6 not only from regularity theories, but also from counterfactual analyses of causation like Lewis (1973b). It is the problem of (symmetric) causal overdetermination cruelly, but standardly, exemplified by the firing squad. Prima facie, cases of causal overdetermination are clearly possible. But as far as I see, they are a great mystery, if not an impossibility for all realistic accounts of causation; it seems that the only thing the realist can do is to explain them away: either by observing that one of the two causal chains from the two apparently overdetermining causes to the effect has not been completed so that one of the two causes is in fact preempted; or by observing that there is an intermediate event (a Bunzl event, as Lewis 1986d, p. 208, calls it) which causes the effect and which is jointly caused by the two apparently overdetermining causes so that the two causes in fact jointly cause the effect.[33]

For Definition 6, however, there is no mystery at all. The two overdetermining causes may be simply conceived as additional causes; each of the two is an additional cause of the effect in the presence of the other one. The crucial difference is that additional causes cannot be defined within a counterfactual approach, let alone a regularity theory. Something true can counterfactually be still true or not true, but not more or less true. But something conditionally believed can be believed more or less firmly under different conditions.[34] Of course, I do not claim that this simple remark solves the problem of causal overdetermination. What it does is first to do justice to the prima facie existence of causal overdetermination and secondly to locate the problem; it arises when we try to objectivize our epistemically relativized causal picture, because there is no realistic counterpart to additional causes as defined above.

So much for the partial account of causation we need. It is easily extended to a partial account of explanation. I shall not comment on explanation of laws and theories and on whether there is non-causal explanation.[35] But concerning causal explanation, it seems unassailable to say that getting an explanation for *B* is learning a cause of *B* and having an explanation for *B* is knowing a cause of *B*.[36] The problem is only that this statement is unhelpful as long as one does not have an account of causation or tries to explain causation by explaining explanation. But this is not our problem, and thus we may immediately turn this informal statement into a formal definition.

---

[33] See the enlightening discussions of Bunzl (1979) and Lewis (1986d, pp. 193–212).

[34] This, by the way, is also a point of difference between Gärdenfors' belief revision model and mine. Gärdenfors has plain conditional belief, but not more or less firm conditional belief and therefore nothing like additional reasons and causes. Cf. Gärdenfors (1988, ch. 3 and 9).

[35] However, I tend to join Lewis (1986d, pp. 221ff.) in thinking that there is no non-causal explanation of singular facts.

[36] This is essentially also what Lewis (1986d, pp. 217ff.) maintains, though he points out that knowing a cause of an event is not the only way of having information about the causation of that event. I neglect here the other ways.

Knowing some fact to be a cause at least involves believing this fact to be a cause. And believing A to be a cause of B according to a NCF κ means believing the actual world to be among the worlds in which A is a cause of B relative to κ. Since only direct causes have been formally defined, this leads to

Definition 7: Let i, j, A, and B be as in the previous definition. A's *range* $C_{A,B}$ *of directly causing B relative to* the NCF κ or its associated belief function β is defined as the $I_{<j}$-$\{i\}$-measurable set of all $\omega \in \Omega$ such that $\beta(B \mid A \cap {}^{\omega}(I_{<j}\text{-}\{i\}) > \beta(B \mid \bar{A} \cap {}^{\omega}(I_{<j}\text{-}\{i\})$. Hence $A \cap B \cap C_{A,B}$ is the set of all $\omega \in \Omega$ such that A is a direct cause of B in ω relative to κ or β.[37] A's *range* ${}^{s}C_{A,B}$ or ${}^{n}C_{A,B}$ *of*, respectively, *sufficiently or necessarily directly causing B relative to* κ or β is defined accordingly. Then, *A causally explains B* (*as necessary, as possible*) *relative to* κ or β iff $\beta(A \cap B \cap C_{A,B}) > 0$ ($\beta(A \cap B \cap {}^{s}C_{A,B}) > 0$, ($\beta(A \cap B \cap {}^{n}C_{A,B}) > 0$).[38]

The only deviation of Definition 7 from its informal statement is that knowledge of a cause has been weakened to belief in a cause. This corresponds to the old debate whether explanation requires true or only accepted antecedents. I think there are both usages; '*B* is explained by *A*' may be factive or not according to whether it is taken as the passive of the apparently factive '*A* explains *B*' or as an ellipsis of the apparently non-factive '*B* is explained by *A* by some explainee'. Since I have always talked only of belief and not of knowledge, I settle for the weaker version. I do not see that our topic is seriously affected by this issue. In particular, I do not see that this issue drives a wedge between explanation and understanding, as Lambert (1988, pp. 308–310) and (1991, pp. 138f.) argues. Understanding as well can be taken factively or non-factively, and it seems only fair that, when assessing the relation between explanation and understanding, only the corresponding interpretations are compared.

9.4 Reason and Truth

The first task of giving a partial account of explanation need not be developed further.[39] The next task is to give an independent account of understanding or, rather, of some not too bad substitute thereof. I approach this task by discussing three

---

[37] In sect. 4 of my (1990a) I have called $C_{A,B}$ the actually relevant circumstances of (the direct causal relation between) A and B in the widest sense.

[38] The need to consider explanations by additional or weak causes will not arise; thus I did not formally introduce them.

[39] It would be useful to extend my comparative remarks about causation to some remarks about how Definition 7 relates to other accounts of explanation; but this is beyond the scope of this paper. Just this much: my account seems to me to fit van Fraassen's theory in sect. 5.4 of his (1980) insofar as Definition 7 tries to say more about van Fraassen's relation of explanatory relevance for the case of direct causal explanation.

principles of increasing strength which I take to be basic principles of coherence, believability, and truth.

Let's start with a simple question: If *B* is a contingent proposition, is there a reason for *B*? Trivially, yes. There always are deductive reasons; each non-empty subset of *B* is a sufficient, and each non-tautological superset a necessary, deductive reason for *B*. So, the question should rather be whether there is an *inductive*, i.e. non-deductive reason for *B*. Or, put in another way, if *B* is a contingent *j*-proposition for some $j \in I$, is there a *I*-{*j*}-proposition which is a reason and thus an inductive reason for or, for that matter, against *B*?[40] Not necessarily, of course. There may be variables which are independent of all other variables in the given frame *I* relative to the given belief function β; and since *I* may be an arbitrary collection of variables, such counter-instances ensue naturally.

Consider now an extension *I*′ of the frame *I* and an extension β′ of β for *I*′. There are many such extensions of β, and, trivially, there exists an extension of β according to which there is a *I*′-{*j*}-measurable reason for *B*. Thus, we should, more precisely, consider *the* extension β′ of β as determined by some unspecified epistemic subject with a belief function covering also *I*′-propositions. Is there a *I*′-{*j*}-measurable reason for *B* according to β′? Again, not necessarily. The case of *I*′ is not different from the case of *I*.

But now consider all extensions *I*′ of *I* and the appertaining belief functions β′ which are within the subject's range. Should then an inductive reason for or against *B* come to the fore? Once more, not necessarily; but that would be a grave matter. It would mean that the subject could not learn anything whatsoever about *B*; wherever it looked, it could not find the slightest hint concerning *B; B* would be outside its world of experience, outside its bounds of sense.

It may of course happen that a proposition is beyond a subject's present grasp. This may change; an individual's inductive scheme as well as that of a society or of mankind keeps evolving.[41] It may even be that a proposition is forever beyond the grasp of an individual or of actual mankind. But these are all accidental limitations. My real concern is the status of the *j*-proposition *B* with respect to all possible extensions of the frame *I* whatsoever and the appertaining belief functions to which a subject would extend its belief function β, if it came to consider these extensions of *I*. Is it still conceivable that in this sense no extension of *I* contains an inductive reason for *B*? Now, finally, it seems plausible to say no. Otherwise, there would be no way whatsoever to reason or to learn anything about *B*, not because of accidental limitations, but due to the inherent structure of the all-inclusive inductive scheme underlying all these extensions of β; *B* would be literally senseless, unreasonable.

---

[40] It would not be reasonable to ask without restrictions on measurability whether there is an inductive reason for any contingent proposition *B* not of the form {ω} or Ω – {ω}, because the answer turns out to be yes whenever Ω has more than four members.

[41] However, I don't know of any theory about the evolution of inductive schemes, i.e. about the change of belief functions, probability measures, or whatever *for changing frames*, except of conceiving it as generated by an underlying, more inclusive inductive scheme.

I have referred to all possible extensions of some initial frame and inductive scheme. But it is simpler to refer instead to the universal frame I^* comprising all variables whatsoever, to the set Ω^* of possible worlds generated by I^*, and to a universal belief function β^* for I^*. It may seem earthlier to talk only of extensions. But the set of all extensions is not earthlier than its union; both are philosophical fiction. Talking of I^*, Ω^*, and β^* is just much less clumsy than quantifying over extensions.

I^*, Ω^*, and β^* are what, in a loosened usage of Kantian and Peircean terms, has been called regulative ideas, ideal limits of inquiry, etc. The question whether one can legitimately and sensibly appeal to such limit concepts is certainly pressing. Here I just follow all those who do so. And I take it that, insofar our epistemic activities may at all be described by frames and belief functions, we conceive these activities as embeddable into the universal frame I^* and a universal belief function β^* and that we consider this embeddability as a fundamental requirement of consistency.[42]

What we have arrived at, then, is a first plausible principle of coherence:

(PCo1) For any $j \in I^*$ and any contingent j-proposition B there is a I^*-$\{j\}$-measurable reason for B relative to β^*.

PCo1 may be taken as a condition on β^*, on how β^* has to connect propositions. But it may also be conceived as a condition on I^* (and the generated Ω^*) saying that no logically simple proposition exists unless appropriately connected by β^*. The best is to view PCo1 as what it is, as a condition simultaneously on I^* *and* β^*.

PCo1 is, of course, akin to the positivists' verifiability principle and other criteria of empirical significance. But PCo1 is a weak version, because it requires at best confirmability and not verifiability and because it does not refer to a directly verifiable basis, to evidential certainty, and the like. And PCo1 is unambiguous about the nature of the required ability of confirmation. This ability is not to be taken as restricted by our sensory outfit; PCo1 does not refer to any specific senses. It is not restricted by limited computing capacities; β^* will not be computationally manageable, anyway. It is not restricted by our spatiotemporally and causally limited access to facts. This ability is constituted exclusively by the inherent structure of the limiting inductive scheme and thus of the actual inductive schemes approaching it.

Given the above explication of direct causation, PCo1 is, by the way, tantamount to the following weak principle of causality:

(PCa1) For any $j \in I^*$ and any contingent j-proposition B there is a direct cause or a direct effect of B in some world $\omega \in \Omega^*$ relative to β^*.

At least the equivalence of PCo1 and PCa1 holds, if I^* is linearly and discretely ordered by time.[43] Note also that the reference to I^* and β^* eliminates the frame-relativity of that explication, but not its epistemological involvement.

---

[42] Cf. also the quite similar remarks of Ellis (1979, pp. 9ff.) about what he calls the ideal of completability.

[43] This is easily proved on the basis of two properties of conditional independence between sets of variables which are stated as assertion (7) in my (1990b) and proved as Theorems 11 and 13 in my (1988) [cf. here pp. 35f.]. A probabilistic counterpart of the present claim, or rather a considerable generalization thereof, is proved as Theorem 5 in my (1980).

PCo1 is symmetric with respect to positive and negative relevance; whenever a proposition is a reason for B, its negation is a reason against B. This symmetry will break in the next step when we consider *true* propositions; truth is biased towards positive relevance.

We have first to introduce another limit concept: the actual world taken not as a spatiotemporally maximally inclusive thing, but as everything that is the case. We naturally assume that among all the possible worlds in Ω^* exactly one is actual; let's call it α^*. Thus, a proposition A is true (absolutely, not relative to a model or a world) iff A is true in α^*, i.e. iff $\alpha^* \in A$.

The question now is this: Suppose that the contingent j-proposition B is true. PCo1 asserts that there are I^*-$\{j\}$-measurable reasons for B. But will there be a true reason among them? Let's look at the question in a more earthly setting of a small frame I, the small actual world α (which is the restriction of α^* to I), and a subject's belief function β for I. Within this setting, the answer may certainly be no. If so, however, the truth would be undetectable, unbelievable for the subject within this setting. If it believed only truths, it would have no reason for believing B; and if it has any reason for believing B, then only by believing some I-$\{j\}$-propositions which are false. This situation is not critical by itself, but it again becomes more and more critical when it does not change as larger and larger extensions of I are considered. And relative to I^*, α^*, and β^*, finally, this situation seems absurd; all true evidence which could conceivably be brought to bear on B would univocally speak against B and for $\overline{B}$, though B is true and $\overline{B}$ false. Thus it seems plausible to answer the initial question in the affirmative.

This can be stated as a second principle of coherence:

(PCo2) For any $j \in I^*$ and any contingent j-proposition B with $\alpha^* \in B$ there is a I^*-$\{j\}$-measurable proposition A with $\alpha^* \in A$ which is a reason for B relative to β^*.

Briefly: for each singular truth there is a true inductive reason. Of course, PCo2 implies PCo1.

In Peirce-Putnamian terms one might say that PCo2 is part of the assertion that the epistemically ideal theory cannot be false. The ideal theory has, of course, recourse to all true evidence; and in a case violating PCo2 the ideal theory would have to falsely affirm $\overline{B}$ on the basis of that evidence and the universal inductive scheme β^*. PCo2 prevents this and thus captures at least one aspect of Putnam's internal realism.[44]

Indeed, PCo2 fits well under the heading 'coherence theory of truth'. The theoretical standing of the coherence theory is not exactly brilliant, because of difficulties in saying precisely what coherence is. Explications in deductive terms, say as consistency or deducibility, were precise, but unprofitable; and other, more interesting explications were always vague. A noticeable exception is Rescher (1973) and (1985); but I find his underlying theory of plausibility indexing not fully satisfying. Here,

[44] Which is the basic theme of Putnam's recent work; cf., e.g., the introduction and ch. 4 of Putnam (1983a).

coherence is construed as inductive coherence as constituted by positive relevance relative to a belief function. PCo2 is thus one way of saying that truth must cohere. Of course, a workable theory of induction or belief revision for plain belief is vital to this construal.

PCo2 does certainly not yield a definition of truth. For propositions, being true is defined as having α^* as a member; and for sentences, Tarski's truth definition may need an underpinning by a theory of reference, as called for by Field (1972), but as a definition it does not need a coherentist supplement. PCo2 also does not yield a criterion of truth; it is of little help in determining the truth of B because it is kind of circular in requiring true reasons for B and because it does not tell what to do in the case of conflicting reasons. In fact, PCo2 is not a condition on truth alone; it must again be viewed as a condition on I^*, α^*, *and* β^*, on how truth and reason relate in the universal frame.

There is also a principle of causality associated with PCo2:

(PCa2) For any $j \in I^*$ and any j-proposition B with $\alpha^* \in B$ there is a direct cause or a direct effect of B in α^* relative to β^*.

Briefly: each singular fact has a direct cause or a direct effect in the actual world. This principle of causality is, of course, much stronger and much more interesting than PCa1. PCa2 is even stronger than PCo2; the former implies the latter, but not vice versa.[45] It would be nice to find a plausible principle of coherence entailing PCa2; so far I have not succeeded.

There are, however, plausible strengthenings of PCo2. One of them is my next goal.

PCo2 asserts that a true I^*-$\{j\}$-measurable reason A may be found for the contingent true j-proposition B. Now imagine that a piece C of true information is received and that A is then no longer a reason for B, i.e. A is not a reason for B conditional on C. This is not impossible; if A is positively relevant to B given one condition, A may be positively, negatively, or not relevant to B given another condition. And it is not excluded by PCo2. But this seems an implausible way to satisfy the plausible PCo2.

This opens up a new kind of question: How does the relevance of some truth to B evolve in the infinite process of acquiring more and more true information? Formally, everything is possible. The relevance may (a) vacillate for some (or no) time and then forever stay on the positive side, or (b) vacillate for some (or no) time and then forever stay on the non-positive side, or (c) vacillate forever. A truth of kind (b) is a very casual kind of reason for B, if at all, and one of kind (c) an odd and deeply undecided kind.

---

[45] *Proof*: Let the i-proposition A be a direct cause of B in α^* relative to β^*. If $D = {}^{\alpha^*}(I_{<j\text{-}}\{i\})$, this says that $\beta^*(B \mid A \cap D) > \beta^*(B \mid \bar{A} \cap D)$. Let $E_1 = A \cap D$, $E_2 = \bar{A} \cap D$, $E_3 = A \cap \bar{D}$, $E_4 = \bar{A} \cap \bar{D}$, and $E = \bigcup\{E_r \mid \beta^*(B \mid E_1) \geq \beta^*(B \mid E_1)$. The law of disjunctive conditions (after Definition 3) immediately implies that $\beta^*(B \mid E) > \beta^*(B \mid \bar{E})$, i.e. that E is a (unconditional) reason for B relative to β^*. The same reasoning applies if B has a direct effect in α^* instead of a direct cause.

Is it conceivable that all true reasons for *B* one finds after some true information or other turn out to be of these unreliable kinds (b) and (c)? Formally, there are again three ways how this might happen. First, it might be that true reasons for *B* run out after sufficient true information. This case definitely violates the basic idea of PCo2 that in the limit all truth must be believable. Secondly, it might be that at infinitely many stages of the acquisition of true information there are true reasons for *B* and at infinitely many other stages there are no true reasons for *B*. This case again violates the basic idea. As often as one gains confidence in *B*, one loses it; one can never hold it fast. Thirdly, it might be that after some true information there always are true reasons for *B*, though different ones at each subsequent stage of the process. This case seems to be compatible with the basic idea, but it is still strange. Each time when asked why one believes *B* one has to withdraw the previous answer and to give another one; and this continues forever. This does not seem to be an acceptable process of truth tracking.

I therefore conclude that there should be at least one reason for *B* of the reliable kind (a); I shall call it an *ultimately stable* reason. This is the key concept of the following considerations; it is more precisely defined thus:

Definition 8: Let $\omega \in \Omega$ and $A, B, C \subseteq \Omega$. Then A is a ω-*stable* (*sufficient, necessary*) *reason for B within C relative* to a belief function β for I (or the associated NCF κ) iff $\omega \in A \cap B$, $\omega \in C$, $A \cap C \neq \emptyset \neq \bar{A} \cap C$, and A is a (sufficient, necessary) reason for B relative to β conditional on each $D \subseteq C$ with $\omega \in D$ and $A \cap D \neq \emptyset \neq \bar{A} \cap D$. A is an *ultimately* ω-*stable* (*sufficient, necessary*) *reason for B relative to* β iff A is so within some condition. The set of all $\omega \in \Omega$ such that A is an ultimately ω-stable (sufficient, necessary) reason for B is called A's *range of being an ultimately stable* (*sufficient, necessary*) *reason for B* and denoted by $S_{A,B}$ (${}^sS_{A,B}$, ${}^nS_{A,B}$).

Note that the truth of A and B in ω is made a defining characteristic of A's being an ultimately stable reason for B. Note also that, if A is an ultimately ω-stable reason for B, so is B for A.

In these terms, then, I have just argued for a third principle of coherence:

(PCo3) For any $j \in I^*$ and any contingent j-proposition B with $\alpha^* \in B$ there is a I^*- $\{j\}$-measurable, ultimately α^*-stable reason for B relative to β^*.

Briefly: for each singular truth there is an ultimately stable inductive reason. If there are reasons with stronger than ultimate α^*-stability, say, with α^*-stability within Ω, all the better. But such stronger forms of stability do not seem to be required in PCo3 on coherentistic grounds. Still, PCo3 implies PCo2.[46]

---

[46] This is so because, as the proof in the previous note shows, there is an unconditional reason for *B*, whenever there is a conditional reason for *B*.

9.5 Explanations and Stable Reasons

Now I can finally offer my substitute for (scientific) understanding: it is knowing ultimately stable reasons. I do not want to defend this as an explication of the complex notion of understanding. But what has been said in this context about understanding is captured fairly well by my proposal; and knowing ultimately stable reasons is epistemologically significant in its own right. Let me explain.

What is meant by knowing an ultimately stable reason A for B? Not only that one knows A and A is in fact an ultimately stable reason for B, but also that one knows A to be so, i.e. that one knows A's range $S_{A,B}$ of being an ultimately stable reason for B to obtain. As in the case of explanation, there is a factive and a non-factive understanding of understanding, and as in the former case I deal only with the latter, in order to be able to confine myself to belief and to be silent about knowledge. To be precise, then, A's being believed to be an ultimately stable reason for B relative to β simply comes to $\beta(S_{A,B}) > 0$.

The significance of believing in ultimately stable reasons is this: When one believes A and B to be true, one thinks that A and B are part of, fit into, α^* in some way or other. But one may do so as a mere recorder of facts without any understanding of what is going on, without any grasp of *how* A and B fit into α^*. And one may, adhering to PCo3, simply proclaim that it should be possible to find an ultimately α^*-stable reason for B. When one believes $S_{A,B}$ to be true, however, one does not only believe A and B, and one does not merely postulate an ultimately α^*-stable reason for B. Rather, one thinks to know a particular one, namely A. And one has a partial grasp of how A and B fit into α^*, namely as one element of coherence, as one coherent link among many others which have to exist in α^*. Thus, for the believer of $S_{A,B}$ A and B better qualify as part of the final truth than for the believer of A and B alone.

How else is understanding characterized? Lambert (1991), p. 129, says that "the metaphor of 'fitting into', and its stylistic variants such as 'incorporated into' or 'integrated into', seem especially germane vis à vis scientific understanding as it relates to scientific explanation" and quotes a number of important authors using this metaphor. For him, then, "the sense of scientific understanding relevant to scientific explanation may be characterized as an answer to the question 'How does state-of-affairs *S* fit into theory *T*?' " (p. 130), where, as he goes on to explain, "fit into" may mean various things.

Similarly, Friedman (1974) and Kitcher (1981), again adducing a number of witnesses, take unification as the key concept. Friedman (1974, p. 15) explicitly claims:

> [T]his is the essence of scientific explanation – science increases our understanding of the world by reducing the total number of independent phenomena that we have to accept as ultimate or given. A world with fewer independent phenomena is, other things being equal, more comprehensible than one with more.

And he goes on to say more precisely how he understands independence or independent acceptability.

These seem to be appropriate ways of talking also about ultimately stable reasons; indeed, I myself slipped into these ways three paragraphs ago. Of course, the metaphors take on different senses with the different authors. But this is not an unhappy homonymy which hides incomparable interests. On the contrary, I think, there is one common idea which is vague and allows of various explications, and there is less a disagreement about its content, but rather a common need in surveying this idea and tracing fruitful explications. Here, in any case, fit and unification, like coherence, are construed as inductive fit and unification as constituted by (conditional) positive relevance relative to a belief function.

On a strategic level, the main difference between the papers referred to and the present proposal is that there fit and unification are somehow construed as relations between facts or phenomena and theories, whereas here they are construed as a relation between facts and inductive schemes. Talking of theories is certainly closer to scientific practice, but talking of inductive schemes is nearer to epistemological theory. Is there a substantial difference? This is unclear as long as the relation between theories and inductive schemes is not made clear. Without doubt there is a close relation, and it is incumbent on me to say how theories are implicitly contained in inductive schemes; I shall not attempt to do so here. Conversely, however, there is an urgent need to say how inductive schemes are implicitly contained in scientific theories; I am convinced that mere reference to theories is not helpful for all the topics here addressed as long as theories are conceived as something modally inert, e.g. as sets of extensional sentences or extensional models.[47]

These remarks also suggest an answer to the question on which Salmon (1978) hangs his discussion. Salmon asks on p. 684:

> Suppose you had achieved the epistemic status of Laplace's demon … who knows all of nature's regularities, and the precise state of the universe … at some particular moment. … you would be able to predict any future occurrence, and you would be able to retrodict any past event. Given this sort of apparent omniscience, would your scientific knowledge be complete …? Laplace asked no more of his demon; should we place further demands upon ourselves?

In the sequel Salmon explains what Laplace's demon lacks. From the present point of view, omniscience – whether it is direct as presumably that of God or inferred from a complete set of axioms as that of Laplace's demon – is neither an ideal nor a counterfactual epistemological possibility for us. The reason is not that it is impossible on various scores to know so much. The reason is rather that we could not merely know everything; having an inductive scheme, proceeding inductively in the broad sense here always referred to is an essential and indispensible feature of our epistemic constitution which would not fade by approaching omniscience. Laplace's demon is indeed granted too little; it would not know what to believe, if

[47] On this score, then, the sentential and Sneed's and Stegmüller's structuralist view of theories seem equally insufficient. This insufficiency is also felt, for instance, by Kitcher (1981), when he associates explanatory stores of argument patterns with scientific theories. Cf. also Mühlhölzer (1989, ch. 6).

it were to discover that it is wrong. We would know, even while approaching omniscience. If I am right, all the other things which the demon is held to be wanting in this discussion including those mentioned by Salmon (1978, p. 701) result from this central lack.[48]

Having thus shed some light on the epistemological locus of stable reasons, I can finally turn to the object of my paper: the relation between explanations and ultimately stable reasons. Though the definitions of $C_{A,B}$ (Definition 7) and of $S_{A,B}$ (Definition 8) look quite similar, this relation is not straightforward. The main difference is that direct causes are characterized by conditionalization on the whole past of the effect, whereas ultimately stable reasons are characterized by conditionalization on many and finally all other truths, whether past or future. This prevents a direct comparison. There is help, however: just restrict all the coherentistic considerations about the *j*-proposition *B* in the previous section to the past of *B*. This move brings easy success, indeed too easy, and therefore two disappointments. I shall first state in precisely what the move and its success consist, before explaining what the disappointments are and what might be done about them.

The move is simple: Restate PCo1 as saying that for any $j \in I^*$ and any contingent *j*-proposition *B* there is an $I^*_{<j}$-measurable reason for *B* relative to β^*. This is equivalent to a modified PCa1 saying that for any such *j* and *B* there is a direct cause of *B* in some world $\omega \in \Omega$ relative to β^*. Change PCa2 and PCo2 in the same manner; the former is again implied by the latter.[49] Modify finally Definition 8: Define *A* to be a ω, *j-past-stable* (*sufficient, necessary*) *reason for* β *within C relative to* β by additionally requiring *C* to be $I^*_{<j}$-measurable and by requiring *A* to be a (sufficient, necessary) reason for *B* relative to β conditional only on each $I^*_{<j}$-measurable $D \subseteq C$ with $\omega \in D$ and $A \cap D \neq \emptyset \neq \bar{A} \cap D$; define accordingly *A*'s being an *ultimately* ω, *j-past-stable* (*sufficient, necessary*) *reason for B* and *A*'s *range of being an ultimately j-past-stable* (*sufficient, necessary*) *reason for B*; and denote this range by $S_{j;A,B}$ (${}^sS_{j;A,B}$, ${}^nS_{j;A,B}$). PCo3 may then be reformulated correspondingly.

After this modification the comparison is immediate: If *A* is an *i*-proposition and *B* a *j*-proposition with $i < j$ and if *i* is a binary variable,[50] then $A \cap B \cap C_{A,B} = S_{j;A,B}$ ($A \cap B \cap {}^sC_{A,B} = {}^sS_{j;A,B}$, $A \cap B \cap {}^nC_{A,B} = {}^nS_{j;A,B}$) and thus *A* causally explains *B* (as necessary, as possible) relative to β if and only if *A* is believed to be a *j*-past-stable (sufficient, necessary) reason for *B* relative to β. For proof one has only to look at the definitions and to observe first that ${}^\omega(I_{<j}\text{-}\{i\})$ is *the* smallest $I_{<j}$-proposition *C* with $\omega \in \Omega$ and $A \cap C \neq \emptyset \neq \bar{A} \cap C$, if *i* is binary, and secondly that being an ultimately ω, *j*-past-stable reason only requires being a reason conditional on this smallest proposition *C*.

---

[48] Of course, the demon has other epistemological defects as well. For instance, it may be one of the two gods of Lewis (1979b, pp. 520f.) unable to localize itself. But this is obviously another kind of defect.

[49] The proofs given in the notes 41 and 43 also apply to these claims.

[50] This means that Ω_i has only two elements. This premise is technically required and I am not sure about the best way to get rid of it.

Hence the justification of explanation I propose runs as follows: On the one hand, there is the explication of direct causation and consequently of causal explanation in Section 9.3. On the other hand, there are the independent coherentistic considerations of Section 9.4 which suggest that truth is tied to ultimately stable reasons, as stated in PCo3, and that believing in ultimately stable reasons is thus an indispensible ingredient of having a true world picture. And, as has turned out now, it is explanations and only explanations which provide these ingredients, at least if the relation of being a reason is considered only with respect to pairs of logically simple propositions about single variables and if the coherentistic considerations are restricted to the past of the later proposition of such a pair.

Somehow, however, the last step appears too trivial. It falls short of the expectations I have created in two respects.

One disappointment is that in the short proof of the identity of $A \cap B \cap C_{A,B}$ and $S_{j;A,B}$ being an ultimately stable reason takes on an unexpectedly weak sense. According to my definition, being an ultimately ω(j-past)-stable reason boils down to being a reason conditional on, sloppily put, all the rest of the truth in (the j-past of) ω. But according to my motivation in the previous section, the idea rather was that an ultimately ω-stable reason is a reason after some finite information true in ω and stays a reason after all further information true in ω.

This deficiency can be cleared, however, because the cause specified in a causal explanation is in fact a reason which is stable within the cause's range and not only from some remote point onward. More precisely, the following assertion holds true: If A is an i-proposition and B a j-proposition with $i < j$, then A is a sufficient reason for B conditional on each non-empty, $I_{<j}$-$\{i\}$-measurable $D \subseteq {}^{s}C_{A,B}$ relative to β.[51] Hence, for each $\omega \in A \cap B \cap {}^{s}C_{A,B}$, A is a ω, j-past-stable sufficient reason for B not only ultimately, but within no less than ${}^{s}C_{A,B}$. The assertion with 'necessary' and '${}^{n}C_{A,B}$' replacing 'sufficient' and '${}^{s}C_{A,B}$' holds correspondingly.[52] However, the assertion fails to generally hold for reasons and direct causes simpliciter.[53]

Do explanations also provide unconditional reasons? Under mild assumptions yes, provided only sufficient or necessary reasons are considered. More precisely, the following assertion holds true: If A causally explains B as necessary relative to

---

[51] *Proof*: For each $\omega \in {}^{s}C_{A,B}$ we have $\kappa(\bar{B} \mid A \cap {}^{\omega}(I_{<j-}\{i\}) > 0$ and $\kappa(\bar{B} \mid \bar{A} \cap {}^{\omega}(I_{<j-}\{j\}) = 0$. Trivially, each $I_{<j-}\{j\}$-measurable $D \subseteq {}^{s}C_{A,B}$ is equal to $\bigcup\{{}^{\omega}(I_{<j} - \{i\} \mid \omega \in D\}$. Therefore, the law of disjunctive conditions (after Definition 3) implies the assertion that for each such D $\kappa(\bar{B} \mid A \cap D) > 0$ and $\kappa(\bar{B} \mid \bar{A} \cap D) = 0$.

[52] At this point it is particularly clear that my analysis of explanation is closely related to Hempel's requirement of maximal specificity (cf. Hempel, 1965, pp. 397ff.) and to Skyrms' notion of resiliency (cf. Skyrms, 1980, parts IA and IID).

[53] The failure of the analogous probabilistic assertion is related to Simpson's paradox. Cf. also my (1990a), pp. 128.

β and $\beta({}^{s}C_{A,B} \mid \bar{A}) \geq 0$, then A is a sufficient reason for B relative to β[54]; and if A causally explains B as possible relative to β and $\beta(B - {}^{n}C_{A,B} \mid \bar{A}) < 0$, then A is a necessary reason for B relative to β.[55] This assertion, or at least its 'sufficient'-part, very much resembles the thesis "that an adequate answer to an explanation-seeking why-question is always also a potential answer to the corresponding epistemic why-question"[56] and may thus be taken as a proof thereof. The additional premise of the 'sufficient'-part that $\beta({}^{s}C_{A,B} \mid \bar{A}) \geq 0$ will usually be satisfied, I think; and one might argue that it is just this premise which is violated in alleged counter-examples to that thesis.[57]

The other disappointment is the restriction of the coherentistic considerations about the *j*-proposition B to the past of B in the way specified above. This is disappointing because thus restricted these considerations lose much of their persuasiveness. I have great confidence in PCo1-3 as I have stated them in the previous section; but I do not know how to convincingly argue for PCo1-3 as modified in this section. The modified PCo1-3 (and in particular the principles of causality associated with them) still look desirable, but it is not clear why they should be necessary on coherentistic grounds. This is a gap in my argument.

Perhaps, however, this unsupported restriction of the coherentistic considerations is not really necessary. How is it possible that the *i*-proposition A is a direct cause of the *j*-proposition B in ω and thus an ultimately ω, *j*-past-stable reason for B, but not an ultimately ω-stable reason for B? The only possibility is that some true information about the future of B turns the positive relevance of A for B given the rest of the past of B into irrelevance or negative relevance. But there is something odd about this possibility. Consider a simple formal example:

Let ω, A, B, and C be such that A precedes B, B precedes C, and $\{\omega\} = A \cap B \cap C$. Now suppose on the one hand that A is a sufficient reason for B and thus also a sufficient direct cause of B in the small world ω, and on the other hand that A is a necessary reason for $\bar{B}$ given C and thus not an ultimately ω-stable reason for B. These assumptions imply: First, $\beta(C \mid A) < 0$; thus A and C cannot both believed to

[54] Proof: Let ${}^{s}C_{A,B}$ be abbreviated by C. It was just shown that A is a sufficient reason for B conditional on C, i.e. that (a) $\kappa(\bar{B} \mid A \cap C) > 0$ and (b) $\kappa(\bar{B} \mid \bar{A} \cap C) = 0$. Since A causally explains B as necessary, A, B, and C are believed; this implies $\kappa(A) = 0$ and $\kappa(\bar{C}) > 0$; hence $\kappa(A \cap \bar{C}) > 0$ and (c) $\kappa(\bar{C} \mid A) = \kappa(A \cap \bar{C}) - \kappa(A) > 0$. And the additional premise says that (d) $\kappa(C \mid \bar{A}) = 0$. Now, (a) implies $\kappa(\bar{B} \cap C \mid A) > 0$, and (c) implies $\kappa(\bar{B} \cap \bar{C} \mid A) > 0$; therefore $\kappa(\bar{B} \mid A) > 0$. Moreover, (b) and (d) imply $\kappa(\bar{B} \cap C \mid \bar{A}) = 0$ and thus $\kappa(\bar{B} \mid \bar{A}) = 0$.

[55] Proof: Let ${}^{n}C_{A,B}$ be abbreviated by C. It was just shown that A is a necessary reason for B conditional on C, i.e. that (a) $\kappa(B \mid A \cap C) = 0$ and $\kappa(B \mid \bar{A} \cap C) > 0$, hence (b) $\kappa(B \cap C \mid \bar{A}) > 0$. Since A causally explains B as possible, A, B, and C are believed; thus $\kappa(A \cap C) = 0$ which implies (c) $\kappa(C \mid A) = 0$. And the additional premise says that (d) $\kappa(B \cap \bar{C} \mid \bar{A}) > 0$. Now, (a) and (c) imply $\kappa(B \cap C \mid A) = 0$ and thus $\kappa(B \mid A) = 0$. Moreover, (b) and (d) imply $\kappa(B \mid \bar{A}) > 0$.

[56] Hempel (1965, pp. 368). This is the part of the thesis of the structural identity of explanation and prediction which Hempel (1965, pp. 364ff.) endorses.

[57] I have in mind Michael Scriven's examples of the jealous murderer and the collapsing bridge discussed in Hempel (1965, pp. 370ff.).

be true, and A is at best a weak reason for C. Secondly, C is a necessary reason for $\overline{B}$ given A and, because of symmetry, B a reason for $\overline{C}$ given A.[58] Hence, C very badly fits A and B under these assumptions.

Similar assumptions create similar oddities. This suggests a general conclusion which looks at least plausible: If for all $j \in I$ true j-propositions cohere with all past truths, then, for any $i \in I$, a true i-proposition coheres with all other truths, because it coheres with all past truths, as just stated, and also with true j-propositions for all $j > i$, since coherence is symmetric. In this way general coherence with the past seems to imply general coherence with past *and* future. If this is true, the above restriction of the coherentistic considerations would, after all, not really be a restriction. However, this is only a vague conjecture, neither precisely stated nor proved.

If the presented line of reasoning is correct, we ask 'why?', we search for explanations because this is one and, in a way, the only way of finding coherent truth and, insofar as truth must be believable and coherent, the only way of finding truth. Why search for truth? Here I cannot think of any further theoretical justification; to some extent we seem to be intrinsically curious beings. Papers must end, justifications presumably, too. But the present one does not end here; there is beautiful further justification for the search for truth of a practical, decision-theoretic kind.[59]

[58] For proof note that (a) $\kappa(\overline{B} \mid A) = \min [\kappa(\overline{B} \mid A \cap C) + \kappa(C \mid A), \kappa(\overline{B} \mid A \cap \overline{C}) + \kappa(\overline{C} \mid A)]$. We have assumed (b) $\kappa(\overline{B} \mid A) > 0$ and (c) $\kappa(\overline{B} \mid A \cap C) = 0$. All three immediately imply the first claim $\kappa(C \mid A) > 0$. According to the law of negation (after Definition 1), the latter entails $\kappa(\overline{C} \mid A) = 0$; this and (a) in turn entail $\kappa(\overline{B} \mid A \cap \overline{C}) > 0$; and this and (c) say that C is a necessary reason for $\overline{B}$ given A.

[59] I refer to the observation in Savage (1954, sect. 7.3) that the expected utility of free information is always non-negative, and to the strong generalizations offered by Skyrms (1990, ch. 4). A different generalization to free memory may be found in Spohn (1978, sect. 4.4).

Chapter 10
Two Coherence Principles[†1]

10.1 Introduction[1]

The purpose of this paper is twofold. On the one hand, it is a self-contained continuation of Spohn (1991) [here: ch. 9]. I studied there the relation between three principles of coherence and two versions of the principle of causality, thereby transferring the plausibility of the former onto the latter. Ever since then, I have wondered what more can be done to defend the coherence principles than simply appeal to their plausibility. This paper tries to give an answer which, however, is partial since I shall discuss only one of the old coherence principles.

On the other hand, a more important purpose interfered. Everyone engaged in the epistemological issue of foundationalism versus coherentism will grant that the notion of coherence is in bad shape. Since pondering the second of the present coherence principles, I thought that it offers a nice explication of the notion of coherence, which I have not found in the literature, which is perfectly precise and theoretically fruitful, and which therefore deserves to be presented. In view of the richness of the notion of coherence it would be silly to claim that this is *the* explication of the notion. The intent of this paper is rather to make this explication attractive by briefly relating it to other conceptions of coherence, by explaining the epistemological picture behind it, and by showing how one might argue for the associated principles.

The plan of the paper is this: Section 10.2 introduces some of the basics of epistemology, in particular the notion of a reason which is essential for the rest of the paper. Section 10.3 goes on to explain the two coherence principles which are the

---

[†1] This paper was originally published in: *Erkenntnis* 50 (1999) 155–175.

[1] I am grateful to the members of our Forschergruppe, Wolfgang Benkewitz, Ulf Friedrichsdorf, Ulrike Haas-Spohn, Volker Halbach, and Erik Olsson, for various suggestions and criticisms helping to improve the paper, to Jeffrey Knight for improving my English, and to an anonymous referee for further helpful suggestions.

subject of this paper and depicts their epistemological setting. Sections 10.4–10.7 finally offer four attempts to further ground these principles, the results of which are, briefly, that it is neither enumerative induction, nor the nature of propositions as objects of belief, nor consciousness, but rather an even more fundamental principle of rationality and an elementary theory of perception which entail these principles.

A final warning: In the course of the paper I shall make many claims which may be formally elaborated within the theory of ranking functions.[2] Here, however, I mostly dispense with formal details. This has obvious advantages. One of them is that I am not immediately committed to all the assumptions built in into the theory of ranking functions and can try instead to be more general. Thus I indicate, in an informal way only, which features of doxastic modelling are needed for the reasoning at hand. However, it may not always be clear to what extent I have avoided falling back on the features of ranking functions. Opacities of this kind belong to the drawbacks of informality which, I hope, do not outweigh the advantages.

10.2 Reasons

It seems uncontroversial to me that any kind of formal epistemology must represent a doxastic state by a function β with at least the following three features:

First, β must be defined on some set of propositions, where propositions, just by definition, are to be appropriate objects of belief. For the time being we may leave the exact nature of propositions an open question, which, of course, is much discussed; I shall only make the minimal assumption that they have Boolean structure.

Second, β must allow for degrees of belief, i.e., the range of β has to be some (usually linearly) ordered set of degrees. This condition is almost trivial in view of the fact that 1 (= belief), −1 (= disbelief), and 0 (= neutrality) also form such a set of degrees, indeed the minimal one.

Third, β must allow for conditionalization, i.e., it must assign conditional degrees of belief in some substantial, reasonable way. I am not sure how to strictly prove this, but any account of the dynamics of doxastic states I know of assumes conditional degrees of belief, and I have no idea what an alternative account could look like.

These three features immediately yield a most natural notion of confirmation, justification, or reason: A proposition A confirms, supports, or is *a reason for* a proposition B relative to a doxastic state β iff A strengthens the belief in B, i.e., if

[2] Introduced in Spohn (1988) [here: ch. 1] (where I still called them ordinal conditional functions). Ranking functions are particularly suited for more formal accounts of the present discussion, because they include a straightforward notion of belief – a point which has always been difficult for the probabilist.

the belief in B given A is stronger than given non-A.[3] We may thus define a bit more formally:

A is a reason for B given C (relative to β) iff $\beta(B \mid A \wedge C) > \beta(B \mid \neg A \wedge C)$
A is irrelevant to B given C iff $\beta(B \mid A \wedge C) = \beta(B \mid \neg A \wedge C)$ and
A is a counter-reason to B given C iff $\beta(B \mid A \wedge C) < \beta(B \mid \neg A \wedge C)$

The unconditional relations are defined by reference to the tautological condition; thus *A is a reason for B* (relative to β) iff $\beta(B \mid A) > \beta(B \mid \neg A)$. Hence, being a reason is nothing but positive relevance, and being a counter-reason is nothing but negative relevance – an old idea which reaches back at least to the discussion between Carnap and Popper about confirmation.

Which properties does the reason relation have? It follows trivially (assuming that $\neg\neg A$ is the same proposition as A) that

A is a reason for B given C iff $\neg A$ is a counter-reason to B given C.

All other properties of the reason relation depend on specific assumptions about β. The most common and useful choice is, of course, to conceive β as a probability measure. Then we obtain a reason relation which is symmetric and embraces logical consequence:

A is a reason for B given C iff B is a reason for A given C; and
if B is logically implied by A, then A is a reason for B (and vice versa), provided neither A nor B has an extreme probability.

We get many other important properties in addition, which, however, will not be relevant in the sequel. Moreover, it is worth mentioning that this probabilistic reason relation is *not* transitive.

Exactly the same properties result if we conceive β to be a ranking function. It would be interesting to find out about the properties of the reason relation if β is conceived as in the AGM-theory, as an entrenchment relation, for instance (cf. Gärdenfors 1988 or Rott 2001), as a Dempster-Shafer belief function (cf. Shafer 1976), etc. I believe, though, that the behavior of the reason relation turns out to be most satisfying relative to probability measures and ranking functions. There is no space to look closer into this issue; but I indeed think that this behavior is an unduly neglected adequacy criterion for formal representations of doxastic states.

This paper will be entirely based on the reason relation of positive relevance. It is obvious that this will bias the paper from the beginning. Are there not many other reason relations or similar notions around? So why use this one? This is a large question, but to attempt a lump-sum answer: It is my impression that those engaged in the epistemological issues I am going to address usually operate with a reason relation too vague to allow any rigorous theorizing and that alternative formal

[3] Why "given non-A" rather than "given nothing"? If we interpret β in the most familiar way as a probability measure, the two alternatives are equivalent as long as the relevant conditional probabilities are defined. However, if we interpret β, e.g., as a Popper measure or as a ranking function, a simple reflection shows my alternative to be preferable.

reason relations are less suited for these issues. A better answer, however, would first grant that no explication of the reason relation is to be expected to dominate all others and then provide an extended argument comparing the virtues of the theories built around the various explications – a task too large for a small paper. In a way, however, this paper may be seen as part of such an argument.[4] In any case, I shall simply proceed with positive relevance.

10.3 Two Coherence Principles

Since logical entailment abounds among propositions, the more embracive positive relevance does so as well. Therefore it will be most crucial to observe how much of positive relevance there is beyond logical entailment. To this end we must give a bit more structure to the propositions. I shall assume that we can discern atomic propositions and that these atomic propositions are logically independent. Or to be a bit more specific: I assume a Boolean algebra of propositions as it is usually constructed in probability theory or, e.g., in Carnap's latest inductive logic (1971/80). This construction starts from a set of *variables* (not in the logical sense, but in the sense of stochastic variables). Each variable can take values from a certain range; in the simplest case it is a yes/no variable ranging over $\{0,1\}$. A *possible world* or a *possible course of events* is a function specifying a value for each variable; this is the value the variable takes in this world or course of events. A *proposition* is any set of possible courses of events. Let U denote the set of all variables, and for $V \subseteq U$ let $P(V)$ denote the set of all propositions over V; thus, $A \in P(V)$ iff A does not discriminate outside V, i.e. iff for any world w in A all worlds differing from w only outside of V are also in A. Then, a proposition A is *atomic* iff it is about a single variable, i.e., if there is a variable X such that $A \in P(\{X\})$; thus, atomic propositions concerning different variables are logically independent. Finally, a proposition is *a posteriori* iff it is neither empty (a priori false) nor identical with the set of all worlds (a priori true).

How should positive relevance spread over the set of propositions? It is impossible to say. If U is some gerrymander, a subject's beliefs concerning U may take any form whatsoever. However, if U is the set of *all* variables within the grasp of a subject's doxastic state β (certainly an ill-defined set), we have more definite expectations. One plausible expectation is stated in the *special coherence principle*:

> For any variable X and any a posteriori proposition $A \in P(X)$ there is a proposition $B \in P(U - \{X\})$ such that B is a reason for A (relative to β).

Thus the special coherence principle says that there is some inductive support for each atomic a posteriori proposition or, more simply, that no variable is independent from all others.

---

[4] Other papers of mine (Spohn 1991, 1997c, 1997/98) [here: chs. 9, 12, and 11] may be seen as further parts of such an argument. Spohn (1997/98, sect. 2) [here: sect. 11.2] in particular, contains some remarks comparing positive relevance with other reason relations.

I refer to Spohn (1991) [here: ch. 9] for one way of expanding and strengthening the special coherence principle.[5] Here I shall take another way leading to an explication of coherence. The idea is simply that the special principle looks just as plausible when we replace the single variable X by some arbitrary set of variables. Then we get the much stronger *general coherence principle*:

For any proper subset $V \subset U$ and any a posteriori proposition $A \in P(V)$ there is a proposition $B \in P(U - V)$ such that B is a reason for A (relative to β).

Thus the general coherence principle says that the set of all variables does not fall into independent parts. Or in graph-theoretic terms: If one represents the (conditional) dependencies and independencies given by the doxastic state β in a so-called Bayesian network,[6] the general principle requires that this network is a connected graph which cannot be separated into unconnected parts. Or to be a bit more pompous: The general principle really affirms something like the unity of science, the unity of our empirical world picture.

So far, I have only claimed that these principles are plausible; in the subsequent sections we shall have to inquire into what the deeper truth behind them might be. However, let me first ask what these principles have to do with coherence. The answer is simple; the general principle *defines* coherence:

A doxastic state β is *coherent* iff β satisfies the general coherence principle.

Coherence *is* connectedness, integratedness. This explication is as precise and clear as the underlying reason relation; it thus compares favorably with most alternative offers.

However, what we really like to know is, of course, how the explication and the principles relate to coherence as it figures in the debate between coherentism and its alternatives, or, for short, in the "knowledge debate" (since the alternatives have arisen in the quest for the nature of knowledge). So let me introduce four rough characters: the foundationalist, the coherentist, the externalist, and the (formal) belief theorist, for want of a better term. The former three are the well-known archetypes in the knowledge debate. The primary epistemological interests of the last, however, do not lie in this debate. They are, rather, to build formal models of the statics and the dynamics of doxastic states, to develop their theory, and to somehow justify the assumptions built in into the models as rational. Which stance, if any, in the knowledge

---

[5] The explication of causation defended in Spohn (1991) [here: ch. 9] entails that the special coherence principle is equivalent to a very weak principle of causality which says that each atomic proposition has a cause or an effect in some possible world. Moreover, I present there two strengthenings of the special coherence principle, one entailing and the other being entailed by a weak principle of causality saying that each atomic fact has a cause or an effect in the *actual* world.

[6] This is a directed acyclic graph the nodes of which represent variables and the vertices of which represent conditional independencies between variables obtaining according to β insofar as *all* these independencies can be read off from the vertices by help of the so-called criterion of d-separability; cf. Pearl (1988, sect. 3.3). The theory of Bayesian nets is an utterly useful tool for the epistemologist, not only because of its graphical qualities; however, it is applicable only where conditional independence behaves as in probability measures or in ranking functions.

debate is thereby entailed is only a secondary question. The attitude Carnap finally took towards inductive logic (cf. his 1971/80) is certainly prototypical, belief revision theory and probabilistic epistemology are carried out in the same spirit, and I consider myself to be a formal belief theorist in this sense as well.

There are various agreements and disagreements among these characters. All of them have some notion of the reason relation. However, the foundationalist, the coherentist, and the externalist diverge on the properties of the reason relation in well-known ways. The belief theorist is certainly an internalist; I do not know of any belief theorist providing theoretical means for allowing external facts to be reasons for or to justify beliefs. Whether he sides with the foundationalist or the coherentist will, however, depend on his doxastic model.[7] For instance, if he takes the reason relation to be symmetric, as I did above, he thereby opposes the foundationalist who insists that basic beliefs are reasons for other beliefs, but cannot have reasons outside themselves.

There is a much deeper disagreement, though. Those engaged in the knowledge debate assume that there is not only the binary relation of one belief being a reason for another, but also a unary predicate (or quantity) of a belief being justified or warranted (to a certain degree). To put it graphically, the common picture[8] is this: The binary reason relation provides a network of channels between its relata, the thickness of which governs how much of the viscous quantity called *degree of warrant* can flow through them. By itself, however, the network is empty. It still needs to be filled with this quantity. Now the disagreement starts. The foundationalist thinks that this quantity is created in what he calls basic beliefs and then flows to the other beliefs. The externalist seeks the source of this quantity in appropriately related external facts. The coherentist either says that this quantity is bestowed on a belief in virtue of its relational coherence with all other beliefs,[9] or that this quantity is created by the network itself according to its degree of intrinsic coherence and then distributes differentially among its nodes.[10] It is clear that many mixtures are conceivable, and have indeed been suggested.

Now, the deep schism is that the belief theorist does not at all know what to make of this picture. It is hardly explicable for him and, what is worse, he has no use for it. Not that his theory of doxastic states would be complete; but a theory of warrant is not among the things he is missing. There is overwhelming evidence that the theory of belief contents requires much more sophistication. He may strive for

---

[7] And on his explication of the reason relation – he need not adopt my above proposal.

[8] It may be explicitly found in BonJour (1985, sect. 5.2) or in Plantinga (1993, ch. 4). In fact, it is built in into the set-up of the justification trilemma which drives the knowledge debate and according to which one can choose only between three unpalatable alternatives: infinite justificatory regress, circular justification, or stopping justification at some unjustified or obscurely self-justifying point.

[9] This is, roughly, the version of Lehrer (1990, pp. 147ff.).

[10] This is pure coherentism as explained by BonJour (1985, sect. 5.2) and amended later on. Plantinga (1993, p. 78) criticizes this version as pure magic; indeed it looks like *creatio ex nihilo*.

more realism by considering other kinds of degrees of belief, probability intervals for instance, instead of point probabilities, or by adding a badly needed theory of computational management of doxastic states. The theory about a priori states is severely underdeveloped in my view. The theory of doxastic changes does not say much about non experiential changes, for instance conceptual change. The input theory of observation and experience could certainly be more detailed; and the output theory of action and behavior need not stick to decision theory. Such are the tasks for the belief theorist to complete his theory (all of which are belabored, of course). As far as I know, however, the knowledge debate has not advanced any good reason for the belief theorist to think that he needs to add a theory of warrant as well. In a way, this is not surprising since knowledge is simply not a relevant topic for the belief theorist and since the notion of justification or warrant plays its primary role precisely in the difference between true belief and knowledge.[11]

In ch. 6 on Bayesian coherentism Plantinga (1993) arrives at the same conclusion, suggesting that it is a defect of the Bayesian, or the belief theorist in general, that he is unhelpful to the knowledge debate. This is only half of the truth, however. The concern should really be mutual. Of course the belief theorist should be deeply worried about the fact that he cannot, and does not want to, say much about the notion of warrant which seems to arise so naturally and is taken so seriously by many serious philosophers. Conversely, however, the knowledge debate should be deeply worried about the fact that the notion of warrant is apparently unimportant to a large part of epistemology and to equally many equally serious philosophers. The schism is unbridged.[12]

I am explaining all this because it clearly entails that whenever a belief theorist like me starts using the terms so central to the knowledge debate, he is bound to stand crossways to that debate. The conclusion I draw from this situation is this: If the belief theorist has complete ways of theorizing, or ways to complete theorizing, without referring to the knowledge debate, this is so either because that debate is really immaterial or because it is somehow implicit in his theorizing. Since I cannot believe the former, I try to verify the latter. This is how my efforts here should be seen.

For instance, defining the reason relation as I did above is something the pure belief theorist need not do; it is merely an attempt to approach the knowledge debate. Likewise, I might progress from the binary relation to the unary predicate by saying that a belief is justified iff the balance of reasons is in its favor. However, this is no more than an insubstantial metaphor so far. The belief theorist does not have the idea of an active weighing of reasons which results in a justified belief. Rather, in his rationalized picture, a doxastic state *eo ipso* satisfies the basic laws of his doxastic model (e.g., the axioms of probability), and hence each proposition is automatically in balance, so to speak, within a doxastic state: it could not be

[11] One should note that doubts about the role of justification have also been articulated within the knowledge debate; cf. von Kutschera (1982, ch. 1) or Sartwell (1992).

[12] This schism seemed to me, on reflection, to be at the center of the conference whose results are published here. Perhaps the conference has at least spanned a rope between the sides.

believed to any other degree without violating these laws (without violating, e.g., coherence in the probabilistic sense).[13]

Finally, the belief theorist has certainly great difficulties in understanding the notion of coherence, as it figures in the knowledge debate, in a warrant-creating or warrant-conferring role (cf., however, Olsson 1999). Moreover, he certainly cannot make sense of measuring coherence by measuring (probabilistic) inconsistencies (unless he resorts to something like paraconsistent logic). However, he has no difficulties in understanding the standard examples of consistent but incoherent doxastic states which simply consist in a set of unconnected or independent beliefs.[14] Connectedness and dependence is precisely what the reason relation creates. Hence, this aspect of the notion of coherence is most adequately captured by the general coherence principle. And as such it should also be of interest within the knowledge debate.[15]

Having thus roughly clarified the setting within which the above explication of coherence is placed, I can finally turn to the main purpose of this paper, i.e., to considering on which grounds doxastic states should satisfy these coherence principles.

10.4 Justifying the Coherence Principles via Enumerative Induction?

Let me first briefly look into the relation between the coherence principles and inductive logic. Indeed, this is the only place, as far as I know, where similar relevance principles are stated.

The most important and most convincing one is the principle of positive instantial relevance (cf. Carnap 1971, sect. 13), which is the probabilistic analogue to enumerative induction and says, roughly, that the fact that one individual has a certain attribute makes it likelier that another individual has this attribute as well. This clearly entails the special coherence principle, provided that the set U of variables has an appropriate structure.[16] However, positive instantial relevance is silent on the general principle, because it does not say anything about the relation between different attributes.

---

[13] The metaphor would be more substantial if it would be possible to reconstruct the degree of belief in a proposition from the strengths of the reason relations in which it stands. However, it is easy to see that this is not possible for my above reason relation and doxastic states conceived as probability measures or ranking functions. It might be worthwhile investigating which stronger assumptions allow such reconstruction.

[14] See also the coherence conditions (3) and (4) in BonJour (1985, p. 98).

[15] In Spohn (1991, sect. 5) [here: sect. 9.5] I try to argue that this kind of coherence is closely related to explanatory coherence.

[16] The structure is appropriate if the variables are constructed from attributes, relations, or magnitudes and objects such that each attribute etc. figures in more than one variable. This condition is certainly satisfied if U is the set of variables within the grasp of a given subject, and indeed implied by what Evans (1982, sect. 4.3) calls the generality constraint.

Such relations are rather specified in Carnap's theory of the analogy influence (cf. Carnap 1980, sect. 16f.). However, it is not at all clear whether Carnap's full inductive logic would satisfy the general coherence principle. This would depend on whether all attributes are integrated in one attribute space and, if not, whether any relations between different attribute spaces are specified, and how precisely the analogy influence spreads within one attribute space. Moreover, it must be admitted that this theory of analogy has been put forward quite tentatively and that it has not met many friends in the last decades; without further scrutiny no strong case can be built on it. It therefore seems advisable to look for other ways of justifying the coherence principles.

10.5 Justifying the Coherence Principles via the Essence of Propositions?

The next possible answer, though much deeper, will also be considered only very briefly. First, equating propositions with sentence meanings seems quite innocent. What precisely meanings are is, however, an inexhaustible topic. One view, which is still popular in the wake of the verifiability theory of meaning, is to construe sentence meanings or propositions not as truth conditions, but rather as assertibility, justifiability, or acceptability conditions of sentences. There are many places in the philosophy of this century where such a view is suggested. Properly understood, this approach takes the reason relations which a proposition bears to other propositions as *individuating* this proposition,[17] though this is rarely endorsed in an explicit way.

This definition of propositions entails the special coherence principle: there can be at most one exception, i.e. at most one proposition which stands in no reason relations whatsoever. Despite my sympathies for such ideas, I think that this justification of the coherence principle is at least doubtful. My concerns are fourfold.

First, I do not know of any satisfying formal implementation of the idea. The proponents of acceptability conditions are usually stuck in metaphorical descriptions, and as far as I know, the formal literature does not address the question. If the individuation of propositions is aided by the logical relations between them, it becomes trivial because each proposition is uniquely characterized by the set of its logical consequences. However, if the undertaking is restricted to the reason relations as explained above, I do not know how it might be accomplished, how, for instance, the Boolean structure of propositions might be generated. As long as this technical task is not achieved,[18] this justification of the coherence principles does not work.

[17] A nice parallel would be Davidson (1969) who individuated events via the causal relations they bear to other events.

[18] I know of two attempts which get close to what would be needed, namely the ingenious proposal of Popper (1934/69, Neuer Anhang *IV) to extract the Boolean structure of propositions from the properties of conditional probabilities and the construction of Gärdenfors (1988, ch. 6) which achieves the same by starting from the properties of the dynamics of belief.

Second, this definition of propositions can avoid outright circularity only by claiming a thoroughly holistic conception of sentence meanings or propositions. To maintain such a holism is certainly difficult in view of the large and on-going philosophical debate about it.[19]

A third and related concern is that there are competing accounts of propositions which do not seem worse: for instance, the account which defines, as I did in Section 10.2, propositions as sets of possible worlds or more complex indices, or the account which takes propositions as internally structured, i.e., as somehow composed of properties, relations, and objects by various rules of composition. Thus, before this line of reasoning in favor of the coherence principles can succeed, one would have to engage in intricate arguments showing that the individuation of propositions via justifiability or acceptability conditions is to be preferred to the other ones within the given context. Here one certainly moves on very general and problematic grounds.

Finally, we have the same problem as with Carnap's inductive logic. So far, the proposed strategy does not yield the general coherence principle and I cannot see any feasible strengthening of the strategy which would do so. Hence, success is again incomplete. All this is sufficient reason for looking further.

10.6 Justifying the Coherence Principles via Consciousness?

If the general principle is so recalcitrant, we better face it directly. The general line of reasoning for it seems quite obvious. Suppose my doxastic state violates the general principle and the set of variables within my grasp divides into two independent separate parts. Where am I? Certainly, my self-picture is an indispensible part of my doxastic state,[20] there are a lot of variables about myself. Apparently, these variables cannot belong to both parts, the dividing line cannot cut through myself. Thus they are wholly within one part. But then it is hard to see how the other independent part could be within my reach. My learning seems to be restricted to the part containing me, and I could not come to believe anything about the other part.

This line of reasoning may look promising, but it is a different matter to turn it into a sound argument. Clearly, the suggestion has a Kantian ring. When I just said that at least the propositions about myself must be connected, I should probably have been so cautious to refer only to the propositions concerning my consciousness. And then we seem to be in the vicinity of Kant's profound idea that the "I think" must potentially accompany all my thoughts and ideas, i.e., in the vicinity of the transcendental unity of pure apperception which Kant declares to be the first principle of understanding lying at the base of all our judgments. So, in a nutshell,

[19] Fodor (1987, ch. 3), e.g., offers a most forceful criticism of such holism.

[20] See, e.g., Perry (1979) concerning the irreducibility of attitudes de se.

the suggestion is that we may somehow derive the connectedness of our empirical beliefs from the unity of consciousness. However, closer inspection fails to confirm this; we rather encounter a class of propositions which must be exempt from the coherence principles: facts of consciousness are not within the field of the reason relation. This is the consequence of the following considerations.[21]

The suggestion from Kant is that the relevant sort of facts of consciousness are propositions about one's own beliefs; in a sense, I simply know what I do, and do not, believe. However, it would be intuitively very strange to defend, justify, or reason for one's beliefs with the help of such knowledge. Suppose someone claims: "Clinton will resign before the end of the year," and when asked for his reasons he responds: "I believe so." Then he has certainly given no reason at all, even if the answer is, unnaturally, interpreted not as the affirmation of the original claim, but as an expression of a second-order belief. Believing to believe that *A* is somehow tantamount to believing that *A*, and therefore the former cannot be used in reasoning for the latter.[22]

This intuition should be substantiated, though. This is done by Benkewitz (forthcoming, sect. 5.3), in an extended argument. Instead of adapting this argument to the present purposes,[23] however, I shall try to confirm a fairly common thought which runs as follows: Facts of consciousness are maximally certain and, generally, maximally certain propositions cannot have, or be, reasons.

Let me start with the latter claim. Why can maximally certain propositions not have, or be, reasons? Observe first that, if *A* is maximally certain, it is so under any conditions; this is so at least if doxastic states satisfy an analogue to the formula of total probability, i.e., if the degree of belief in a certain proposition is in some sense a weighted mixture of the conditional degrees of beliefs of that proposition under mutually disjoint and jointly exhaustive conditions. This observation entails that no proposition can be a reason, in my sense, for a maximally certain proposition. If one further accepts the symmetry of the reason relation, then this in turn entails that maximal certainties cannot be reasons for other propositions either. But one may also argue that a maximal certainty cannot be a reason for other propositions because relative to the negation of a maximal certainty, to which the minimal degree of belief should be assigned, no conditional degrees of belief can be defined. For, if such

---

[21] This is not intended to disprove Kant, of course, since I shall not be concerned with the special role of "I" which is so important for Kant. However, my implication certainly is that whatever kind of unity is generated by the special role of "I", it is not the unity in terms of the reason relation.

[22] Because of this I wondered about the account of observation in BonJour (1985, ch. 6) for which this kind of reasoning is essential (and hence I tried in Spohn 1997/98 [here: ch. 11] to give a coherentist account of observation without alluding to second-order beliefs). That second-order beliefs find no place in the reason relation is reflected in BonJour's work also in the role which is played by his Doxastic Presumption, which is special since he admits that there is no further justification for the beliefs about one's own beliefs.

[23] Benkewitz argues for the more consequential thesis that in an important sense a subject cannot causally explain its own present beliefs, and it would require some explanation to show how the present thesis is implicitly contained in that argument. I am grateful to Wolfgang Benkewitz for alerting me to assertions of this kind.

conditional degrees of belief were non-trivially explained, i.e. in such a way that they may have different values, this would entail an impossible splitting-up of the minimal degree of belief into several different degrees. This reasoning establishes a large class of exceptions to the coherence principles, namely the set of maximally certain propositions all of which cannot engage into reason relations.

The next question is: Which propositions are maximally certain? There seem to be two kinds. The first kind consists of propositions which are a priori in the sense of being necessarily believed in any doxastic state capable of grasping them. All analytic propositions, like "bachelors are unmarried" or "5 + 7 = 12", are a priori. But there also are Kripkean non-analytic propositions a priori like "I exist," "I am here now," "the *F* is an *F*"(provided that "the *F*" is read referentially), and reduction sentences for dispositions (cf. Spohn 1997c, [here: ch. 12]) – a class of propositions which strongly recommends itself for further investigation. Still, the fact that such a priori propositions do not fall under the scope of the coherence principles is no cause for worry. The coherence principles are designed for empirical beliefs a posteriori, and thus this kind of exception is easily tolerable.

Besides, however, it is usually held that there is a second class of propositions which are maximally certain, namely, facts of consciousness. These comprise facts about my perceptual or experiential state such as "I am now appeared redly" (to use Chisholm's phrase) or "I am in pain now", facts about my present propositional attitudes like "I think that *A*", "I believe that *A*", "I desire *A* to be the case", or "I intend to do *A*", and maybe other kinds of facts. If these propositions are maximally certain, the strategy presently considered apparently fails.

Why, though, should we think of facts of consciousness as maximally certain? We might try to elaborate one of the following two argument sketches. Both proceed from the following starting point: What precisely are facts of consciousness? We have listed examples, but a general explication would be better. The following applies to the examples and seems generally adequate: *A* is a *fact of consciousness* iff *A* is true and necessarily equivalent with, i.e., the same proposition as the proposition that I (presently) believe that *A*.[24,†2] Moreover, it seems that in this case such necessity is a priori and hence that the two propositions are even analytically equivalent. I am well aware that in giving this explication I am opening a Pandora's box; but for the present purpose let us neglect this and just look what follows from it.

The first argument sketch is this: To believe something presumably means to believe it at least to a certain degree (analogously, to be tall for a man means something like to be taller than, say, 6'4").[25] Hence, if *A* is a fact of consciousness, it is the same as believing *A* at least to a certain degree. Believing *A* in a specific, sufficiently large degree would then be something stronger, and something different, for each different degree. But if *A* is the same as believing *A* there seems to be no room for such varying degrees of belief in *A*. This suggests that there is no proper degree

[24] Thus, facts of consciousness are the same as what Benkewitz (1999, sect. 5) calls internal contents (as opposed to external contents of beliefs).

[†2] In Spohn (2005d) I have more fully developed this analysis.

[25] This idea and its vagueness is propounded by Hunter (1996).

of belief for A, only an improper one, so to speak; and the only improper degree of belief (which is sufficiently large) is the one expressing maximal certainty.

The other argument sketch is this: We have already seen above that doxastic states cannot be conditionalized with respect to negations of maximally certain propositions. Likewise, it looks strange and even seems impossible – though I have no further argument for this – to conditionalize a doxastic state with respect to something which denies that very state. According to my explication of facts of consciousness, however, which declares such a fact to be part of a doxastic state, we would try to do exactly this if we try to conditionalize a doxastic state with respect to the negation of a fact of consciousness. Hence, if such conditionalization does not make sense, the above explication of the reason relation does not apply to facts of consciousness; that is, facts of consciousness cannot be reasons for other propositions.

So whether we are content with declaring that facts of consciousness are maximally certain or add one of the further arguments, the conclusion is in any case that such facts are not in the field of the reason relation and that this attempt, at least, was not the right way to get help from Kantian insights. Still, one may wonder about this conclusion. It seemed to be generally agreed that the foundationalist is right insofar as the basic beliefs he postulates have at least some justifying force; the question was rather whether all justification ultimately reduces to them and whether they are really foundational in the sense of having no justification outside themselves. Moreover, conscious phenomenal or experiential states (or the identical beliefs in them) appeared to be first-rate candidates for such basic beliefs. This appearance is false, however, if my conclusion is right. But how could it then be so plausible? Let me close this section with offering two brief thoughts for reconciliation.

First, phenomenal facts of consciousness are really quite special and can only be expressed by phrases like "it looks now *thus* to me", accompanied by a deferred ostension to my present phenomenal experience. Propositions like the one that the tomato in front of me looks red to me, or even that I am appeared redly now, may also seem to be facts of consciousness. But they are not, they are subtly different; and the subtle difference suffices to make them unexceptional and to integrate them into the circle of reason. So, they may well serve as a substitute offer to the foundationalist.[26]

Second, one must pay close attention to the dynamics of the reason relation. Doxastic states change and positive relevance changes with them. Consider the proposition that I shall be in such and such a conscious phenomenal or doxastic state in an hour. There is no problem for this proposition to be a reason for, and to find reason in, other propositions. An hour later, I am in such and such a conscious state and thus believe it to obtain with maximal certainty or in a way excluding it from the reason relation.[27] Still an hour later, my doxastic state will have changed again. Then I believe that I was in such and such a conscious state an hour ago – in a less than maximal

[26] This is more fully argued in Spohn (1997/98) [here: ch. 11]. However, I argue at the same time that these propositions are not basic in the foundationalist's strict sense.

[27] Strictly speaking, it is not the same proposition as before which I believe then, because the temporal index has shifted. However, being precise about this would only enforce my point.

degree, however, not because I have learnt new things in between, but simply because the conscious state has turned into a less than maximally certain recollection which is again justificatorily related to other propositions in both ways. Hence, even the phenomenal proposition is within the circle of reason for most of the time; it jumps out of the circle only during the dazzling moment of conscious experience.

10.7 Justifying the Coherence Principles via a Theory of Perception

Should we conclude therefore that the line of reasoning sketched at the beginning of the previous section fails? No. I suggest that we stay away from that dazzling moment and replace the subject's consciousness by his beliefs about an arbitrary perceiver who may be a third person or he himself at another time. Thereby we can turn the rough sketch into a more rigorous argument proceeding in seven steps. The first six steps deal with the special coherence principle. A simple further step will finally carry us to the general principle.

(1) The argument must start somewhere. I propose the following principle of rationality: A subject should have a variable degree of belief in any a posteriori proposition within his grasp. That is, if the subject believes in such a proposition to a certain degree, there should be a possible dynamics which leads him to believe that proposition to some other degree.

This sounds almost tautological. Recall that a priori propositions were defined above as propositions necessarily believed in any doxastic state (capable of grasping them). Hence, a posteriori propositions may or may not be believed or, more generally, may have varying degrees of belief in different doxastic states. This is weaker than the rationality principle; the different states need not be connected by a possible doxastic dynamics. Still, the principle thereby appears evident. If, by definition, varying attitudes are possible towards an a posteriori proposition, one should not be so dogmatic to fix one's attitude once and for all.

(2) Now let A be a proposition about a single variable which does not belong to the exceptions already admitted, i.e., which is a posteriori and not a fact of consciousness, and which is thus believed to a non-extremal degree. How can this degree change? Mainly by obtaining reasons for or against A, that is, by coming to believe or, more generally, by changing the degree of belief in other propositions which are positively or negatively relevant to A so that the belief in A changes its degree as well. Now, if A would violate the special coherence principle, there would be nothing that counts for or against it, there would be no way to change the degree of belief in A – in contradiction to the rationality postulate in (1).[28]

[28] Of course, we always suffer from a large and grave practical inaccessibility of reasons, simply because our experience is so restricted in space and time. The case at hand is worse, however; there, the non-existence of reasons would be irrevocably fixed in the internal structure of the doxastic state.

(3) The proof in (2) leaves a gap, however. The degree of belief in A may also change directly, not mediated by changes concerning other propositions. Indeed, the foundationalist will point out that this is the case with basic propositions as he conceives them, namely, as propositions which do not find any reasons outside themselves and are thus defined to violate the special coherence principle *without* necessarily being facts of consciousness. One may rejoin that this definition is empty because basic propositions are certainly used as reasons for other propositions, and the symmetry of the reason relation then entails that these allegedly basic propositions have reasons as well. However, this rejoinder has two shortcomings. First, nothing has been said so far to exclude the strange case of a basic proposition which is not good for justifying anything else; and secondly, the symmetry of the reason relation is, of course, something the foundationalist cannot accept. So, the proof in (2) needs some amendment.

(4) To this end we should first ask: What are the basic propositions in the foundationalist's sense? There is no perfect agreement, as far as I see, but the usual answer is that basic propositions are *directly perceived* propositions. What, in turn, are these? Some say, or think they are forced to say, that directly perceived propositions are facts of consciousness having purely phenomenal qualities as their contents. However, directly perceived propositions then reduce to a class of exceptions which we have already seen not to serve the foundationalist's purposes. So we may dismiss this reduction.

There is a more fruitful notion of direct perception according to which other propositions can be directly perceived as well. It runs as follows: Let us first assume that we can distinguish doxastic changes caused by perception from other doxastic changes (due to new concepts, drugs, forgetfulness, etc.). Changes through perception are usually accounted for by rules of conditionalization.[29] Now it is easy to check that these rules have the following property: Given the prior and the posterior doxastic state, and given that the change from the former to the latter was governed by a rule of conditionalization, the minimal set of propositions relative to which conditionalization was applied is uniquely determined; we may call this minimal set the *source* of the change. It seems then appropriate to say that the proposition(s) directly perceived in a perception is (are) just the proposition(s) in the source of the change brought about by the perception.

It must be emphasized that it is possible, but certainly exceptional that facts of consciousness are directly perceived in this sense. Usually, directly perceived propositions are public and in principle perceivable for many observers. Moreover, directly perceived propositions then stay firmly within the circle of reason; there is no need to exempt them from the circle. Their distinctive role rather lies, according to the account given, in the role they play in doxastic changes.

---

[29] In probabilistic terms these rules are simple conditionalization and generalized conditionalization as introduced by Jeffrey (1965, ch. 11). These rules can also be stated in terms of ranking functions; cf. Spohn (1988, sect. 5) [here: sect. 1.5].

(5) The next step is to introduce the following standard theory of perception: If *x* directly perceives that *A* (and if *A* is not a fact of consciousness), then *A* is a cause of the fact that *x* believes (more firmly than before) that *A*. Despite many disagreements concerning the theory of perception this seems to be one uncontested corner-stone.

A more contested question is, among others, whether this theory can be turned into an analysis of (direct) perception. The answer must be negative, it seems; there are certainly many propositions not directly perceived (or not perceived at all) for which this causal relation also obtains. People have tended then to require that it is this causal relation which must be somehow direct. However, this only leads to completely assimilating directly perceived propositions to facts of consciousness. But this seems wrong: the directness does not lie in the causal relation, but in the kind of belief change, as is also expressed in the familiar assertion that the directly perceived is non-inferentially known.[30]

(6) Now we may finally close the gap left in (2) and noticed in (3). The gap was that *A* may also be a basic proposition, i.e., a directly perceivable proposition which is not a fact of consciousness, which may directly change its degree of belief, and which thus appears to have no reason. This appearance is, however, refuted in five steps. First, since *A* is directly perceivable, it is possible that some observer *x* directly perceives that *A*. Suppose, secondly, that I believe in the above uncontested theory of perception. Then I believe that, given that *x* perceives that *A, A* is a cause of *x*'s belief that *A* (where, however, we should exclude the case where the perception is my own present one). Thirdly, we may assume that whenever I believe that *B* is a cause of *C*, then *B* is also a reason for me for *C* (in the sense defined above), and vice versa.[31] This entails, fourthly, that under the condition that *x* perceives that *A, A* is a reason for me for assuming that *x* believes that *A*, and vice versa. If *A* is far-fetched, this condition will be far-fetched, too. Still, it is a posteriori and its falsity not maximally certain. Then, fifthly, some further mild assumptions[32] will turn the conditional reason relation into an unconditional one. Hence, the special coherence principle holds even for all directly perceivable or basic propositions.

How did I thereby avoid the two shortcomings noted at the end of (3)? First, I refuted the strange case of a basic proposition which is not a reason for anything else by specifying for each basic proposition another proposition for which it is a reason. And second, I think the foundationalist can concede that an effect is a reason to infer the cause, just as the cause is a reason to infer the

[30] Cf., e.g., Armstrong (1968, p. 234). I am well aware that steps (4) and (5) move on highly controversial grounds. However, in pursuit of the argument I want to give it may be legitimate to cut just one aisle through these grounds.

[31] Indeed, this assumption is a theorem of my theory of causation, given some weak restrictions; cf. Spohn (1991, p. 188 and notes 51, 54, and 55) [here: pp. 230f.]. Because of its plausibility I take this theorem rather as confirming that theory.

[32] Cf. again Spohn (1991, p. 188 and notes 54 and 55) [here: pp. 230f.].

effect, i.e., that at least in the case considered the reason relation is indeed symmetric.

(7) This may seem an improperly long-wided argument in favor of a fairly weak principle. The only excuse I have for proposing it is that I see no other argument extending to the general coherence principle as well. But now the extension is straightforward.

Consider any partition $\{V, U - V\}$ of the set U of all variables, and let $S = P(V)$ be the set of all propositions over V and $T = P(U - V)$ the set of all propositions over $U - V$. Then one possible case is that S contains all directly perceivable propositions and T none. In this case, however, the reason relation must relate S and T. Otherwise, nothing whatsoever could be found out about propositions in T, nothing could change my degree of belief in propositions in T. Again this contradicts the rationality postulate stated in (1). The same holds for the case where T contains all directly perceivable propositions and S none.

The final case is where both S and T contain directly perceivable propositions. Hence assume that $A \in S$ and $B \in T$ are directly perceivable. According to the above theory of perception, A is a cause of the fact that a given perceiver x believes that A, and B is a cause of the fact that x believes that B. Then, the trick goes, x also believes that $A \wedge B$. Both A and B are then partial causes of x's belief that $A \wedge B$.[33] Hence, if I believe in this theory of perception, then, as in (5), A as well as B are reasons for me for the proposition that x believes that $A \wedge B$, and vice versa.

Where, however, is the proposition that x believes that $A \wedge B$? It may not be totally clear which variables describe the doxastic state of x (at a certain time t). Let us try the two most plausible proposals. The most coarse-grained procedure would be to assume a single variable with a rich range consisting of all possible states x might be in (at t). But then both A and B are reasons to assume x to be in a certain doxastic state. There is thus at least one reason relation between S and T, since this rich variable must be either in V or in $U - V$. The most fine-grained procedure would be to assume a separate variable for each proposition taking all the possible degrees of belief of x (at t) as possible values. Then it is the variable for x's degree of belief in $A \wedge B$ which must be either in V or in $U - V$. And again there must be a proposition in S and another in T which are related by the reason relation. This finishes my proof of the general coherence principle.

Let me briefly sum up: I hope to have made clear the relevance of the two coherence principles discussed here and thus also the relevance of providing some argument for them. Clearly, I have offered only an argument sketch; but I believe that the steps and premises I have suppressed do not invalidate my argument. There were, however, a number of important premises. Some of them were linguistic, consisting in the explications of crucial notions I have used in the course of the argument. But there was also a substantial premise, namely, the rationality principle stated in (1). Moreover, I have introduced two assumptions. First, the proof of the special

---

[33] I use here "partial cause" for emphasis and not as a new term. Here, as in every-day language, "cause" always means "partial cause".

coherence principle assumed that the subject believes in the theory of perception mentioned in (5). Second, the extension to the general coherence principle additionally relied on (the subject's belief in) the capability of an arbitrary perceiver to form conjunctive beliefs. In this way the line of reasoning envisaged at the beginning of Section 10.6 and modified at the beginning of this section could be made to work. Whether this is a trivial or a significant result I do not dare to assess.

Chapter 11
How to Understand the Foundations of Empirical Belief in a Coherentist Way†

11.1 Introduction

The discussion between foundationalism and coherentism has been around for a long time, but for about two decades it has, in a way, become more serious than before, currently forming one of the central epistemological issues. It starts from the well-known justification trilemma which runs as follows.

Any rational subject is concerned with having rational or justified beliefs. Apparently, the only way to justify beliefs is to justify them with other beliefs which are in turn justified. This sounds obvious, but it immediately generates the trilemma: The claim that justifying beliefs have to be justified in turn triggers a regress leaving two unappealing options. Either the regress continues endlessly, in which case no one has any idea how the infinite regress could build up any justificatory force; or the regress turns back on itself, but then it seems puzzling how this circularity can avoid being vicious. Still, it is this second option coherentists venture to defend. There is a third option, namely to deny the claim generating the regress and to maintain that there are basic beliefs having justificatory force without requiring justification for themselves. This is the foundationalists' position which differentiates according to the kind of beliefs held as basic; the most usual variant is to take our perceptual or observational beliefs as basic, at least as far as our empirical beliefs are concerned.

I said that the present discussion is, in a way, more serious than before. This is so because the possibility of such basic beliefs, which many held to be obvious, has become more and more doubtful, and because coherentism has only recently found more precise non-metaphorical formulations which can escape the most obvious objections. In any case, I felt strongly confirmed in my coherentist prejudices by

† This paper was originally published in the *Proceedings of the Aristotelian Society, New Series* 98 (1997/98) 23–40. It is reprinted here with kind permission of the Aristotelian Society.

BonJour (1985) and others.[1] On the other hand, it became increasingly clear that the coherentists only have a chance to convince the foundationalists if they are able to provide a compelling account of the special role of those beliefs which the foundationalists erroneously describe as basic in their special sense. I am not fully satisfied by the existing attempts to do so, and therefore I would like here to add another attempt.

This will require two preparatory explanations. The first relates to my major discontent with the whole discussion, i.e. with the fact that the relevant epistemological notions such as justification, coherence, being a reason for, etc. usually remain relatively vague. BonJour (1985) excuses himself by pointing out that the clarification of these notions is not the particular task of the coherentist. This is certainly correct. Still, the discussion would be greatly helped, I find, if it were based on precise models of our doxastic constitution which captured at least the most relevant aspects. My main motivation for this paper is that I believe myself to be in possession of such a model, though this is not the place to introduce it. Instead, as a first preliminary I will briefly sketch the epistemological outlines of this model.

The second preliminary will be concerned with how I intend to account for the epistemological role of dispositional concepts within this model. The assumption that the whole world is in principle disposed to appear to us in perception will then immediately lead to what I have to offer as a coherentist account of observation.

11.2 Belief, Belief Change, Reasons, and Apriority

Epistemology has two parts: a theory of knowledge and a theory of belief. I am concerned with the latter which is certainly more basic because doxastic notions play a crucial role in the theory of knowledge.

What would a doxastic model, a theory of belief have to accomplish? Primarily, it would have to account for the statics and the dynamics of doxastic states; and it would have to do so not as a merely empirical theory, but from the perspective of a theory of rationality which leads a characteristic normative and empirical double life. The static part describes doxastic states as they rationally are at a given time; and the dynamic part describes how doxastic states rationally change over time.

Probability theory yields one very powerful model. It represents rational doxastic states as probability measures, the rational change of which is described by various rules, for instance, by the old and simple rule of conditionalization or by van Fraassen's very general reflexion principle.[2] The theory of ranking functions[3] which I developed 15 years ago yields another powerful model. Ranking functions behave

[1] Even though BonJour (1997) seems to turn away from coherentism.

[2] Cf. van Fraassen (1984) or Hild (1998b).

[3] Cf. Spohn (1988) and (1991) [here: chs. 1 and 9]. There I clumsily called these functions ordinal or natural conditional functions. Goldszmidt and Pearl (1992) introduced the term 'ranking functions', a terminology I like much better.

very much like probability measures in surprisingly many ways. They are less well suited than the latter in some important respects, but have one big and consequential advantage: They allow for a natural notion of plain belief which is difficult to capture within probability theory (as the famous lottery paradox makes clear). The notion of plain belief is also extensively dealt with in the belief systems of the AGM theory,[4] but I believe ranking functions exhibit a more general and satisfying dynamics than AGM belief systems.

In any case, whatever the precise theory, it seems that its dynamics cannot be stated without introducing both something like conditional doxastic states and something like degrees of belief (which need not be probabilities). This gives rise to a perfectly natural notion of reasons, i.e. of one proposition or belief content being a reason for another relative to a given doxastic state. Intuitively, what a reason would do if received is simply to strengthen the belief in for what it is a reason. In formal terms, this means that the proposition A is a reason for the proposition B in a given doxastic state just in case the conditional degree of belief in B given A is higher than that given non-A. In other words, the reason relation is just positive relevance.

Since this notion is of central importance, one must be aware of the fact that people talk of many different justificatory relations.

There is, first and most importantly, deduction, i.e. the notion that the premises of a deductive argument are reasons for the conclusion. Positive relevance embraces this notion; a premise is positively relevant to its deductive conclusions. However, positive relevance also admits inductive, non-deductive reasoning. And fortunately so; it seems fairly clear that deductive reasoning alone is insufficient for the justification of empirical beliefs.

There is, secondly, a causal notion according to which the reasons for a belief are simply those beliefs (or possibly other items) causing its acquisition and maintenance. This diverges from positive relevance in two respects. The two notions differ in their objects. Positive relevance is a relation between belief contents,[5] whereas the causal notion is a relation between belief state tokens (and possibly other items, thus opening the externalist strategy of seeking justification from outside). Moreover, they refer in different ways to the dynamics of belief. As explained, positive relevance is related to the rational dynamics of belief which actualizes itself in rational subjects, whereas the causal notion refers to the actual dynamics of belief, the rationalization of which still needs to be explained. These remarks also indicate how closely related the two notions are.

Thirdly, there are computational notions of reasons formalized in various kinds of calculi. They emphasize the process character of reasoning. They may have advantages, for instance, in explaining how mathematical assertions may be justified. They fall, however, on the other side of a fundamental chasm in the theory of belief. There are semantic theories of belief which seem unable to cope with what

[4] Cf. e.g., Gärdenfors (1988) or Gärdenfors and Rott (1995).

[5] Merely for stylistic variance I shall also speak of propositions or even of facts, though these terms have other uses as well.

Stalnaker (1984) calls the deduction problem, and there are computational theories of belief which are problematic in many other ways. Computational notions of reasons inherit these problems. By contrast, positive relevance, as explained, falls under the scope of semantic theories by conceiving of belief contents in a purely semantic way and not as syntactically structured.[6]

The multiplicity of concepts is certainly a main source of unclarity in epistemological discussions. The notion of coherence, or degrees of coherence, makes matters worse, insofar as its relation to the conceptions of reasons just mentioned is quite unclear in turn. The prominent notion of explanatory coherence is, however, well in line with my preference for positive relevance. If my argument in Spohn (1991) [here: ch. 9] is sound, the search for explanations is tantamount to the search for positively relevant reasons in a very specific sense.

Anyway, this brief discussion indicates why I think that the notion of positive relevance which embraces deductive and inductive reasons is the most appropriate for discussing empirical belief. I shall henceforth always refer to positive relevance when talking of reasons, of support, or of justification.

This move has grave consequences which are succinctly epitomized in the following observation: According to the deductive and the causal conception the reason relation is transitive, but not symmetric; and the same holds for most computational notions (though this depends on the specific calculus). In sharp contrast to this, positive relevance is symmetric, but not transitive! This already settles the dispute, in a way, for coherentism and against foundationalism, since it immediately opts for the circular dissolution of the justification trilemma and denies basic justificatory propositions, i.e., propositions which are reasons without having reasons.[7] Of course, a main task of this paper will be to make this conclusion credible.

Another central notion immediately springs from considering the dynamics of belief, the notion of apriority. It takes on two forms both of which will play an important role later.

In one sense, ‘a priori’ means ‘unrevisable’; apriority accrues to those beliefs, or generally to those features of doxastic states, which are unrevisable and hence necessarily and always present in doxastic states. The beliefs that I exist now, or that if *p* then *p*, are *a priori* in this sense. It is important to note that unrevisable beliefs like these cannot enter the reason relation. Nothing can change the status of an unrevisable belief, hence there cannot be any reasons for or against them, and since the reason relation is symmetric, they cannot be reasons for or against other beliefs.

In another sense, ‘a priori’ means ‘initial’. In this sense, apriority accrues to those doxastic states or parts thereof which initially obtain with respect to a given subject matter, i.e. before having any experience about it. This notion is not

---

[6] I have more fully discussed this chasm in Spohn (1997a). My main reason for sticking to semantic theories is that only they seem capable of capturing the normative aspect of theories of rationality.

[7] If relevance would be transitive as well as reflexive and symmetric, it would be an equivalence relation with either one equivalence class – in which case it would be absurdly universal – or several equivalence classes – in which case it would badly fail to yield coherence in any reasonable sense. So, it had better not be transitive, and it is not.

unproblematic,[8] but not useless, either. A priori probabilities – for instance, an equal distribution over the possible results of a throw of a die – exemplify this kind of apriority; default assumptions as studied in default logic may also be taken as an example. Clearly, what is initially present need not be forever, it is revisable or defeasible. Hence, the second sense of apriority is much weaker than the first.

Both notions have a rich history, as indicated by the examples. Recent discussions have focussed on *a priori* justification as the key notion since it seems to provide the only route to *a priori* beliefs. How does this relate to my notions? On the one hand, if *a priori* justification is to justify *a priori* beliefs, it can do so only in a computational sense. So, these discussions belong to another field. On the other hand, *a priori* justification is, in a way, easily subsumed under my notions. I have deliberately applied apriority to features of doxastic states in general. Thus, also justificatory or positive relevance relations can obtain *a priori* in each of the two senses; for instance, a premise is unrevisably positively relevant to its deductive consequences, and in Carnap's inductive logic certain initial positive relevancies had a central place. Indeed, such initial positive relevancies will play a crucial role in the sequel.

11.3 Dispositions and Reduction Sentences

Thus armed let me turn first to a topic *prima facie* unrelated: dispositions. We all know well enough what a disposition like solubility is:

(1) x is soluble if and only if x would dissolve if it were placed in water.

Being unsure, however, of the truth conditions of subjunctives logical positivists resorted to explaining dispositions with the help of reduction sentences, i.e., sentences of the form:

(2) If x is placed into water, then x dissolves if and only if it is soluble.

The logical empiricists at first thought reduction sentences were analytic. But they are not, as the case of dispositions with two or more characteristic manifestations made clear; a pair of reduction sentences may have synthetic consequences. Indeed, reduction sentences are, strictly speaking, false. They hold only *ceteris paribus*: the presence or absence of the characteristic manifestation is not a sure sign of the presence or absence of the disposition. So, (2) should be reformulated as:

[8] The main difficulty is this: Either, one takes 'initial' in an absolute sense in which it becomes something like 'innate'. But then it is quite obscure whether what is innate can be described in doxastic terms, e.g., as innate concepts. Or one relativizes 'initial' to a given subject matter (as I have implicitly done). But then one needs concepts for structuring the subject matter at hand, concepts which are to be acquired only through experience, and so the problem arises how to separate experience which is allowed to inform a (relatively) initial doxastic state from experience which turns the state into an *a posteriori* state.

(3) If *x* is placed in water and normal conditions obtain, then *x* dissolves if and only if it is soluble.

Indeed, the reference to normal conditions seems ubiquitous.[9] But what are they? We have to investigate, describe, and list them, but apparently they are only extensionally equivalent to such a descriptive list. Are they so defined as to make (3) true? Again no. The literal understanding is the best: The normal conditions are those conditions relevant to the case at hand which normally, or usually, obtain in our environment.

However, the real force of the reference to normal conditions emerges only when we place them into an epistemological perspective. Then it appears to be an *a priori* default assumption that normal conditions obtain, and the conditionals appear to express justificatory relations. In this way (3) turns into:

(4) Given that *x* is placed in water, the fact that *x* is soluble is an *a priori* reason for assuming that *x* dissolves (and vice versa).

Of course, the reason relation in (4) is not fixed forever. New facts can turn up in a given case on the basis of which solubility is not a reason or even a counter-reason for dissolving, and vice versa. For instance, it may turn out that the pot of water is already saturated with the stuff in question, or is exposed to unusual pressures, or is influenced by electromagnetic fields which hinder or further the dissolution. The space of further reasons, counter-reasons, and relevant conditions is to be explored only by empirical research. Still, this consideration already provides a more informative understanding of normal conditions: *they are just those conditions under which the reason relation (4) continues to hold.* To find out what they actually are is the task of an empirical investigation which ends with the required descriptive list, while being constrained precisely by the epistemological role of the normal conditions just given.

These observations obviously entail that the 'a priori' in (4) has to be understood in the sense of 'initially'. They also entail that the refined reduction sentence (3) is unrevisably *a priori*: If normal conditions are those confirming the relation between solubility and actual dissolving, any counter-reason to this relation must be an instance of non-normality; hence, (3) cannot turn out to be false. However, this is not to say that (3) is analytic. Following Kripke, I take analyticity to be *a priori* necessity. Thus to find out about the analyticity of (3), one would have to inquire into the metaphysical status of (3), but that would lead us astray. The unrevisability of (3), in turn, entails that the original reduction sentence (2) is *a priori* in the sense of being initially accepted, since we also believe prior to any investigation that normal conditions obtain.[10]

This analysis of solubility may suffice as an illustration of the machinery of reasons, apriority, etc. in a fairly uncontroversial case. The very same considera-

[9] Cf., e.g., Hempel (1988).

[10] All this is more fully explained in Spohn (1997c) [here: ch. 12] where I also consider the metaphysical side of the matter.

tions, however, apply to the far more delicate case of the foundations of empirical knowledge, as I shall argue in what follows.

11.4 A Thesis Concerning the Basis of Empirical Beliefs

As an intermediate step consider briefly secondary qualities. Who would not subscribe to the following assertion?

(5) An object *x* is red if and only if it looks red[11] under normal conditions.

Nevertheless, the status of this assertion is controversial. Does it define 'red'? Is it a necessary truth? I think everything I have said about solubility applies here as well.[12] The core of (5) is, again, *a priori* positive relevance:

(6) The fact that an object *x* looks red is an *a priori* reason for assuming that *x* is red (and vice versa).

Hence, as before, (5) is an unrevisable truth *a priori*, and without reference to normal conditions, it would express a defeasible belief *a priori*. However, (5) as it stands need not be analytic. Again this depends on the resolution of hidden ambiguities.

The next step will not be a surprise. Not only do some objects look coloured to us, the world incessantly appears to us in this and that way, at least as long as our awareness is directed outwardly. Thus we may generalize (6) to the following claim:

(7) The fact that it looks as if *p* is an *a priori* reason for assuming that *p* (and vice versa).

However, this formulation is too imprecise. Our discussion requires a more explicit version:

(8) The fact that it looks to person *x* at time *t* as if *p* is an *a priori* reason for person *y* to assume that *p* (and vice versa, given that *x* observes at *t* the situation in which *p* obtains[13]).

I believe that this claim is universally correct, i.e., correct in all its instantiations. The matter is extremely intricate, however, and my discussion is bound to be

---

[11] Obviously it is dangerous to use the crucial phrase 'looks red' without further comment. The way I understand it here will unfold in the following sections.

[12] In Spohn (1997b) [here: ch. 13] I more fully discuss the epistemological and metaphysical status of the various readings of (5).

[13] The 'given that' clause indicates a conditional reason relation; hence it is still within the scope of this relation. The clause is not really necessary, but it guards the 'vice versa' direction from *prima facie* objections.

incomplete. Before trying to defend (8), let me briefly discuss its general significance for our epistemological concerns.

There is a characteristic indecision among foundationalists when pressed to specify the alleged basis of our empirical beliefs. They oscillate between a physicalistic and a phenomenalistic basis. A physicalistic basis contains such propositions as 'there is a computer on the table in front of me' or 'the pointer points to 2.6'. They provide a common-sense basis, in the sense that they usually need no defense. Doubts concerning basic propositions of this kind are usually not answered by argument, but by the request to look again more carefully. Still, such doubts *are* often legitimate; hence, this kind of basis seems to be neither really certain nor really basic. So foundationalists are driven to a phenomenalistic basis consisting of propositions about sense-data. Though sense-data belong to the more problematic species in the ontological zoo, the intended propositions can be simply expressed in common-sense terms, for example, as 'it looks to me as if there is a computer on the table in front of me' or 'the pointer seems to me to be pointing to 2.6'. This kind of basis seems both really certain and really basic. It is affected, however, by the problem how to build anything on it.

Claim (8) brings the matter into a more plausible perspective, I think. It says how the two alleged bases of the foundationalists are related. It explains why the physicalistic base is not really basic, and how something can be built upon the phenomenalistic base. Because of the symmetry of the reason relation it also does the converse and says how phenomenalistic propositions are not basic, but can have reasons, a point to which I shall have to return. Thus, (8) fits into a thoroughly coherentist picture. The reason relation claimed in (8) provides a pervasive coherentist link as a crucial building block of our empirical world view from which further coherentist links spread to other propositions about the external world more remote from observation.[14] Experience may refine or even replace this building block in particular cases, but it is guaranteed to be initially present by its apriority. All of this is achieved without claiming any absolute certainties where there are none.

Claim (8) can also be viewed as an attempt to answer skepticism,[15] or a least one version of it, by showing that there is an *a priori* argument leading from assertions about our sense impressions to assertions about the external world. Nothing is thereby declared indubitable, and the argument is defeasible. But it is a good argument and generally applicable, and it is not prone to skeptical questions, but only to positive counter-reasons (which the skeptic refuses to provide). Obviously, however, this topic deserves much more scrutiny.

Thus, we have plenty of reasons to wish (8) to be true. Is it really true? Well, let us look at it more closely.

---

[14] The metaphor of spreading is, I find, nicely explicated in the theory of Bayesian nets (cf. Pearl 1988), which works for ranking functions just as well as for probability measures.

[15] This kind of attempt is launched by von Kutschera (1994).

11.5 Defending the Thesis

Five observations concerning (8) seem to be the most relevant.

First, the domain of the propositional variable p in (8) roughly consists of observation sentences such as 'there is a computer on the table in front of me' or 'the pointer points to 2.6'. This does not mean, however, that there is a distinguished observation language ('computer' would not typically belong to it). Indeed, I do not believe in such a language. Hence, in the absence of a more precise theory about the domain of the variable p, we should stick to our ordinary understanding of what can be observed or perceived.

Second, it should be emphasized that assertion (8) seems perfectly reasonable when x and y are different persons. In one direction (8) says that we initially trust the senses of others. If they make us believe, by credible assertions or whatever, that certain things looked so and so to them in a particular way, we also believe that these things were that way. This conclusion can only be obviated by particular counter-reasons.

The same holds for the opposite direction. If p is an observable state of affairs, as just assured, and if the person x is observing the situation in which p obtains, as presupposed in (8), then normally it should look to x as if p. Again, special reasons are required for assuming otherwise.[16]

Third, the case where $x \neq y$ is the epistemologically less exciting one. Only the case where x and y are the same person is relevant to the debate between coherentists and foundationalists. To a large extent, however, this is as unproblematic as the interpersonal case. To see why, let us look more closely at the temporal relations in (8). x's observation in which certain things appear in a particular way to him takes place at a certain time t. However, the *a priori* reason relation asserted in (8) is timeless, it holds for any initial doxastic state. Still, we can apply it to a given time t', since the initial reason relation is maintained at t' if the information available to the doxastic subject up to time t' is not unfavourable.

Again there are two cases: t and t' can be different times or the same time. Now it seems to me that the case where t and t' are different times is like the interpersonal case. If you are reasoning now about the relation between past and future facts and the ways past and future things appear to you, you are in a similar position towards your past or future selves as you are towards other persons. I cannot see a relevant difference.

So the hard case, as to be expected, is the case where x and y are the same person and t and t' the same time. I shall call this the reflexive case of (8). One might argue that the case cannot really occur, because as soon as we start to reason about or from how things appear to us, the appearance is already in the past, and we can

[16] This corresponds to the negative case discussed by BonJour (1985, sect. 6.3), where the subject infers the absence of a given external state of affairs from the absence of the corresponding spontaneous belief.

reason about it only via recollection. However, this argument sounds like a lame excuse. It would be more convincing to face the problematic case, not to deny it.

Fourth, to this end we have to take a closer look at the verb 'look'. How crucially it appears in claim (8) is clear from the fact that it ultimately fixes the domain of the variable *p*. Obviously all and only such *p* for which it makes sense to say that it looks to *x* as if *p* are allowed in (8). However, the treatment of this verb requires considerable delicacy, and here I cannot fully treat it. Let me make just two observations.[17]

On the one hand, the verb 'look' appears at least in three different constructions which are not equivalent. The fact that an object looks red to *x*, for example, is not quite the same as the fact that it looks to *x* as if this object were red. Again, the fact that an object looks like a car to *x* is not quite the same as the fact that it looks to *x* as if this object were a car. Still, in the context of claim (8) these constructions seem to be exchangeable. I do not see which difference it should make to my reasoning to replace 'look as if' by 'look like' or by 'look' followed by an adjective. Hence, my remarks are intended to cover the two latter constructions as well.

On the other hand, the verb 'look' has, according to Chisholm's familiar doctrine, three different readings: the epistemic, the comparative, and the phenomenal reading.[18] I need not decide whether the phenomenal or the comparative reading is more adequate. The important point is that assertion (8) cannot be maintained with the epistemic reading of 'look' in the reflexive case. The reason is this: If the phrase 'looking as if *p*' were defined solely in doxastic terms, as it is in the epistemic reading, then (8) would claim that second-order beliefs are inductive reasons for first-order beliefs, and vice versa. This, however, contradicts the widely accepted reflexion principle of doxastic logic. This principle says that it is logically true that I believe that *p* if and only if I believe that I believe that *p*, and thus it entails that the reason relations between second-order and first-order beliefs are deductive and unrevisable, not defeasible, as required by (8); there is no way to drive any wedge between first-order and second-order beliefs, as it were needed for (8) to be true in the epistemic reading of 'look'.[19] Hence, I have to reject the epistemic reading of 'look' as inappropriate for (8).

This conclusion may sound implausible. However, the impression of implausibility certainly derives from the fact that the verbs 'look', 'appear', and 'seem' superficially seem exchangeable, that their subtle differences, emphasized by philosophers,[20] are blurred in every-day language, and that at least the verb 'seem' has a broad usage in which it expresses in the first person, or describes in the third

---

[17] I have considered the matter more thoroughly in Spohn (1997b) [here: ch. 13].

[18] Cf. Chisholm (1957, ch. 4).

[19] I owe this point to Benkewitz (forthcoming, sect. 5.3). The point also marks a difference to BonJour (1985), who proposes in sect. 6.3 to justify observational beliefs with reference to metabeliefs.

[20] Austin (1962, ch. IV) gives a paradigmatic investigation of the differences.

person, nothing but a tentative or feeble belief. But 'look', or 'sound', *does* have a more narrow usage according to which nothing looks any way to the blind or sounds any way to the deaf, though things may well seem to them to be red or loud or some other way. It is this narrow use, the use according to which it could not look to x as if p unless x has a certain kind of qualia, which is intended in assertion (8).

I have argued so far that the fact that it looks to x as if p is a non-doxastic fact about x, and therefore is suited to enter x's own inductive reasoning. Yet danger threatens from another direction which is dealt with in my *fifth* remark.

It is often said that beliefs about introspective facts like 'this flower appears red to me' are infallible and unrevisable. This was the reason why many sought a phenomenalistic foundation of empirical knowledge. These beliefs are not *a priori*, of course, they do not exist all along. But once they have arisen, they seem unrevisable, at most they may be forgotten. If this were true, claim (8) taken reflexively would be in trouble again, because, as explained in Section 11.2, unrevisable beliefs cannot enter justificatory relations. Hence, claim (8) can fully be maintained only if beliefs about such introspective facts may be mistaken and confirmed or disconfirmed by other beliefs.

Indeed, they can be mistaken for a simple, but general reason. When I come to believe that it looks to me as if p, I subject my sense-impressions to a certain conceptual scheme or linguistic classification, and in this I may err. Austin's well-known example of magenta is a relevant case at hand.[21] But there are more far-fetched and dramatic examples to the same effect. A strong case can be built, I think, that there may be people with inverted qualia: red or reddish things look green or greenish to them, and vice versa. In fact, the hypothesis that such people actually exist has been seriously entertained on scientific grounds.[22] Of course, these pseudonormal people, as they are called, do not realize this. It is very hard (and presently unfeasible) to recognize pseudonormal vision. Hence they *believe* that red things look red to them as to normal people, though red things actually look green to them. Nevertheless, they may find reason to believe in their pseudonormality. This is, after all, a scientific hypothesis confirmable in indirect and complicated ways. Thus someone may indeed learn that the ripe tomato actually looks green to him, though starting with the firm belief that it looks red.

If such examples are telling, the alleged unrevisability of the relevant introspective beliefs is cleared away – even in the seemingly hardest case of beliefs about which color things look to us. So the last obstacle to accepting (8) in the reflexive case seems to be removed, and I conclude that (8) should be endorsed in full generality. This in turn seems to license us to proceed to the favorable and exciting conclusions sketched in Section 11.4.

---

[21] Cf. Austin (1962), pp. 112f.

[22] Nida-Rümelin (1993) and (1996) presents the hypothesis in more detail and thoroughly discusses its philosophical relevance.

11.6 The Foundationalist's Last Resort?

Many things would still need to be said. Some remarks comparing what I have said with other theories would be in order. It would be worthwhile to extend the applications of my notion of *a priori* reasons, as giving it some concrete work to do was, in a way, a major point of the paper. However, the philosophically most significant continuation is perhaps the following, which I would finally at least like to indicate.

The last argument may have raised the suspicion that I have not yet done full justice to foundationalism. The argument used the fact that 'looking red', for instance, is already a linguistic concept controlled by the linguistic community. That is, when I say that something looks red to me, I am *not* necessarily referring to my currently experienced phenomenal quality or kind of quality. Rather, even if my utterance is taken in the phenomenal reading, I am referring to that kind of quality the experience of which most people express by that locution, and which may or may not be the one I am currently experiencing. The question which quality this is, if any, is the source of doubt, reasoning, and error just exploited by my argument.[23]

However, if this observation about the semantics of 'looking red' is correct, it follows that the state of being appeared to *thus* – where the 'thus' is accompanied by a sort of inner pointing – is linguistically ineffable, even if the experienced quality actually is a specific shade of red. Yet the proposition that something looks or sounds *thus* to me can very well be believed. There are pure concepts of phenomenal quality, even if they are ineffable,[24] there is undoubtedly a purely perceptual memory which is not helped by linguistic concepts, and so there are such purely phenomenal beliefs.

Hence, there seems to be a third option for basic beliefs in the foundationalist's sense. There are not only physicalistically basic propositions expressed by observation sentences *p* or phenomenalistically basic propositions expressed by observation reports of the form 'it looks to *x* as if *p*', both of which we have treated from a coherentist point of view. There are also purely phenomenal propositions. Do they save the case for foundationalism? Let us see how the picture changes when we add these purely phenomenal propositions.

---

[23] The assumption that it is one and the same kind of quality which is mostly expressed by "looking red" is the presupposition characteristic of the phenomenal reading. The comparative reading does without it, and the epistemic reading even works in the case of missing qualia. Cf. Spohn (1997b) [here: ch. 13].

[24] This ineffability is nothing mysterious or even impossible. One has to observe here that the concept expressed by a linguistic predicate differs in general from the property denoted by it. Thus, the claim that purely phenomenal concepts are ineffable amounts to the fact that we have no linguistic predicates for expressing these concepts. At the same time, however, these phenomenal concepts *are* phenomenal properties, and as such they may well be, and presumably are, *denoted* by linguistic predicates.

First, it seems clear that propositions of the form 'it looks *thus* to *x*' are positively (or negatively) relevant to propositions of the form 'it looks to *x* as if *p*', and vice versa. That something looks *thus* to me strongly suggests, but as we saw, does not guarantee, for instance, that it looks red to me; and that something looks red to me strongly suggests, but again does not guarantee, that it makes me experience a certain kind of quality.

Moreover, I think that this positive relevance holds *a priori* (if the missing qualia case is excluded). For, I can acquire, for instance, the linguistic concept 'something looks red to *x*' only by associating it with some purely phenomenal concept. The association may turn out to be erroneous, but I have to start with it. Hence, it is defeasibly *a priori*; and it obtains as long as it is not defeated.

This, finally, raises the question whether the positive relevance even holds in the problematic reflexive case. Here I feel I have no choice but to admit an exception. If I attentively look at the scene before me and it looks *thus* to me, I believe at the same time that it looks *thus* to me, and I do not see how this belief could be supported or weakened by any reasons or counter-reasons. Has foundationalism thus got the upper hand at last? I think not, on two scores.

First, the indubitability of the belief that it currently looks *thus* to me is a genuine doxastic singularity. The indubitablity fades as soon as the belief turns into a recollection, and thereby it becomes accessible to doubt and reason. Hence, what I called a lame excuse above is perhaps a good excuse in this case.

Second, even if we grant the possibility of such momentarily indubitable beliefs, it would be a mistake to conclude that our empirical beliefs are ultimately based on them. Introducing the notion of direct perception, I can surely grant that the dynamics of our beliefs is basically driven by what we directly perceive. However, I take this to define what is directly perceived. It does not mean that only such purely phenomenal propositions were the objects of direct perception. On the contrary, very often we do not pay much attention to our phenomenal experience. I see, for instance, that I am standing in front of my car, I act accordingly, and I would have to *reconstruct* how it looked to me. Therefore, the proposition that I am standing in front of my car is what I directly perceive, it is the base or source of the belief change I thereby undergo, and at the same time it is open to reason and counter-reason.

This holds generally: The rules of rational belief change mentioned in Section 11.2 allow us to identify the source or base of each specific change. This source, I propose, consists of the propositions directly perceived, and as my example suggests, these propositions may or may not be purely phenomenal. If this very rough sketch of direct perception can be maintained, the coherentist picture still stands.

However, I am about to open a new and large chapter in the inexhaustible book of epistemology. I should refrain.

Part V
Concepts

Chapter 12
A Priori Reasons: A Fresh Look at Disposition Predicates[†1]

12.1 Introduction

As indicated by the title, this paper can be seen from two perspectives.[1] Good old solubility being my main example, it can be understood as another discussion of disposition predicates, which have been causing so much trouble since the times of logical positivism. More precisely, it can be understood as an attempt to analyse the meaning of disposition predicates in the setting of the new orthodoxy in the philosophy of language that has been initiated by Donnellan's (1966) theory of definite description, Kripke's (1972) theory of proper names, and others, which in my view was perfected by Kaplan's (1977) theory of characters and supplemented by Stalnaker's (1978) variant theory of propositional concepts.[2] This orthodoxy seems to throw new light onto disposition predicates. Since it is around for quite a while I am surprised that I could not find such an attempt being explicitly carried through.

Mainly, however, this paper is a study of apriority. This notion has received at least two markedly distinct meanings in the history of philosophy. On the one hand, it denotes necessary, unrevisable features of doxastic states, i.e., properties that inhere in all doxastic states of doxastic subjects. Beliefs in analytic sentences, in Kant's synthetic sentences a priori or in sentences like "I am here now" count as

[†1] The original paper is German: "Begründungen a priori – oder ein frischer Blick auf Dispositionsprädikate", in: W. Lenzen (ed.), *Das weite Spektrum der Philosophie. Festschrift für Franz von Kutschera*, Berlin: de Gruyter 1997, pp. 323–345. It is translated here with kind permission of the Walter de Gruyter Publishing Company. I am grateful to Ludwig Fahrbach for preparing a first version of this translation.

[1] I am indebted to Ulrike Haas-Spohn and Wolfgang Benkewitz; pondering their ideas and writings gave the main impetus for this article. Furthermore, I would like to thank Hans Rott, Jay Rosenberg and my former colleagues of Bielefeld for the valuable discussions I had with them.

[2] The appropriateness of the label "new orthodoxy in the philosophy of language" is exhibited in Haas-Spohn (1995, in particular ch. 1), and in Spohn (1992/93). The precise sense in which Stalnaker's theory is an important variant of Kaplan's is explicated in Haas-Spohn (1995, sect. 1.4, 2.1, and 3.9).

paradigms of this kind of apriority; their truth is considered to be a priori and unrevisable. On the other hand, the notion refers to features of a doxastic state that it has prior to any information about the part of reality considered by it and which may well change through such information.[3] The most prominent example is provided by a priori probabilities which one sought to characterize by a principle of insufficient reason or by various symmetry or indifference postulates, as most forcefully attempted by Carnap's inductive logic. Of course, such a priori probabilities may and should change through experience or information. I shall continue to label both notions with the word "a priori"; however, no confusion will arise from this.

The two notions of apriority have had a hard time in this century: the first because apriority was merged with analyticity by the influential logical empiricists and could thus not gain an independent role (in my perception this has radically changed only through Kripke 1972); and the second because the discussion on a priori probabilities displayed, rather than solved, the difficulties involved (as I shall indicate, it was only default logic which contributed an important new idea to this topic). Here I shall try to make progress on these notions by providing new clear instances of them and clarifying the relations between them. These new instances will arise in connection with disposition predicates – whence the two facets of this paper.

The plan of the paper thus is: Section 12.2 will introduce some epistemological preliminaries. In Section 12.3, I shall try to succinctly rehearse Kripke's notion of apriority within the Kaplanian setting, something that does not entirely seem to be common ground. Section 12.4 will rehearse the discussion on disposition predicates and reduction sentences. The upshot is Section 12.5 joining the topics of the previous sections and providing, among other things, a new characterization of normal conditions. Section 12.6 is an annex on the metaphysics of dispositions. Finally, Section 12.7 will briefly indicate, as an outlook, that the approach to dispositions developed here may have deep epistemological consequences.

12.2 Beliefs and Reasons

I have just talked of a priori features of doxastic states. This may have been slightly mystifying and needs to be cleared up first: Commonly one speaks of a priori beliefs (or sentences or statements expressing these beliefs). However, it is a feature or a property of a doxastic state that a certain belief is held in it; and if the belief is a priori, so is this feature. Now there surely are many other features of doxastic

[3] The phrase "prior to any information" may seem sufficiently clear, but is beset with notorious difficulties. For instance, the subject must be equipped with concepts for structuring the relevant part of reality in its prior doxastic state, concepts, however, which it cannot have acquired without rich wordly experience. Quine (1969a, p. 86), turns this difficulty into one of his arguments against the analytic/synthetic distinction. However, I hold this phrase to be useful despite such difficulties, as the examples to be given will show.

states. If, e.g., a doxastic state assigns subjective probability 1/6 to a certain outcome of a throw of a die, then this is not what is usually called a belief; but it is a property of this state that may or may not be a priori.

Indeed, it is useful to divide the features of doxastic states into static and dynamic features (as each science does with the objects it deals with): the examples just mentioned, and many more, belong to the statics of a doxastic state which deals with how the state is at a given time. The dynamics of doxastic states, however, is concerned with the way and the rules according to which these states change. These changes have many causes; besides inevitable forgetfulness the most important surely is that one gathers experience and thereby updates one's doxastic state.

It should be clear that here, as usually in philosophical contexts, doxastic states are considered not as a purely empirical phenomenon, but also in the normative perspective of a theory of rationality which tells how doxastic states should reasonably be and change.[4] This makes intelligible why reasons also belong to the dynamic features of doxastic states: Rationally we form those beliefs for which we get sufficient reasons, keep those for which the cluster of reasons and counter-reasons does not relevantly change, and give up those to which sufficient counter-reasons emerge. It is of utmost importance here not to conceive reasons narrowly as deductive reasons. In fact, for almost all our empirical beliefs, in particular those about the future, we only have non-deductive or, as I shall say, inductive reasons; this way of talking refers only to their feature of being non-deductive and is not meant to imply any specific so-called inductive method.

More precisely, I take the relation of one belief (content) or proposition to be a reason for another to be constituted by positive relevance: Talking of reasons makes sense only if beliefs can be conditionalized and come in degrees of firmness[5] – be they modeled as probabilities, as OCF-values as introduced in Spohn (1988) [here: ch. 1] (or ranks, i.e., values of ranking functions, as I prefer to say in the meantime[6]), or in some other way. Then one can express the perfectly natural explication that an assumption or proposition A is a *reason for* an assumption or proposition B in a given doxastic state if and only if A is positively relevant for B, i.e., if B is more firmly believed conditional on A than conditional on non-A.[7] And A is a *reason for B given C* if and only if A is positively relevant for B given C, i.e., if B is more firmly believed conditional on C-and-A than conditional on C-and-non-A. Of course, precise sense is given to this explication only relative to a precise model of

[4] In Spohn (1993b) I tried to explain the peculiar normative-empirical double life of rationality theory.

[5] The former indeed presupposes the latter, if my considerations in Spohn (1988) [here: ch. 1] are correct.

[6] Because Goldszmidt and Pearl (1992) have coined the term "ranking functions" for the OCFs.

[7] The reason relation thus explicated belongs to the dynamics of doxastic states because the conditional degrees of firmness to which the explication alludes are essential for describing the dynamics.

doxastic states; however, we shall get along here without introducing such a model in formal detail.[8]

In the sequel, apriority will be considered with respect not only to static, but also to dynamic features of doxastic states, not only in form of a priori beliefs, but also in form of a priori reasons – where apriority is taken in both senses. I have given examples for a priori beliefs (more will come). A first example for a priori reasons is provided by deductive reasons as given in analytically valid deductions. Later we shall see, however, that a priori reasons are not confined to analytic reasonings, but extend to inductive, non-analytic reasonings – a fact apparently insufficiently recognized by the philosophical community.

12.3 Kant, Kripke, Kaplan and Beliefs A Priori

Let us first take a careful look, though, at beliefs that are a priori in the sense of being unrevisable. In my view, these beliefs should be analysed in the framework of Kaplan (1977), which in turn can best be understood by recognizing how Kripke (1972), and with him major parts of analytic philosophy, did not catch up with, but nevertheless outstripped Kant. This sounds paradoxical, but it is not.

Surely, Kant's introduction and application of his two central dichotomies, the analytic/synthetic distinction and the a priori/a posteriori distinction, was of fundamental philosophical importance. Because of the non-existence of analytic judgments a posteriori these distinctions were not logically independent; nevertheless, and this is their point, they were not identical, either, due to the existence of synthetic judgments a priori. Equally surely, trying to come to terms with Kant's views filled subsequent generations of philosophers with despair.

In contrast to the concept of analyticity which is a key concept of the philosophy of language and therefore took center stage during the linguistic turn in analytic philosophy, the concept of apriority was marginalized, mainly, it seems, because the logical positivists did not know what to do with it and simply identified it with the concept of analyticity.

Some philosophers, analytic or not, may have recognized this defect early on, but it was generally realized only through Kripke's (1972) reinstitution of the a priori as an independent concept. Kripke did not catch up with Kant because his examples of synthetic sentences a priori to be mentioned in a moment are more or less banal. It is still not clear, for instance, how Kant's principle of causality, one of his cardinal synthetic principles a priori, can be tackled in Kripkean terms.[9]

[8] Still, it would be useful with respect to the later parts of my paper to have such a model in mind: either probability theory or the theory of ranking functions of Spohn (1988) [here: ch. 1] both of which allow a precise account of positive relevance.

[9] My own suggestion can be found in Spohn (1991, sect. 4) [here: sect. 9.4], where I deduce some versions of the principle of causality from certain coherence principles that should turn out to be a priori if interpreted in Kaplan's (1977) framework.

Nevertheless, Kripke also outstripped Kant. His resurrection of the a priori was but a by-product of his efforts that mainly concerned the reinstitution of the ontological dimension. The real focus of his lectures was the concept of ontological or metaphysical necessity. Something is ontologically or metaphysically necessary, if it could not be otherwise. For example, there is no way how it could be false that

(1) 2 + 2 = 4, or that
(2) Bachelors are unmarried.

Inasmuch as

(3) Water consists (mainly) of H_2O,

it could not be otherwise; nothing not consisting of H_2O could be water. Since

(4) Hesperus is identical to Phosphorus,

it could not be otherwise; whatever we counterfactually assume about Hesperus we counterfactually assume about Phosphorus; therefore their non-identity could not hold even counterfactually. Since

(5) You, dear reader, are human,

it could not be otherwise; you are necessarily human; if something is not human it could not be identical with you. However, the fact that you are presently reading this paper is not necessarily, but only contingently true; you could find an easier pastime.

On the other hand, a truth is epistemically necessary or a priori according to Kripke if it could not turn out to be otherwise than we assume it to be, i.e., if it is unrevisably and in this sense necessarily believed to be true. 2 + 2 could not turn out to be different from 4, and bachelors could not turn out to be married; hence (1) and (2) are a priori. Yet we could discover even today that water does not consist of H_2O, that Hesperus and Phosphorus are distinct, or that you are not human, but a robot (though we would have to make up fairly fantastic stories displaying these possibilities); hence (3)–(5) are a posteriori.

However, it could not turn out that I am not here. I could not find myself to be at a place now different from the place to which "here" refers when uttered or thought by me now; hence the statement

(6) I am here now

is a priori true. The same applies to the statements

(7) I exist now, and
(8) I am thinking now.

(6)–(8) exemplify contingent propositions a priori; they might be called Cartesian examples.

With the help of these notions, Kripke finally defines analyticity: a truth is analytic just in case it is both metaphysically necessary and a priori; more precisely, a truth is analytic iff its metaphysical necessity is a priori. Thus, 2 + 2 being 4 is analytic as is the bachelors' being unmarried, whereas all my other examples constitute synthetic propositions.

So far, Kripke confirms Kant who also did not recognize analytic a posteriori propositions and who also recognized the other three possible combinations. Nevertheless, Kripke and Kant differ fundamentally. The difference does not lie so much in their accounts of apriority which both conceive in a similar way: namely that a priori beliefs are held by epistemic subjects solely by virtue of their epistemic nature (although Kripke offers little more than a definition and a couple of intuitive examples, whereas Kant builds up an elaborate theory of judgments). The difference rather lies in Kripke's notion of metaphysical necessity which at bottom amounts to nothing else than a revival of Aristotelian essentialism and which is foreign to Kant. A symptom of this is that while Kripke is able to reproduce Kant's twofold dichotomy with his new necessary/contingent distinction and the shared a priori/a posteriori distinction, these distinctions turn out to be logically independent: there are a posteriori necessities such as (3), and there are a priori contingencies such as (6).

In order to appreciate how radical this upheaval was one has to recognize how the revival of apriority could emerge as a by-product of the revival of metaphysical necessity. This is due to the obstinate tendency to view the concept of meaning from an exclusively or predominantly epistemological perspective. This tendency was dominant in the 17th and 18th century in which theories of meaning were treated as mere appendages to epistemology. It can also be found at many places in this century, e.g. in the logical positivists' verifiability theory of meaning or in the tradition of use theories of meaning which fall victim to similar distortions. This tendency inevitably has the consequence of blurring the difference between analyticity and apriority. Only if one clearly recognizes the ontological and epistemological double dimension of the concept of meaning, as Kripke did by insisting on the notion of metaphysical necessity, can the concept of apriority regain its independence. These remarks apply to Kant as well because he regarded ontology (with the exception of the ineffable Ding-an-sich) as thoroughly interwoven with epistemology. This is the basic reason for Kripke's outstripping of Kant.

Still, the two-dimensionality of the concept of meaning only reached maturity with Kaplan (1977). Although Kripke was able to show that intensions as developed by Carnap (1947) in his intensional semantics are suited to capture the modality of metaphysical necessity, he could not offer a comparable theoretical framework for capturing apriority. This was provided only by Kaplan's (1977) theory of indexicals and demonstratives – a fact that has not been fully recognized until now because the amount of indexicality in natural language which can be accounted by this theory has been grossly underestimated by most philosophers and even by Kaplan himself.[10]

[10] Kaplan's own underestimation shows up in section XXII of Kaplan (1977) where he draws the negative, but, according to Haas-Spohn (1995, ch. 4), wrong conclusion that his theory is unable to account for the cognitive significance of proper names. The general underestimation shows up for instance in the fact that, as far as I know, Haas-Spohn (1995, ch. 3), is the first attempt to interpret the hidden indexicality of natural kind terms described by Putnam (1975) within the framework of Kaplan.

Kripke's dichotomies indeed agree perfectly with Kaplan's theory. According to Kaplan, the extension of any linguistic expression in principle exhibits a twofold dependency: it depends, as Kaplan puts it, on the context of utterance, and it depends on the world of evaluation or on the circumstances[11] of evaluation. The function describing both dependencies of the extension of a given phrase is called its character by Kaplan; for him, characters thus are the proper objects of a recursive semantics of natural language. In some cases these dependencies may run idle, but as already noted, this happens far less often than generally thought.

The twofold dependency is capable of capturing both of Kripke's dichotomies: The utterance of a sentence in a given context is necessarily true iff the sentence's character assigns "true" to it in that context and in every possible world of evaluation. Examples are utterances of (1) and (2) in every context and utterances of (3)–(5) in any context given by our world. Otherwise the utterance of a sentence is either contingent or necessarily false. Furthermore, a sentence is a priori true iff it is true in every context, i.e., if it can be only truly uttered or thought of, as exemplified by (6)–(8); in any other case the sentence is a priori false or a posteriori. Here it must be observed that the world in which a given context is situated is a component of this context; it may be called a context world, which, however, may also function as a world of evaluation. Thus a sentence is true in a given context iff the character of the sentence applied to the context and to its context world as a world of evaluation yields the extension "true".[12] This explication finally entails that a sentence is analytic just in case its character assigns "true" to it in all contexts and all worlds of evaluation. It is this embeddability of Kripke's dichotomies into Kaplan's theory of characters that is guiding me in the sequel.

Are there any further examples of a priori sentences beyond the Cartesian (6)–(8) (or a priori beliefs expressed by those sentences)? Kripke's main example is this:

(9) The standard meter is 1 m long.

Of course, this example has become outdated by modern means of length standardization. Apart from this it is obvious that this sentence is contingent: the standard meter might have been subject to changes in temperature and could thus have had a different length. At the same time it is a priori: whatever the length of the stick, it was defined as being one meter and, hence, could not be found to be different from 1 m.

If we accept this example, as I think we should in principle,[13] it becomes apparent that it is a special case of a general pattern which can be more clearly discerned in the following sentence:

(10) The first to climb Mount Everest first climbed Mount Everest.

---

[11] "Circumstances" is more general than "world", and rightly so. But for our purposes it suffices to conceive of circumstances of evaluation simply as possible worlds.

[12] See, e.g., Zimmermann (1991, sect. 1.2), or Haas-Spohn (1995, sect. 1.2) for a more thorough discussion of these concepts.

[13] Kripke's example involves some complications, though, which have provoked intense discussion (see, e.g., van Brakel 1990) and may therefore be not the best of its kind.

Sentences of this kind are systematically ambiguous. They have a so-called *attributive* (or *derigidified*) reading according to which the example is to be understood in the manner of:

(11) Whoever first climbed Mount Everest, first climbed Mount Everest.

According to Kaplan's theory, the definite description "the first to climb Mount Everest" if read as in (11) refers to the world of evaluation; the predicate "first climbed Mount Everest" refers to the world of evaluation anyway; thus, (11) says that the person who first climbed Mount Everest in the world of evaluation, did this there. Hence (11) is analytic (and thus a priori in a way that is presently of lesser interest).

However, sentences of this kind also have a so-called *referential* (or *rigidified*) reading that I take to be the preferred one; read in this way my example becomes:

(12) The person who actually first climbed Mount Everest first climbed Mount Everest.

According to Kaplan's theory, the definite description in (12) refers, as the word "actually" is meant to indicate, to the context and not to the world of evaluation. Read in this way, the sentence is not necessary; Sir Edmund who first climbed Mount Everest in our context world might have failed to do so in other worlds of evaluation for various reasons. Yet, (12) is a priori: (12) is true in every context; whoever turns out to satisfy the definite description in the context world we know of him in advance that he also satisfies the predicate there. Hence, the referential reading is contingent and a priori.

The discovery of this ambiguity, which is so important because seventy years of best analytic philosophy had been systematically blind for the referential reading, is mainly due to Donnellan (1966), even though he first described it as a merely pragmatic phenomenon. Its interpretation within Kaplan's theory as just outlined[14] seems to me to be commonly accepted, at least in linguistic semantics. Therefore I shall keep using the referential/attributive or rigidified/deregidified distinction in the general sense explained, i.e., in the sense of its projection onto Kaplan's character theory as displayed by (11) and (12). Having become fully aware of it, one finds it everywhere:

It is, for instance, connected with natural kind terms like "tiger", "water", "rose", "aids", "language". Consider the sentence:

(13) Water is the fluid we call "water".

Its behavior is just opposite to that of the "Mount Everest" example. It is analytic if the definite description "the fluid we call 'water' " is read referentially, whereas it is contingent and a priori, if read attributively. The attributive reading does justice to the fact that it is completely external to water how we name it and whether there are any beings around capable of talking about water. Moreover, it expresses our a priori knowledge; water cannot possibly turn out for us to be something different

[14] This interpretation is due to Stalnaker (1970) and Kaplan (1978); see also Heim (1991, sect. 1.3).

from what we call "water". Put more precisely: We may be wrong when taking something to be water and therefore calling it "water"; superficiality may even lead to wide-spread error of this kind. Nevertheless, most of what we, to the best of our knowledge, take to be water and hence call "water" must be water; we simply do not have any other basis for inquiring into the nature of water. Analogous remarks apply to other natural kind terms.[15]

Proper names can be treated in a similar fashion. Consider the sentence:

(14) Saturn is the planet called "Saturn".

If we read the definite description referentially, the sentence is analytic. Read attributively, it is another example of a contingent truth a priori. Again, our naming of the object is completely inessential to the object, but it is not possible that we discover that what we call "Saturn" is not Saturn.[16]

These explanations should by now have sharpened our senses for the phenomenon of the contingent (or synthetic) a priori and should have shown that its occurrence is wide-spread and systematic. Thus prepared, let us now turn to our main subject, namely to disposition predicates and the a priori propositions related with them. As far as I know, it has not yet been explicitly discussed in Kaplan's framework; moreover, it is more complicated and less clear-cut, yet it is full of interesting consequences of its own.

12.4 Disposition Predicates and Reduction Sentences

Disposition predicates like "soluble", "magnetic", "digestible", "red", "obedient" or "intelligent" denote properties true of an object if it exhibits a certain typical behavior in certain typical test situations; for example, an object is water-soluble just in case it would dissolve if it were immersed in water. How this is to be exactly understood this is one of the most notorious questions of 20th century philosophy of science.

In their program the logical positivists thought that it was epistemologically mandatory and hence also possible to explain everything not "given", i.e., predicates the application of which is not decidable by mere observation or sentences that describe unobservable states of affairs, in terms of the given, i.e., in terms of the observation vocabulary. For example, the predicate "water-soluble" does not belong to the observation vocabulary, because one cannot see immediately, but only in the relevant test situation, whether or not an object is water-soluble. Let us

---

[15] This thought, by the way, seems to lead to a direct justification of something like the principle of charity that Davidson (1984) takes to be axiomatic in his many papers about the theory of interpretation. Thus, the fact that most of our (reference-fixing) beliefs about a certain subject matter must be true need not be seen as a fact enforced by an external interpretive perspective, but can be directly attributed to the inner functioning of the terms.

[16] Both types of examples, which are in fact much subtler than displayed here are discussed at length by Haas-Spohn (1995, chs. 3 and 4).

assume for the sake of our example that "is immersed in water" and "dissolves" are observation predicates. If I now define:

(15) x is water-soluble just in case: if x were immersed in water, then x would dissolve.

did I not give the desired explication of water-solubility? No, at least not in a manner acceptable to logical positivists. The reason is that the subjunctive conditional in the definiens of (15) is not yet explained; hence it is not yet explained how the truth condition of "x is water-soluble" is determined by the truth conditions of "x is immersed in water" and "x dissolves". No other more acceptable explication was found; thus, it turned out that the program of the logical positivists could not be executed.[17]

Thereafter, a less demanding program was tried; the search for definitions of disposition predicates was replaced by the statement of so-called reduction sentences. Consider:

(16) If x is immersed in water, then x dissolves if and only if x is water-soluble.

Here, this is the decisive advantage, all if-then-constructions are to be construed in the truth-functional sense of classical propositional logic; no unexplained subjunctive "if-then" appears. Moreover, thus understood, the reduction sentence is apparently true. However, this is its decisive disadvantage, it only yields a conditional definition of water-solubility, i.e., no definition at all. Logical empiricism which succeeded logical positivism tried to turn this into a virtue by reducing the demands on reduction: what is not given needs to be only partially interpreted by the given, i.e., non-observation statements are somehow to be connected to observable facts in such a way that they acquire empirical relevance for observation statements, thus making them testable and confirmable by the latter. Indeed, any empirically-minded scientist would subscribe such vaguer assertions without hesitation.

A prominent example for partial interpretations is given precisely by the reduction sentences for disposition sentences. Due to the validity of the reduction sentence (16) claims about the water-solubility of an object become empirically relevant and testable; you only have to put the object into water. Moreover, the validity of the reduction sentence does not derive from empirical facts; for, if the reduction sentence itself would be something in need of testing or confirmation, the partial interpretation would become most puzzling again. Rather, its validity seems to flow directly from our understanding of the predicates involved; i.e., the reduction sentence is analytically true as it should be for a partial interpretation.

However, it soon became clear that this cannot be the whole truth. One point is the existence of so-called multi-track dispositions that have different characteristic manifestations in different situations. The predicate "magnetic" served as an example. Different pieces of magnetic material show different characteristic behavior if brought near compass-needles, iron cuttings, or coils; therefore "magnetic" is characterized by several reduction sentences of the form (16). If put together,

[17] Cf. Carnap (1936/37) or von Kutschera (1972, pp. 264–269).

however, these sentences have synthetic consequences, e.g., that anything showing a given behavior near compass-needles also shows a given behavior near coils. Hence, not all of them can be analytic; and since they have an equal claim to analyticity, none of them is analytic.[18]

Even worse is the fact that reduction sentences of the form (16) for single-track disposition predicates such as "water-soluble" are, strictly speaking, false. The reason is obvious; we can easily think of circumstances in which an object is immersed in water, but does not dissolve, and vice versa. The object might be soluble, but not dissolve, because the water is already saturated with the kind of stuff the object consists of; an exceptionally strong current in the water might lead to the dissolution of an insoluble object; and there are many more bizarre physical possibilities and impossibilities that may uncouple disposition and manifestation.[19] Hence, the reduction sentence needs a little softening:

(17) If x is immersed in water and normal conditions obtain, then x dissolves if and only if x is water-soluble.

It is as common as it is obscure to say that the reduction sentence is thereby subject to a ceteris-paribus-clause. Some[20] even think that the whole of empirical science is to be put under reservations of this kind. Hence, in order to reach a metaphysical as well as an epistemological assessment of (17) we obviously have to gain a better understanding of our vague talk of normal conditions.

12.5 Normal Conditions and A Priori Reasons

To begin with, it is patently the task of the empirical sciences not to be content with the vague reference to normal conditions, but to inquire them in more detail (and also to examine deviant conditions). Their inquiry will yield a possibly very long list of explicitly stated, specific conditions. But precisely because the inquiry is empirical the equivalence between this list and our talk of normal conditions cannot be analytic.

Another suggestion would be to claim that the normal conditions in (17) are by definition those conditions under which all and only water-soluble objects immersed in water dissolve. Reduction sentences such as (17) would then simply be analytic. As such this result is not wrong, as we shall see in the next section; but it would be wrong to derive it from the suggested definition which is too generous insofar as it

[18] Cf. Carnap (1936/37) or Stegmüller (1970, sec. IV.1.c).

[19] This was already noticed by Carnap (1956) – which is why he did no longer introduce disposition predicates via reduction sentences of the form (16), but instead developed what became influential as the received view of scientific theories; cf. also Stegmüller (1970, sect. IV.1.d), von Kutschera (1972, sect. 3.3), or Suppe (1977).

[20] E.g. Hempel (1988) who prefers to speak of provisos.

allows conditions to be counted as normal conditions that are completely crazy and make (17) true by chance.

The third idea is the most literal; according to it normal conditions, though vague, are exactly those conditions that normally, usually, mostly obtain. However, this formulation can be read either referentially or attributively. Which reading is appropriate?

According to the attributive reading normality is to be assessed in the world of evaluation; conditions are normal if they usually obtain in that world. However, the extension of "water-soluble" then varies to an unexpectedly high degree for different worlds of evaluation. If, for example, the world is such that strong currents usually flow through water there, then only those objects that dissolve in water of that kind are water-soluble; and again there are more bizarre variations. As I said, this is unexpected. Intuitively I would think that because in our world sugar usually dissolves in water, other-worldly, possibly numerically distinct occurrences of sugar still belong to the extension of "water-soluble", whatever the conditions of water usually are in that other world.

The deeper reason for this intuition is this: Normal conditions are not those usually obtaining in the entire universe, but those encountered in the small space-time region inhabited by us. This reference to the here and now of us, a big old language community, shows that the normal conditions of the context world, not those of the world of evaluation are the right ones. Otherwise, it would not make sense to ask what would be normal in this sense in a world of evaluation where humans or even the earth do not exist. If nevertheless the extension of "water-soluble" in such a world is to be neither empty nor undefined, as it certainly should be, then this can only mean that normality is to be grounded in our context world.

This entails that our talk of normal conditions must be understood in a referential way; normal conditions are those conditions normally, usually, mostly encountered in the context world – whatever it is. Hence, when empirical scientists of our world put together the above-mentioned list of explicitly specified conditions, this list, if complete and correct, is necessarily equivalent, in the metaphysical sense, with the normal conditions (a fact concealed by the inevitable vagueness of "normally", "usually", and "mostly").

This clarifies the epistemological status of the modified reduction sentence (17). If normal conditions are understood referentially, (17) is a priori in the sense of unrevisability; it cannot turn out to be false. For, wherever in the context world an object being immersed in water dissolves despite being insoluble or does not dissolve despite being soluble, normal conditions obviously do not obtain (which is not to say, as already indicated, that normal conditions would be defined in this way); this much we know even if our knowledge about the actual normal conditions is poor.

This result, however, does not yet completely capture the epistemological role of the talk of normal conditions. Up to now talk of normal conditions has remained disturbingly vague, as has proposition (17). How, then, is it possible that something as vague as (17) is known a priori?

This knowledge seems to me to be grounded in an a priori justificatory relation – which leads us full circle back to Section 12.2. Instead of saying that the reduction sentence (17) with its reference to normal conditions is known a priori, I propose the following reformulation which avoids referring to normal conditions:

(18) Given that *x* is immersed in water, the fact that *x* is water-soluble is an a priori reason for assuming that *x* dissolves, and vice versa.

I already explained in Section 12.2 how talk of conditional reasons is to be understood, namely as positive relevance under the given condition. But how is the "a priori" to be understood?

Apparently, the relation of being a reason a priori or of positive relevance a priori is not fixed forever. On the contrary, this is the point, relevant data can always emerge conditional on which the water-solubility of an object is not a reason for, and maybe even a reason against, its dissolution if immersed in water. The space of further reasons, counter-reasons and relevant conditions can only be unraveled by further inquiry. The relation of positive relevance being a priori therefore signifies that it only obtains initially as long as nothing else is known; we thus deal here with the second sense of the a priori explained in the first section.

From (18) it can be inferred that (16) is also a priori in this sense.[21] This yields the remarkable contrast that the reduction sentence is a priori in the sense of unrevisability if read as (17) and a priori in the other sense if read as (16). This contrast allows us to improve our understanding of normal conditions: they are precisely those conditions under which the a priori reason can be maintained and is thus confirmed. No claim is made about what the normal conditions actually consist in; it is up to science to find out about them. Yet their epistemological role is thereby sufficiently explained.

Finally, this account explains the apriority of (16) and (17). The assumption that normal conditions obtain is the a priori default assumption with which we start; hence, (16) is also a priori in this sense.[22] Furthermore, if the normal conditions

---

[21] This inference can be rendered precise in the setting of the theory of ranking functions; see Spohn (1991) [here: ch. 9].

[22] Goldszmidt and Pearl (1992) explain in which way the doxastic models referred to in footnote 8 (to which I have implicitly alluded all the time) are closely connected to so-called default logic.

are those under which the positive relevance stated in (18) continues to hold then no further conditions can invalidate this positive relevance; hence, (17) is unrevisable.[23,†2]

12.6 The Categorical Base of a Disposition

The epistemological considerations of the last section have been my main concern; they leave open, however, the ontological status of the reduction sentence (17). For the sake of completeness I want to address this issue, too, though I shall not reach a definite conclusion.

We have first to understand what the (categorical) base of a disposition is: It consists just in the intrinsic properties of an object that are responsible for the relevant behavior of the object.[24] In the case of water-solubility the base consists in the bonds holding together the molecules which an object is made of; these bonds have to be sufficiently tight to form a solid object, but not so tight as to withstand to the attack of H_2O dipoles. The digestibility of food is based on a very complex biochemical structure. The assumption that intelligence is simply based on a rich connectivity of the brain would be neuro-physiologically very naïve. Indeed I do not want to assert that we can identify a base for every disposition; for example, it is hard to see what the base of obedience should be.

---

[23] The account also explains why normal conditions should not be defined so as to make (17) true. There may be strange conditions under which the positive relevance of (18) does not hold any more. But then there may additionally be still stranger compensatory conditions under which the positive relevance of (18) is restored. According to the definitory approach this pair of conditions would belong to the normal conditions. Not so according to my account, and rightly so, it seems to me.

[†2] In Spohn (2003a, p. 177) I added footnote 16 at this point that seems worth appending here as well:

How (18) develops from (15) via (16) and (17) should have been sufficiently clear. However, it has turned out (cf. Martin 1994) that (15) is the wrong starting point. A disposition can be *finkish*, i.e., it can vanish just when and because it is put into a test situation in which it should prove itself. In such cases an object has a disposition even though the corresponding conditional (15) is false. (Cf. Lewis 1997 for more complicated cases.) Does this impair (18)? I don't think so:

We have to distinguish two cases. The first is that the test situation in which the disposition is finkish is rare. In this case (18) holds true, and we have to learn that the finkish situation belongs to exceptional circumstances in which the disposition is present, though the manifestation fails.

The other case is that the disposition of an object to show response R is always or mostly finkish in the test situations in which it might prove. In this case, it is perhaps not so clear whether the object really has the disposition to show response R (though Lewis 1997, pp. 147f., thinks so). Certainly, the whole arrangement does not have the disposition to display R (even if the object as such has the disposition which is then thwarted by the rest of the arrangement). But then (18) rightly applies again when its x is taken to refer to the whole arrangement.

[24] Cf., e.g., Armstrong (1968, pp. 85–88) or Prior et al. (1982).

The crucial question is now: what is the relationship between a disposition and its base? The simplest answer is to claim that they are identical, i.e., they are the same property.[25] Just as we should say that being water is the same property as consisting of H_2O, we might say that being water-soluble is the same property as having certain inter-molecular bonds; identities such as these are always metaphysically necessary.

Prior et al. (1982, sect. II) disagree (and Lewis 1986c, pp. 223f., partially concurs). One of their arguments (p. 254) is to insist that dispositions are defined by propositions such as (15). This does not solve our problem, however, which precisely consisted in finding out how best to understand (15).

Their first argument is that there conceivably are many different mechanisms showing the relevant behavior under suitable conditions. In this case, something would be water-soluble not by having the one and only base of water-solubility, but by having one of many bases. We have to distinguish two problems here:

First, the argument suggests the possibility that if we examine all water-soluble objects in our world we may find two or more causal mechanisms leading to their dissolution, which are so different that they cannot be captured by a unified description. Water-solubility then has two or more different bases in our world and can therefore not be identified with any one of them. However, this kind of problem affects any essentialist theory of identity. Surely, water in our world could have been found to be occurring in two different forms, e.g. as H_2O and as XYZ; Putnam (1975, p. 241) mentions the nice example of jade which indeed occurs in two different forms. In this case it is plausible to assert that jade essentially has one or the other form; and it certainly remains a vague matter to determine which number of different forms makes this response implausible. This does not yet decide whether the same position is acceptable in the case of water-solubility. Still, the problem is of a more general kind, not one of dispositions in particular.

So, let us put aside this problem by assuming that a disposition has only one base in each world. According to the argument of Prior et al. it is still possible, however, that the bases differ in different worlds. In this case, we can still maintain that "*x* is water-soluble" means the same as "*x* has the basis of water-solubility". But this is ambiguous in a manner well-known by now, the definite description "the base of water-solubility" can be read attributively or referentially.[26] Since the reference to normal conditions was already to be read referentially, one could think that this is appropriate now as well. I am not sure, though:

Consider a piece of sugar. In our world it is water-soluble. Now, move it into another world of evaluation. It does not change internally; and the same normal conditions obtain as do in our world. Yet, the piece of sugar does not dissolve in water in that world; somehow that world is governed by different causal laws, and

---

[25] Armstrong already argues for such a realistic understanding of dispositions; see his (1968, pp. 85–88).

[26] Prior et al. (1982) do not use the framework of Kaplan, but some of their formulations suggest that they have this ambiguity in mind, too.

hence different objects with a different internal make-up dissolve in water. Should we now say that the piece of sugar is water-soluble in that world as well, in spite of its not dissolving? This would be the referential reading. However, in this case the attributive reading seems to me to be more plausible; according to it the piece of sugar is water-soluble in our world, but not in that other world.

This response seems appropriate for one-track dispositions; dispositions of this kind seem to be especially tightly bound up with their characteristic manifestations. But other examples may be different. If we perform the same thought experiment with multi-track disposition predicates, e.g. with "magnetic", we might rather come to the conclusion that an object which is magnetic in our world is magnetic in this queer world of evaluation as well. The same is true for predicates whose dispositional character is less obvious, e.g. for the predicate "red". On the whole, the situation seems to me to be quite indeterminate.

This finally affects the ontological status of the modified reduction sentence (17). If "water-soluble" is read attributively as indicated, (17) is necessarily true, i.e., true in all worlds of evaluation. Hence, having already recognized (17) to be a priori, (17) turns out to be analytic. If, on the other hand, "water-soluble" is read referentially, which, as just remarked, may be appropriate for other disposition predicates, then (17) would be only true in all worlds allowed by our physical laws, but not necessarily true; thus, reduction sentences would turn out to be another example for contingent truths a priori.

12.7 Outlook

If my analysis of dispositions is reasonable and gives the notion of reasons a priori an essential role to play, then major epistemological consequences, in particular concerning coherentism and skepticism, seem to be entailed. As a kind of appendix, I want to briefly indicate how these consequences might come about.

To begin with, it is clear that secondary qualities are dispositions as well, i.e., dispositions to affect our senses and our minds in a particular way under suitable circumstances; indeed, I listed "red" as one of my examples. Hence, all of what I have said in Sections 12.4 and 12.5 also applies to secondary qualities; instead of the reduction sentence (16) I could equally well have discussed the reduction sentence:

(19) If I look at an object, then it is red if and only if it looks red to me.[27]

If (18) is a good analysis of what is intended by (16), then we may also proceed from (19) to:

---

[27] This would not have been advisable, though, since this example is burdened by particular problems. I say more about these problems in Spohn (1997b) [here: ch. 13] where I have more extensively discussed the epistemological and the ontological status of (19) within the same theoretical framework.

(20) Given that I look at an object, the fact that it looks red to me is an a priori reason for assuming that it is red, and vice versa.

Now, not only do some objects look colored to us, the whole world has the ever-lasting disposition to appear to us in this way and that way. This suggests a generalization of (20):

(21) Given that someone looks at the situation in which *p* might realize, the fact that it looks to him as if *p* is an a priori reason for others as well as for himself to assume that *p*, and vice versa.

It is a huge step from (20) to (21), but a promising one. If (21) is true it provides a coherentist link between a physicalistic and phenomenalistic base of empirical knowledge entailing that none of the two is really basic in the foundationalist sense; and it provides an a priori, though defeasible argument against skeptical doubts. Indeed, I believe (21) to be universally true.[28] However, it is all too obvious that it takes a long way to defend this in all its details.[29]

[28] In the attempt to guard against skepticism von Kutschera (1994) tried to establish analytic relationships between phenomenalistic and physicalistic statements. I could not see any such analytic relations. However, after replacing them by a priori positive relevance von Kutschera's claims suddenly appeared very plausible. And so I came to ponder (21).

[29] In Spohn (1997/98) [here: ch. 11] I attempt to more extensively provide this defense.

Chapter 13
The Character of Color Terms: A Materialist View[†,*]

This paper investigates the character of predicates like:

(A) $\lambda x(x \text{ is red})$ and
(B) $\lambda xy(x \text{ appears red to } y)$

where x stands for a visible object and y for a perceiving subject (the reference to a time may be neglected).[1] I take here "character" in the sense of Kaplan (1977) as substantiated by Haas-Spohn (1995, 1997). The point of using Kaplan's framework is simple, but of utmost importance: it provides a scheme for clearly separating epistemological and metaphysical issues, for specifying how the two domains are related, and for connecting them to questions concerning meaning where confusions are often only duplicated. All this is achieved by it better than by any alternative I know of.[2]

Therefore using this framework seems especially relevant to color talk where metaphysical and epistemological issues are more difficult to tell apart or may even

---

[†] This paper was originally published in: W. Künne, A. Newen, M. Anduschus (eds.), *Direct Reference, Indexicality, and Propositional Attitudes*, Stanford: CSLI Publications, 1997, pp. 351–379. It is reprinted here with kind permission of CSLI Publications.

* I am very much indebted to Wolfgang Benkewitz, Martine Nida-Rümelin, and Ulrike Haas-Spohn; to a large extent the ideas of this paper have emerged in long lasting discussions with them. I am also indebted to Galen Straws on for various helpful remarks concerning style and content. The research was supported by grant No. Sp 279/4-2 of the Deutsche Forschungsgemeinschaft.

[1] It may seem excessively correct to use λ-abstraction here. But I do so only because it will help me later on to avoid awkward English.

[2] The credit equally goes to Stalnaker. In fact, the epistemological usefulness of the framework stands out much more clearly in his work; cf. in particular Stalnaker (1978, 1987). Despite their mutual claims of distinctness the work of Kaplan and that of Stalnaker are so closely related that I feel justified in speaking of one framework; for the precise nature of this relation cf. Haas-Spohn (1995, sects. 2.1 and 3.9).

seem to coincide.[3] And it should help in particular with my more specific goal, namely to clarify the epistemological and metaphysical status of such statements as:

(1) x is red if and only if x would appear red to most English speaking people under normal conditions.
(2) x appears red to y if and only if x (appropriately) causes y to be in a neural state of the kind N.
(3) x is red if and only if the reflectance spectrum of the surface of x is of the kind R.

Indeed, I shall argue that (1) is analytic only in one reading and merely a priori in another reading. Moreover, I shall argue that after having set aside epistemological worries there is no good reason why one should not be able to be metaphysically conservative and to believe that (2) and (3) are necessarily, though a posteriori true for some N and some R, i.e., to sustain physicalism concerning colors and a type-type identity theory concerning color experiences; this is why I have characterized my views in the title as materialistic. Those who share these views anyway might still find it interesting to see how they fit into a broader theoretical framework; and those who oppose these views have to face the whole framework which appears to be successful on other scores. In any case, the framework should help both sides to more easily locate and clarify their divergence. Indeed, it was the main intention of this paper and the twin paper by Nida-Rümelin (1997) to exemplify this potential of clarification.

The paper starts with a presupposition and will then present six claims, the last three being the ones about (1)–(3) I have just indicated.

What I presuppose is simply the general adequacy and power of the framework of Kaplan and Stalnaker; I briefly recall its essentials as I use it here.[4] According to this framework, the right way of doing semantics for a given natural language is to recursively specify the character of all well-formed expressions of that language. The character of an expression is a function which assigns to each context the intension the expression has in that context, where the intension is a function from the set of index worlds or, more generally, from the set of indices into the set of categorically appropriate extensions. Thus, if a possible utterance of an expression is defined to be just that expression in a possible context, then the character of that expression may be represented by a two-dimensional scheme the rows of which show the intensions of all of its possible utterances.

There is wide agreement that intensions are suited for treating metaphysical modalities, in particular metaphysical necessity, but also counterfactuals, causation, and so forth. However, the two-dimensional scheme is also capable, though this is

[3] Almog (1981) is carried by the same enthusiasm concerning this framework. It is the only example I know of which explicitly takes this approach to analyzing color talk. But we differ in details, as will be seen below; moreover, in (1984) Almog withdrew his theory presented in (1981) and developed a new one without saying, however, how it applies to color talk.

[4] Cf. also Haas-Spohn (1997).

less accepted,[5] to generally account for epistemological modalities, apriority, linguistically expressible belief, and so on. How does it do that?

A preliminary point is that each context determines its associated index.[6] Thus, each possible utterance of an expression as such has not only an intension, but also an extension, namely the value of the intension at the associated index; in particular, each utterance of a sentence has a truth value. Following Stalnaker, I call the function which assigns to each context the extension an expression has in that context the diagonal of the expression; this function is, so to speak, the diagonal of the two-dimensional scheme that represents the character of the expression. It is this diagonal which does the epistemological job. The general reason is Stalnaker's, and it is very simple. Namely, whenever a speaker utters a sentence or a hearer hears one, they are not fully informed about the actual context; but in any case the speaker believes she says something which is true in the context and the hearer, if he accepts the utterance, believes he hears something which is true in the context. Thus, their epistemic attitudes are directed to possible contexts, that is, have sets of possible contexts (or the corresponding indicator functions) as their objects,[7] and it is the diagonal of the uttered sentence which represents their belief. Clearly, the belief expressed by speaking and acquired by listening is a belief *de dicto*. Consequently – and this is important – the diagonals of sentences are more specifically to be taken to represent the corresponding beliefs *de dicto* (cf., however, footnote 15 below).[8]

If this epistemological account is to work, a crucial hypothesis is required to hold: There is a stock of philosophical arguments showing that the intensions of sentences are (almost) never the objects of belief.[9] But for context-independent sentences having the same intension in every context the diagonal essentially coincides with the intension; and clearly, sentences built from context-independent expressions are in turn context-independent. Therefore, if the diagonal is to perform its epistemological job, most expressions must be context-dependent.

---

[5] Kaplan, for instance, does not fully believe in it; cf. his skeptical remarks in (1977, sect. XXII).

[6] If an index consists only of an index world, the index associated with a context is just the world of that context. The same holds for less simply conceived indices – as long as for each index parameter there is a corresponding context parameter (that this is so is a substantial semantic claim).

[7] This idea is briefly indicated in Lewis (1983, p. 230), and further developed in Haas-Spohn (1995, ch. 2), and Spohn (1997a).

[8] As will become clear, this marks a basic difference between this paper and Nida-Rümelin (1997). Nida-Rümelin holds that in the special case of utterances of sentences like "the sky is blue" the normal speaker expresses the phenomenal, as she calls it, as well as the non-phenomenal belief that the sky is blue; this agrees with her diverging explanation of the character of color terms. By contrast, I think that also in this special case the belief primarily expressed is only the belief *de dicto* (which roughly, though not fully corresponds to what she calls the non-phenomenal belief), and that the ascription of any further beliefs to speakers on the basis of their utterances is licensed only by additional background assumptions which may or may not hold.

[9] The best known references are, of course, Putnam (1975), Kripke (1979), and Burge (1979).

To arrive at the same conclusion in a slightly different way: A sentence is a priori (true) if and only if its diagonal assigns truth to all possible contexts; if neither it nor its negation is a priori true, it is a posteriori or informative. Clearly, a context-independent sentence which is necessary in one context must also be a priori and in fact analytic, i.e. true in all contexts and at all indices.[10] However, many necessary sentences are informative and not a priori, let alone analytic. Therefore, again, most expressions must be context-dependent.

Hence, for this hypothesis to hold true usual context-dependence as in indexicals and demonstratives is not enough. One has to interpret Putnam's hidden indexicality of natural kind terms as dependence on the context world as it is here understood,[11] and one has to find context-dependence in other predicates, for instance in Burge's examples, and in names as well.[12] For the same reason it will be crucial to find out whether the color terms (A) and (B) are hidden indexicals, i.e. dependent on the context world; this is the only way to tell whether their metaphysics and their epistemology can be treated distinctly in our framework.

So, how then do we determine the character of a given expression? First, we find out what is known a priori about the extension of the expression; in principle, we can do this with good old Cartesian methodical doubt. Having done this, we know the diagonal of the expression; we thus have one entry in each row of the two-dimensional scheme. From that entry we project the entire row, that is an extension for all of the other indices. The vehicle for doing this is what may be called the essentiality convention pertinent to the expression. This convention specifies for each context what is essential for the extension of the expression and thus allows to project it onto other indices. It must indeed be assumed that the linguistic community has such an essentiality convention for each of its referring expressions.[13]

A final preliminary point: My epistemological talk is quite loose in an important respect. Usually, belief, apriority, informativity, etc. are notions applying to individual subjects; something is believed by, or is informative to, a given individual. On the other hand, I have explained a character to be that of a given natural

---

[10] Or, equivalently, a sentence is analytic iff its necessity is a priori. This is Kripke's notion of analyticity in (1972).

[11] How this may be done is explained in Haas-Spohn (1997).

[12] Kaplan was skeptical of the generality of the epistemological strategy (which he had invented for demonstratives) precisely because he denied the context-dependence of names. And Almog withdrew his (1981) precisely because he had there misidentified the context-dependence of names; cf. Almog (1984, pp. 10f.).

[13] For details see Haas-Spohn (1995, sect. 3.5). However, the point is easily explained with Putnam's "water"-example: It is a convention of the English speaking community that "water" is a natural kind term denoting a substance, if there is a single substance underlying most of what we call "water", or any mixture of a few substances, if there are few substances underlying most of our "water"-paradigms, or anything sharing certain superficial characteristics, if no underlying physical structure can be found. This is the English essentiality convention for "water" as Putnam (1975) describes it; and the context world then tells which of the possible cases for which the convention is prepared becomes relevant and thus what is water in other possible index worlds.

language (or, more precisely, of a given and maybe changing state of that language). This entails that all the epistemological notions just derived from the character must be taken as applying to the given linguistic community as a whole and not to any of its subjects; the a priori is that of the linguistic community; informativity is measured by communal standards; etc. Such communal epistemic states are certainly a vague matter, but not worse than meanings and languages; and when talking about the latter, we certainly cannot avoid talking of the former.

There is a certain tension between the individual and the communal notions. Indeed, the tension is irreducible, since I take the communal epistemic state not as a kind of average of all the individual epistemic states or as something like Putnam's stereotype, which may be assumed to be embodied in most or all competent individuals, but rather as a kind of sum of the individual states, as consisting of what is recognized by the community as the best knowledge available to it, which need not be embodied in any individual. If, nevertheless, one wants to stick to the sketched framework, the conclusion is that it has to be doubled, i.e. to be developed on a communal as well as on an individual level, including an explanation of how the two levels relate.[14]

However, all this seems unnecessarily complicated for the present purpose. Therefore I will be deliberately sloppy concerning the two levels, or, rather, my account will explicitly refer to the communal level while pretending – although this is, strictly speaking, false – that it equally applies to the individual level.[15] It seems to me that my account is not essentially affected by this sloppiness; but this is a claim I do not attempt to prove here (even though Nida-Rümelin 1997 may throw doubt on it).

So much about the framework I am presupposing. How does all of this apply to color talk? I shall unfold this in a series of claims:

Claim 1: Color terms like (A) $\lambda x(x \text{ is red})$ are hidden indexicals.

This looks implausible. Our standard example for a hidden indexical is "water", and at first sight "red" seems to be quite different from "water". We all might say to the very best of our knowledge: "This is water", and we might still be wrong, because the alleged sample of water may differ in essential aspects from other samples; water has a hidden nature. On the other hand, if we all say to the best of our knowledge: "This is red", then that object is red. There seems to be no hidden nature to be found in red things which would separate between genuine redness and fake redness.

However, this is not quite true. Though redness seems to have an overt nature, it does not show it under any circumstances. One's individual color judgment can be

[14] This is elaborated in Haas-Spohn (1995, sects. 3.8–3.9); it is here where the crucial difference between Kaplan and Stalnaker unfolds.

[15] In particular, this remark modifies my claim that the diagonal of a sentence represents the corresponding belief de dicto. This is correct only if "belief de dicto" is taken in the unusual communal sense; the beliefs which individual speakers express by utterances are, strictly speaking, not these diagonals. What they do express can be correctly accounted for in the just mentioned doubling of Kaplan's framework.

mistaken; and there is in principle also the possibility of collective error. The light may be strange; there is a whole set of optical tricks and delusions; there is collective madness; and so forth. Thus, the colors show their seemingly overt nature only under normal conditions, and the point is that these normal conditions have a hidden nature. This is most easily and clearly demonstrated, with a familiar type of argument, for the normal conditions concerning illumination.

There is not only daylight and twilight, but also twinlight. Twinlight looks as white and bright as daylight, and under twinlight all the things familiar to us look the very same color as under daylight. Thus, without modern physics we could not tell apart daylight and twinlight, and perhaps even present physics does not yet help. Now imagine that in some possible world there is a kind of objects which we have not encountered so far; let us call them modaleons. In daylight modaleons look deep blue, in twinlight they look glaring red.

In contrast to what Nida-Rümelin (1997) prefers from her point of view, it would not be appropriate, I think, to say that modaleons change color when the index world changes normal light. When talking counterfactually about changing light we would not say, for instance, that sun-flowers would be orange if a huge red filter were fixed between the sun and the earth; rather we would say that they look orange under these circumstances, though they still are yellow. Similarly, we would say that modaleons, which are actually blue, would still be blue, but look red if the world were filled with twinlight.

Consider now different context worlds with different normal light; for all we know the context world we live in may be filled with daylight or with twinlight. If the foregoing is granted, then the modaleon case clearly shows the extension of color terms to vary with the context world. Viewed from a context world filled with daylight, modaleons are blue, whichever index world they inhabit; viewed from another context world filled with twinlight, however, modaleons are not blue, but red. So, this example shows the hidden nature at least of the normal lighting conditions and thus at the same time the context-dependence of the predicate $\lambda x(x \text{ is red})$.

This remote reason for the context-dependence of color predicates of the type (A) vanishes, if we turn to color predicates of the type (B); how things look to us at a given moment does no longer depend on such normal conditions. Thus, we might expect that terms of type (B) are not context-dependent; this would also conform to the traditional view that we cannot be mistaken about which color something looks to us at a given moment. But contrary to this I contend:

Claim 2: Color predicates like (B) $\lambda xy(x \text{ appears red to } y)$ are hidden indexicals.

The reason is basically that there are what I take to be clear cases falling under the heading "inverted qualia". For better explanation I have to introduce a very coarse piece of current color perception theory. As is well known, the human retina contains a lot of cones each of which is equipped with one of three kinds of pigments. All three pigments are sensitive to large parts of the visible spectrum, but in varying degrees. The maximal sensitivity of the pigments lies, respectively, in the red, the green, and the blue segment of the spectrum. So, the pigments are called R-, G-, and B-pigments; and accordingly, the cones containing them are called R-,

G-, and B-cones. A decisive link between the activity of the cones triggered by the incoming light and the color sensation is now provided by the so-called opponent process theory. According to this theory, the activity of the R- and the G-cones is compared closely behind the retina. The more the activity of the R-cones outweighs that of the G-cones, the more reddish is the color impression; and vice versa. Moreover, the activity of the R- and the G-cones is summed up and compared with the one of the B-cones. Again the more the sum outweighs the activity of the B-cones, the more yellowish is the impression; and the more the activity of the B-cones preponderates, the more bluish is the impression.[16] It is important not to get confused here about the classifications underlying the labels R, G, and B. The pigments so labelled are classified according to their chemistry.[17] By contrast, the opponent process theory offers a functional criterion for classifying cones as R-, G-, and B-cones; they are so classified according to their subsequent wiring.[18]

One of the many explanatory achievements of the opponent process theory is that it can explain dichromatism or red-green blindness. The explanation is simply that for some reason both the R- and the G-cones contain the same pigments so that their activity is always the same and no impression tends to be reddish or greenish.[19]

Now, Piantanida (1974) had a special hypothesis about dichromatism. Obviously, red-green blindness may come in two forms; either the R-pigments are contained also in the G-cones, or the G-pigments are contained also in the R-cones. Piantanida conjectured, very roughly,[20] first that both forms are due to genetic defects, secondly that these defects are located on different genes and are thus statistically independent, and thirdly that there is consequently a slight chance of suffering from both defects. For male persons this chance is about 1.4 per thousand. Would such a male be color-blind? No; his discriminatory powers are precisely as fine-grained as ours, only his reddish and greenish impressions are reversed. Such persons are called pseudonormal. Obviously, it is very difficult, if not impossible without violating bodily integrity to find out about pseudonormality, even for the pseudonormals themselves. But perhaps you, dear reader, are one of those! It is not so unlikely; for instance, about 58,000 of the 40 million male Germans would be pseudonormal, if Piantanida is right!

I do not know the scientific fate of Piantanida's hypotheses, and I cannot assess their scientific plausibility. But clearly, they make perfect sense, they are testable,

---

[16] The details are quite complicated, however, and empirical research is extremely difficult; cf., e.g., Boynton (1979, chs. 7 and 8).

[17] In fact, there occur not only the three normal forms, but also a number of chemical variations; cf. Boynton (1979, ch. 10).

[18] Due to their symmetrical role the issue of distinguishing R- and G-cones is quite subtle, however; cf. Nida-Rümelin (1997) for more detailed considerations.

[19] Note that this explanation presupposes the independence of the classifications of pigments and cones which I have just stated.

[20] For details, see also Boynton (1979, pp. 351–358).

and they might well turn out to be true.[21] The crucial point is how we should talk about pseudonormals. I find it very clear that the right way to talk about them is just as I did, namely that their reddish and greenish sensations are reversed; thus, red peppers look green to them and green peppers look red to them. I would not know how to conclusively refute those who refuse to talk that way, but it will become still clearer in the course of the paper that this is indeed a meaningful way of talking.[22] One may also sense an ambiguity and think that it is equally appropriate to say that red peppers look red to pseudonormals, that is, look as red things look to them. I shall discuss this alleged ambiguity in a moment; but the primary sense of "looks", and the one I am presently referring to, is the one in which red peppers look green to pseudonormals.

Now I am finally prepared to explain the context-dependence of the term (B) $\lambda xy(x$ looks or appears red to $y)$. Take a situation in which someone with G-pigment in his R-cones and R-pigment in his G-cones looks at a ripe tomato. Viewed from our actual context world where most English speaking people have R-pigment in their R-cones and G-pigment in their G-cones, that person has a deviant color perception, and the situation must be described as one in which the ripe tomato appears green to him. Viewed from a context world, however, in which most English speaking people have their pigments reversed,[23] that person is perfectly normal; and the situation must be described as one in which the tomato appears red to him. Thus, to conclude, the truth value of "that tomato appears red to this person" as applied to one and the same situation varies with the context – whence the context-dependence of appearance terms.

Is that meant to say that you may be mistaken when you, well educated, fully attentive, and absolutely sincere, as you are, say: "This tomato looks red to me"? Yes, precisely. Unbeknownst to you, you may be pseudonormal, and your utterance may thus be false. The point of the argument is simply that the application of $\lambda xy(x$ appears red to $y)$ is relative to a standard of normal vision, that the context world sets this normality standard, that the nature of this standard is unknown, and that no one knows for sure whether he conforms to that standard or not.

This seems to make the doubtful presupposition that there is a standard of normal vision. Is it not possible that Piantanida's statistics is wrong and that, say, a third of the population is pseudonormal? Surely; in fact, if one looks at perception experiments, one sees a surprisingly large variation in human color perception.[24] But I do

---

[21] Hilbert (1987, p. 92), seems to be the first to have mentioned pseudonormality in the philosophical literature; but apparently only Nida-Rümelin (1993, ch. 4 and 1996) fully realized its philosophical significance.

[22] The point is more fully argued by Nida-Rümelin (1996).

[23] Clearly, this is a possible context world. Which kind of biochemical substance is in which kind of so-and-so connected cones of most English speaking people is a contingent matter about which we need not have any knowledge.

[24] Cf. Boynton (1979, ch. 10), and Hardin (1988).

not need this presupposition, just as Putnam need not presuppose that water, or jade, for that matter, is just one substance.[25] On the contrary, our essentiality convention for appearance terms responds flexibly to various empirical findings.

For further explanation, I would like to relate this point to the familiar view due to Chisholm (1957, ch. 4), that appearance terms have three different readings, a phenomenal, a comparative, and an epistemic reading. This is my

Claim 3: The phenomenal, the comparative, and the epistemic interpretation of λxy(*x* appears red to *y*) are not three different readings; they rather reflect the context-dependence of this term by being appropriate in three different kinds of contexts.

Let me briefly recall these three interpretations:

(E) According to the epistemic interpretation, "*x* appears red to *y*" says as much as "in the absence of counter-evidence, *y*'s encounter with *x* tends to produce *y*'s belief that *x* is red".[26]
(C) According to the comparative interpretation, "*x* appears red to *y*" means "*x* looks to *y* in the way red things usually look to *y*".
(P) For the phenomenal interpretation, finally, there is no such paraphrase; there "*x* appears red to *y*" holds only if *y* has a specific common type of qualitative experience.

We have seen[27] that according to our essentiality convention for "water" the essential properties for being water depend on the actual properties of the many "water"-paradigms we have in the context world – whence the context-dependence of "water". The very same is true of "appearing red", as these three interpretations reflect:

Imagine *case 1* which I take to be actually obtaining: In this case there are few people with deviant perceptual capacities; there are few color-blinds and few or no pseudonormals. There may be variations; the sensitivity of the pigments may slightly differ in different people; the neurons comparing the activities of the cones may not respond in a completely uniform way; and so on. But on the whole most people have a roughly equal functional and physiological arrangement of the visual apparatus including higher brain regions. In that case, we would apply $\lambda y \bigvee x$(*x* appears red to *y*)[28] only to those normal people whose visual system is in a certain state; we could apply it also to some deviant people, if their deviation is as simple

[25] Cf. Putnam (1975, pp. 239–241).

[26] Or in Pitcher's more careful words: "*y* causally-receives, by means of using his eyes in the standard visual way, the (perceptual) belief, or an inclination to have the (perceptual) belief, or a suppressed inclination to have a (perceptual) belief, that *x* is red"; cf. Pitcher (1971, pp. 85–95).

[27] Cf. footnote 13 above.

[28] This is to replace the awkward colloquial phrase "is appeared red to" introduced by Chisholm (1957, p. 62), by a less awkward formal phrase.

as that of pseudonormals. But we would not further extend the application. In that case, i.e. in such context worlds, the appropriate interpretation of $\lambda xy(x$ appears red to $y)$ is the phenomenal one in which it involves a particular phenomenal quality.

Now compare this with *case 2*. Its simplest version is that there are so many pseudonormal persons that they cannot be dismissed as deviant; there are just two normal kinds of visual systems. In that case, each group can claim with equal right that ripe tomatoes, for instance, look red to its members; it would have no point if the members of either of the two groups insisted that tomatoes look red only to them. Thus, $\lambda xy(x$ appears red to $y)$ does not involve a certain phenomenal quality in this case. This is particularly clear from the fact that in this version objects appearing red to one group produce the same phenomenal quality as objects appearing green to the other group. It is still clearer in cases where there are many human visual systems which even the most advanced future science is unable to match; then the phenomenal qualities experienced by our fellows would be just as foreign to us as those of the bat. Still, color talk miraculously runs as smooth as it does. So, these would be cases or contexts in which the comparative interpretation of $\lambda xy(x$ appears red to $y)$ is appropriate; $\lambda y \bigvee x(x$ appears red to $y)$ would then be applicable to all beings having qualitative experiences which somehow enable them to discriminate and classify red things as we do, even though this ability would remain mysterious.[29]

There is the even less demanding *case 3*, the absent qualia case. It seems perfectly imaginable that some individuals behave in the very same way as we do without having any phenomenal experience at all. Why should computers be able to pass the Turing test only if they have built in sensations? Think also of such things as blind-sight where people with a specific brain damage behave towards objects similarly as normal people do, but are unable to report any conscious visual experience.[30] If this is imaginable, it might turn out to hold in the context world.

[29] It may be that I have overestimated human uniformity and that human vision is so varied as to rather fall under Case 2; this is an empirical question (possibly undecidable due to vagueness). However, Strawson (1989) argues, I understand, that Case 2 yields the appropriate description of the meaning of "red" in any case. This is the main point where I do not agree. In an important argument (sect. 6) he considers Monet and Renoir color vision (which is analogous to normal and pseudonormal vision) and asks whether the meaning of "red" changes when English gets smoothly translated into the language of a population with Renoir vision (or when the share of Renoir vision among English speaking people slowly increases from 0 or 1 to 99 or 100 percent). His answer is: surely not; and the reason seems to be that there cannot be meaning changes which nobody noticed. However, if meanings are explicated as characters there can be unnoticed meaning changes, as is carefully explained in Haas-Spohn (1997, sect. IV). Think again of "water" (which is less confusing than "red") and of Putnam's twin earth: It makes a difference whether we travel there before or after being able to distinguish between H_20 and XYZ. If we travel there after having this ability, XYZ never gets into the extension of English "water". But if we travel there before (and do this very often and develop a close interchange with twin earth), then the character and indeed the extension of English "water" has changed; at the outset XYZ did not belong to it and later on it does. Strawson apparently does not observe this difference.

[30] Cf., e.g., Weiskrantz (1980).

You are presumably quite sure that you have phenomenal experiences. But perhaps you are one of the very few gifted people; the normal case may be to have no visual experiences at all, but to talk as if one had some. But we would still have our usual color talk. And we would still have beliefs; in some mysterious way our beliefs are pushed this way and that way by our encounters with the things in the world. This then is the way things appear to us. So, in this extreme case at least, only the epistemic interpretation of $\lambda xy(x$ appears red to $y)$ seems appropriate.

So, what is essential for $\lambda xy(x$ appears red to $y)$ depends, according to our linguistic essentiality convention, on how the context turns out to be; and the three interpretations just mark three significantly different kinds of contexts. They are thus integrated into a single character of $\lambda xy(x$ appears red to $y)$.

Why then did they appear to be three different readings and thus to uncover an ambiguity? The reason, it seems to me, is that "appear" and "look" are conjoined not only with "red", but with many other phrases as well. In fact, the usual claim associated with these readings is that the scheme "looking *F*" (and not its instantiation "looking red") has three different readings, depending on what is taken as *F*; and this claim is usually accompanied by quite determinate opinions concerning which reading is appropriate for which kind of *F*. My claim 3 interprets this determinateness as a (maybe unreflected) certainty about the actual context world and the interpretation of "looking *F*" pertinent to it.

For instance, it seems very likely that we live in a context world where "appearing red" carries the phenomenal interpretation. Again, circumstances seem to be such that a comparative interpretation is most appropriate for phrases like "appearing square" or "looking like a capital A" which are about simple forms possibly appearing in many different ways. Finally, as things stand, the epistemic interpretation seems applicable not only to perceiving beings without phenomenal experience, but also to us for phrases like "appearing to be a car" where the appearance is phenomenally too complex and varied and best reduced to the proximate epistemic effect.

However, for all these instantiations of "looking or appearing *F*" it seems possible to imagine cases which show the same context-dependence as I have displayed it for "appearing red". Imagine, for instance, beings who have phenomenal experiences, but who see only letters, maybe in Garamond. Thus, if a car is approaching them, they read "car" written in Garamond in the relevant place of their visual field (strangely, these beings are tuned to English); and one may refine the example by giving meaning to the size and color of the letters in their visual field. For such beings a phenomenal interpretation of "appearing to be a car" seems appropriate.[31] Hence we find the three interpretations not only across the various instantiations of the scheme "looking *F*", but indeed within each locution of this type; and this, so I

[31] This example came to my mind when reading Cresswell (1980, pp. 129–131), where he invents similarly weird examples for arguing, contra Jackson, that there is no difference between "looking red" and "looking like a tomato", i.e. that both may be equally given a phenomenal and a comparative reading. This argument further illustrates my present point.

have argued, is better accounted for by giving this locution one context-dependent meaning. Only if one neglects this context-dependence do there seem to be three different readings or meanings.

Let me summarize the point of claim 3 in a somewhat different way: There are two extreme views to be found in the literature. Some think that subjects have certain types of sense impressions, qualitative experiences, or however one may call them, that we can refer to the subjects' having them, maybe even in a direct or rigid way, and that we in fact do so with such expressions like $\lambda y \bigvee x(x$ appears red to $y)$.[32] Perhaps the most famous expression of the opposite view is found in Wittgenstein (1953, sect. 293) where he ponders about how we could talk about the alleged beetles in our boxes when everyone can look only into his or her own box and where he says:

> Here it would be quite possible for everyone to have something different in his box. One might even imagine such a thing constantly changing. – But suppose the word "beetle" had a use in these people's language? – If so, it would not be used as the name of a thing. The thing in the box has no place in the language-game at all; not even as a *something*, for the box might even be empty. – No, one can "divide through" by the thing in the box; it cancels out, whatever it is.

My claim 3 proposes a middle course agreeing, in a way, with both views. Wittgenstein is right: Our communication with the very language and the very color expressions we have obviously works well, however the empirical facts turn out to be and even if our boxes are empty; and a theory of meaning for this language has to do justice to this fact, as my account tries to do.[33] But under happy circumstances, we have the same kind of beetles in our boxes, and then we do talk about our beetles. Surely, it is very likely that the circumstances are happy; confirming this is not impossible, and with such findings as the opponent process theory we are indeed beginning to confirm this.

I turn now to the relation between the two kinds of color predicates, i.e. between (A) $\lambda x(x$ is red) and (B) $\lambda xy(x$ appears red to $y)$. Again, we find two extreme views both of which hold that the one predicate is definable by or reducible to the other. On the one hand, those maintaining an objectivist account of colors insist that $\lambda x(x$ is red) denotes an objective property, i.e. a primary quality of objects by themselves.[34] They may add that $\lambda xy(x$ appears red to $y)$ should be explained by how subjects respond to objects' being red; the comparative and the epistemic interpretation of $\lambda xy(x$ appears red to $y)$ precisely are attempts to give that explanation. But since the phenomenal interpretation turned out to be appropriate at least for some contexts, e.g., those where vision is realized in most humans in a relatively uniform way, this additional claim does not seem defensible. On the other hand, those

---

[32] Cf., e.g., Nida-Rümelin (1997). As she emphasizes in sect. 3, this does not necessarily require to posit objects like impressions, sensations, etc.

[33] This is not to say, however, that my application of Kaplan's framework to color talk would have a specifically Wittgensteinian character.

[34] Cf., e.g., Jackson and Pargetter (1987).

defending a subjectivist account of colors say that the colors of objects are secondary qualities which can only be explained by how the objects appear to us; they would thus cite statement:

(1) x is red if and only if x would appear red to most English speaking people under normal conditions

as a definition or an analysis of being red.[35] In order to assess this, we have to inquire into the modal status of (1).[36] This is the content of my:

Claim 4: (1) is a priori in English[37]; but it is analytic only in one reading and not analytic in another reading.

We find out about the apriority of (1) by evaluating it in each context (and its associated index). There (1) seems to be true; I cannot imagine any context world in which the two sides of the equivalence in (1) would differ in truth value. In any case, (1) is true in all three kinds of contexts for which the three interpretations of λxy(x appears red to y) are, respectively, appropriate. This just reflects the point noted earlier that the colors of things have an overt nature given normal conditions; in no context can an object which appeared red to most of us under normal conditions turn out not to be red, and vice versa.[38]

We find out about the analyticity of (1) by evaluating it in each context and each index; (1) is analytic if and only if it is true in all of them. Consider, for instance, the actual context with a ripe tomato before us and transfer the tomato just as it is into some counterfactual index world i. Thus, this tomato is as red in i as it is here. Now let us assume that most English speaking people in i are pseudonormal, that is, have R-pigment in their G-cones and G-pigment in their R-cones; this is definitely a possible counterfactual supposition. Since I take it that the actual context is one in which pseudonormals are deviant people with red-green-inverted sensations,

[35] This is explicitly done, for instance, by McGinn (1983, pp. 5–14) – with the exception that he would not restrict (1) to English speaking people. This minor difference is cleared up in footnote 37.

[36] In Spohn (1997c) [here: ch. 12] I have again used the Kaplanian framework for a related inquiry into the modal status of reduction sentences for dispositional predicates in general and also into the epistemology and ontology of normal conditions. This inquiry may further illuminate the following discussion.

[37] This explains my restriction of (1) to English speaking people; the primary standard for how "being red" works in English is the English speaking community. The situation changes as soon as there are established translations between English and other languages; then the people speaking these other languages become equally important. Cf. Haas-Spohn (1997, sect. III).

[38] Since I am talking about apriority in English. I am not claiming that (1) is a priori for any subject. This also entails that it would be inappropriate to object that a thing could be red in a context where there are no English speaking people for it to appear red. This is so because the existence of the English speaking community is a priori in English, similarly as my existence is a priori for me. But, of course, the existence of the English speaking community is not analytic in English. This as well as the mere apriority of one reading of (1) are examples showing that apriority in English is an independent notion reducible neither to analyticity nor to subjective apriority. And this in turn shows that such communal epistemic notions are needed; at least there is some work for them to do.

we would then have to say that most English speaking people have red-green-inverted sensations in *i*. So in particular, it is true in the index world *i* that the red tomato appears green to most English speaking people. This amounts to a counter-example to the analyticity of (1).

However, my claim 4 seems to state an ambiguity in (1). Where is it? I have just understood "most English speaking people" in an attributive way by evaluating it at the index world *i* itself. But we can take this phrase also in a referential way as referring to most English people not in the index, but in the context world *c*. The same kind of ambiguity can be found in the phrase "under normal conditions"; it can refer to conditions counting as normal in *i* or counting as normal in *c*.[39] To make this referential reading more explicit, statement (1) thus read says in each context *c*:

(4) *x* is red in the index world *i* if and only if *x* would appear red in *i* to most English speaking people from *c* under conditions normal in *c*.

The above counter-example does not apply to this reading. Indeed, I cannot think of any counterfactual index world *i* for which this reading would not be true. If this is correct, then (1) is analytic in this reading.[40]

There would be no point in arguing which reading is more natural. The important thing to note, I think, is that (1) is a priori in any case and that it is analytic only in the referential reading (of the relevant phrases), but not in its attributive reading.

Does this result support the subjectivist in any way? No. A preliminary point to note is that an analytic equivalence like (1) in its referential reading need not give a definition or analysis; as an analytic truth it just states a certain meaning relation. But since it is not so clear, anyway, what an analysis or a definition (of an already meaningful term of natural language) is beyond an analytic equivalence, we had better concentrate on the subjectivist's claim that colors are secondary qualities, or that predicates of type (A) are dispositional or, more abstractly, relational, i.e. relative to perceiving subjects:

What the subjectivist would need is the necessary or, as McGinn (1983, p. 14), puts it, intrinsic dispositionality or relationality of type (A) terms. This, however, cannot be inferred from my claim 4. λx(x is red) would be necessarily relational in a given context if and only if, viewed from that context, an object could be red in an index world only if it would stand in a certain (maybe only dispositional or counterfactual) relation to other objects in that index world. (1) seems to assert such a thing; but it does so only in its attributive reading which I observed to hold only a priori, i.e., not to be projectible from the context to all indices. Nor does the referential

[39] The distinction between an attributive and a referential use of denoting phrases was originally introduced by Donnellan (1966), however in an apparently different way. By using it as just explained I refer to its standard interpretation within the Kaplanian framework which is to be found in Stalnaker (1970) and Kaplan (1978).

[40] Maund (1986, pp. 173–176) makes a similar point by distinguishing a purely comparative and a referential use of "looks" and arguing that something like (1) is analytic in the first, but synthetic in the second use; however, he does not represent his distinction within the Kaplanian framework.

reading help, despite its analyticity, since it asserts only a certain relation between objects' being red in an index world and the people in the context world. Compare this with λx(x is a mother), the standard example of a necessarily relational predicate. Here, viewed from any context world, someone is a mother in an index world if and only if someone else exists or has existed in that index world who is her child. The analogous assertion for "being red" is simply not licensed by the modal status of (1).[41]

What is licensed by the apriority of the attributive reading of (1) is the conclusion that λx(x is red) is a priori relational. But this is no peculiarity of color terms; according to Haas-Spohn (1997), "water" and other natural kind terms, in fact all hiddenly indexical predicates and thus many predicates which unquestionably denote primary qualities are a priori relational.[42]

But even if the necessary relationality of λx(x is red) cannot be demonstrated, it may actually hold in a given context. Maybe; I was, however, unable to imagine such a context. I shall return to this issue with my claim 6 when I speculate about our actual context world.

So far, I have considered all possible contexts for our two sample predicates (A) and (B). I have, implicitly and explicitly, discussed their diagonal and how to project their extension from contexts to other indices; in this sense I have carried out an epistemological and semantical investigation. After all this it is not so difficult to give a summarizing definition of the character of these predicates.[43] In the

[41] Jackson and Pargetter (1987, pp. 130f.) argue for the same point. They call (1) the dispositional truism and argue that the truism does not justify one in identifying colors with the disposition to look colored; rather, colors should be identified with the categorical base of this disposition. I agree, if one takes the difference between a disposition and its categorical base to be only an epistemological one: The intension of a dispositional predicate and the intension of the predicate describing the categorical base are the same, both predicates denote the same property; but their diagonals and a fortiori their characters are different. With this account of the difference between dispositions and their bases, my argument is the same as theirs. However, Prior et al. (1982), pp. 253ff., give an ontological account of the difference. So, there remains a disagreement. I discuss this disagreement more fully in Spohn (1997c) [here: ch. 12].

[42] Compare also the discussion about the notion of response-dependence and its rigid and its non-rigid interpretation which addresses the very same issues; cf., e.g., Vallentyne (1996). I think this notion nicely fits into the Kaplanian framework; its rigid and its non-rigid interpretation, in particular, corresponds to the referential and the attributive use as explained above.

[43] Applying the general scheme of Haas-Spohn (1995, p.151) to our sample predicates we get: x is red in the context c and the index i iff x shares in i all the properties which, according to the English essentiality convention for "being red", are essential in c for the redness of the objects to which "being red" is typically applied in c by the English speaking community; and x appears red to y in the context c and the index i iff x and y possess in i all the properties and relations which, according to the English essentiality convention for "appearing red", are essential in c for the relation of appearing red between any two objects to which "appearing red" is typically applied in c by the English speaking community. This abstract explanation is neither circular nor badly metalinguistic (cf. Haas-Spohn, 1997 and 1995, sects. 3.4–3.5). But it is less illuminating than the substantiation of its key terms for the case at hand; and I am here rather concerned with the latter.

rest of the paper, however, I shall engage in a metaphysical speculation concerning the intension of these predicates at our actual context world.

I first take up statement:

(2) x appears red to y if and only if x (appropriately) causes y to be in a neural state of the kind N by defending:

Claim 5: For some (possibly disjunctive) kind N of neural states, statement (2) is necessarily true in the actual context.

The parenthetical "appropriately" in statement (2) is to exclude deviant ways of causation which do not count as an object's appearing to a subject; but I am not concerned with spelling out what is to count as appropriate here.[44] Of course, claim 5 similarly holds for other color appearance terms; thus it says in effect that color appearance terms are strongly supervenient on neural state terms, or, what comes to the same, that the property of having-a-so-and-so-colored-sensation is type-type identical with the property of being in a certain (possibly wildly disjunctive) neural state.[45]

Claim 5 consists of two parts. The first is a factual hypothesis, namely that (2) is actually true for some N, or rather that in most of our actual paradigm cases for some object's appearing red to some subject we find in that subject an activation of a certain neural structure or a realization of a certain, possibly very complex neural state. As far as I know, brain research has not come up so far with results disconfirming this hypothesis; but perhaps I am underestimating the complexity and diversity of neurological findings. On the other hand – perhaps I am again underestimating present expertise – we do not seem to have any good theory what that kind N of neural state might be. But this only shows how poor our knowledge is; it does not show the senselessness or illegitimacy of that hypothesis.

The second part of claim 5 is a claim about our essentiality convention for $\lambda xy(x$ appears red to $y)$, namely the claim that, given the factual hypothesis that we find a neural state type N uniformly realized in most of our paradigm cases of $\lambda y \bigvee x(x$ appears red to $y)$, this state type N provides the essence of $\lambda y \bigvee x(x$ appears red to $y)$; that is, we would correctly apply $\lambda y \bigvee x(x$ appears red to $y)$ only to factual and counterfactual cases in which this state type is realized. So, this is rather a linguistic claim about our counterfactual talk. It is to be defended mainly against two doubts.

One doubt is whether, given the factual hypothesis, the essence of $\lambda xy(x$ appears red to $y)$ is really to be conceived so narrowly as to conform to no wider than the phenomenal interpretation. I have briefly discussed this already in case 1 following the three interpretations (E), (C), and (P). One possible ground for abandoning this doubt is how I said we would talk about pseudonormals; when we say that red peppers would appear green to them, we precisely assume the narrow essence. Another

[44] Cf., e.g., Lewis (1980c).

[45] Here I identify the property expressed by a predicate with its intension, so that necessary universal equivalence of two predicates is necessary and sufficient for the identity of the properties expressed. For the equivalence of strong supervenience and type-type identity cf., e.g., Kim (1984, sect. IV).

possible ground is that we say that, strictly speaking, nothing appears green or red to red-green blind people even if they should have other clues for correctly guessing the colors. Still another ground is that we refuse, as I think we should, to carry over human color talk to, say, bees upon finding that bees carve up the space of electromagnetic wave mixtures in quite a different and incomparable way than we do. So this doubt seems unfounded.

The other doubt is whether claim 5 provides a correct understanding of the phenomenal interpretation. One may rather think that it is the phenomenal quality itself which is essential for red appearances, i.e. that, necessarily, x appears red to y if and only if x (appropriately) causes y to have a red-sensation[46]; it would thus be a matter of contingency which kind of brain-states are correlated with red-sensations.[47] I have two reasons for resisting this doubt.

First, if the correlation of neural states with red-sensations is contingent in any case, then conceiving $\lambda xy(x$ appears red to $y)$ as context-dependent and describing this dependence as I did in claim 3 loses its plausibility; it goes together more naturally with the view (endorsed by Nida-Rümelin 1997) that everyone, when claiming that something appears red to him, refers to the kind of phenomenal quality which he is just experiencing and the awareness of which leaves no room for error and thus for hidden indexicality. The consequence of conceiving $\lambda xy(x$ appears red to $y)$ as involving a fixed kind of phenomenal quality in all contexts and indices is, however, that the few pseudonormals, if they exist, always refer to another quality than normal people do, hence use $\lambda xy(x$ appears red to $y)$ with a different meaning (character) and speak, in a sense, a different language. The more varied version of case 2 mentioned after the three interpretations (E), (C), and (P) comes out even worse according to this view; there would be a Babylonian confusion where $\lambda xy(x$ appears red to $y)$ would have many different meanings and people would talk many different languages.

This seems unwarranted to me. I do not know whether cases 1, 2, or 3 obtains (though I have already expressed my prejudice); but in any case I see no reason to assume such a possible multiplicity of languages. For instance, if case 1 should turn out to hold and if some pseudonormals should be identified, my prediction would be that these pseudonormals would not insist to continue speaking as before; they would rather correct themselves and agree to such things as that, strictly speaking, red tomatoes look green to them, i.e. they would submit to common usage. Or, if, to our great surprise, case 2 should turn out to obtain, my prediction is that linguistic practice would not change a bit; after this discovery, all of us would talk of

[46] Here, the unusual locution of having a red-sensation is defined as denoting the property which is caused to apply to a subject by an object iff that object appears red to it; in other words, it denotes the intrinsic, non-relational property which a subject has whenever the relational property $\lambda y \bigvee x(x$ appears red to $y)$ applies to it (and which a subject may also have, as it turns out, without external cause).

[47] Certainly, this better catches the intentions of the adherents of the phenomenal interpretation. Kripke (1972) seems to think so with respect to pains (though not necessarily with respect to colors) (cf. pp. 334ff.). Clearly, Nida-Rümelin (1997) also favors this view.

things appearing red to us as we did before. This does not look like a discovery of many languages where there seemed to be only one.[48]

Maybe, however, the disagreement is not about the context-dependence of $\lambda xy(x$ appears red to $y)$, but only about the essential properties of $\lambda xy(x$ appears red to $y)$ in the presumably obtaining case 1. Then I have a second reason for sticking to claim 5, namely internal realism.

Internal realism, as I understand it, asserts that truth is believable or discoverable; given a correct understanding of the "-able" – this is all-important – I believe that internal realism provides the defensible core of verificationism.[49] Now, it seems to me that internal realism may be strengthened to assert that essences are believable or discoverable. I have no clear argument for this claim[50]; but if so much is granted, my argument can proceed.

Let us imagine that we have investigated vision in human as well as in other sensing beings as completely as possible; for instance, we have constructed fabulous devices with which we can scan brain states in real time. After endless ingenious theorizing and ingenious experimenting we have come up with our final theory about vision, how visual input is processed, how consciousness comes into play, how all this leads to linguistic and other behavioral output, etc.[51] According to internal realism this final theory which cannot be shattered or improved by any further findings is true. The final theory will contain many equivalences of the form (2) all of which are true; an object will appear red to a perceiver if and only if a many-membered chain of events is realized each of which is a necessary and sufficient cause of the later ones. Among all these equivalences there will be one referring to a special neural state type N^* with the further characteristic that, given a subject x is in state N^*, there is no further or overriding reason whatsoever for or against x's having a red-sensation and that, given a subject x is not in state N^*, there is no further or overriding reason whatsoever for or against x's not having a red-sensation; that is, any reason for a divergence between being in state N^* and having a red-sensation would at the same time disconfirm the final theory. But then it would be strange to insist that the essence of having a red-sensation does not consist in the neural state type N^*, but in something else. In any case, no reasons whatsoever

[48] If these predictions would turn out false, however, this might well be reason for me to revise my position.

[49] I interpret the "-able" in the following way: the set of a posteriori truths and our inductive standards (taken in a broad sense) must be such that each truth is inductively supported by other truths (conditionally on arbitrarily many truths) and can thus be believed on true grounds. In Spohn (1991) I formally explicated this idea and proved it to be equivalent, in a way, with the universal feasibility of causal explanation.

[50] A major difficulty is here to adapt all the notions involved in the explication of internal realism to the more sophisticated Kaplanian framework. In Spohn (1991) [here: ch. 9] I have not dealt with this difficulty simply because I was not yet aware of it.

[51] Maybe we even have constructed a transmitter cap and a receiver cap directly connecting two brains, and our final theory says that the human under the receiving cap should experience similar sensations to the being under the transmitting cap.

could be adduced in favor of this, not even by the perceiver herself; and then it is simply false according to the strengthened form of internal realism.[52]

The parallel claim concerning statement:

(3) x is red if and only if the reflectance spectrum of the surface of x is of the kind R

is my final:

Claim 6: For some (possibly disjunctive) kind R of reflectance spectrum, statement (3) is necessarily true in the actual context.

In order to see this, we do not have to do much more than putting together claim 5 and the analytic reading (4) of assertion (1). If we do this we get:

(5) In the actual context it is necessarily true for some neural state type N that x is red if and only if x would cause most of the actual English speaking people under actual normal conditions to be in state N.

Now, there are certainly many ways for people to get into a neural state of kind N and for a given object to bring this about; the actual causal story seems to be a matter of contingent physics and of contingent neurobiology. So how do we get from the above necessary truth in (5) to the necessary truth of (3)? This is achieved by the reference to normal conditions. Recall my speculation about twinlight and the modaleons. Of course, an index world may be filled with twinlight, and because physics is very different there, modaleons there produce state type N in us, that is, they appear red to us. But as I have already argued after claim 1, modaleons, when viewed from the actual context, would not count as red in that index world, but as blue, because under normal conditions such as daylight they would appear blue to us. Similar considerations apply to the normal conditions within the subjects like not being mad or intoxicated, and so on. Thus it is the function of the reference to normal conditions to keep the kind of causal process between visible objects and the observers as it normally is in the actual context world fixed throughout all possible index worlds. This enables us to locate, so to speak, the color of an object with necessity in the object itself; we do not have to settle for merely contingent correlations between the physical properties of an object and its color. And for all we know, it is the reflectance spectrum of the object's surface which is the relevant

[52] Let me clarify the hypothetical and the positive content of the argument: In any case, I think, the final theory will come up with some equivalence of the form "x has a red-sensation iff x is P" with the characteristic just described. My positive claim is then that, according to strengthened internal realism, this P is the essence of having a red-sensation; and my hypothetical claim is that this P will actually turn out to be of the form "being in neural state of type N^*". But the latter seems at least plausible. In any case, if we tend not to leave it open, but to positively assert on the basis of the opponent process theory that green peppers look red and red peppers look green to pseudonormals and to stick to this until receiving counter-evidence, we are on the track of searching for, and being prepared to accept, ever more sophisticated neural conditions for having red-sensations – a track which will eventually lead us to the type N^* required for Claim 5 to be true.

physical property. Of course, the class *R* of reflectance spectra characterizing redness forms an extremely wild and certainly quite vague region in the space of possible reflectance spectra. This is so because the class *R* is specified only in relation to the equally vague neural state type *N* and thus to a very complex biological contingency.[53]

According to claim 6, the nature of being red is hidden and unknown. Did it not seem to be overt? Yes, it seemed so. But then we observed with claim 1 that already the normal conditions have a hidden nature. With claim 2 we realized that the nature of appearing red is even more profoundly hidden. And this entails via the analyticity of (4) that the nature of being red is equally profoundly hidden. The appearance of overtness could be confirmed only under the variant of the phenomenal reading of $\lambda xy(x$ appears red to $y)$ which I have criticized under claim 5.

According to claim 6, moreover, colors are not dispositional properties or, more specifically, secondary qualities of objects, contrary to a familiar view. What claim 6 does, in effect, is simply to identify redness, i.e. the disposition of appearing red with its categorical base. My general presumption is here that many, though probably not all dispositions are such that having the disposition is necessarily, though certainly not analytically equivalent with realizing the categorical base of the disposition. This looks implausible only if one confuses ontology and epistemology. One may say that being red is a dispositional concept, since it is a priori according to claim 4 that red things have the disposition to look red; and it is this disposition which determines for each context which property being red is. But this is an epistemological point which does not entail the ontological point that this property itself is dispositional.[54] The epistemological point is also reflected in the fact that in order to find out about, and succinctly describe, the class *R* of reflectance spectra we have to find out about, and to refer to, the human visual system and possibly to the class *N* of neural states. But again, this does not entail that the property of having a reflectance spectrum belonging to the class *R* would be relational in any way.

To be sure, what claims 5 and 6 say about the actual context world may be far from the truth, and then the a posteriori necessities may be very different; no one can claim certainty about this. But there is at least hope that the context world we live in is so nice as to allow us to stick to the claims 5 and 6 and thus to be metaphysically conservative and parsimonious, even though the epistemological picture I have drawn is much richer.

---

[53] In having this metaphysical position concerning colors I thus join what Hilbert (1987) calls anthropocentric realism. My only disagreement is that I would insist that metamers have the same color, because metamers look to have the same color under normal conditions (bright daylight etc.) and because the real color of an object shows itself only under normal conditions. If a color is thus constituted by the class of its metamers, the class *R* in assertion (3) is, for all we know, bound to be a wild one. To say, as Hilbert (1987, pp. 83f.) does, that only isomers have the same color would mean, I think, to revise ordinary color talk. Whether the revision would be a reasonable one is another question.

[54] Cf. my remarks about necessary and a priori relationality in my discussion of Claim 4.

Chapter 14
Concepts Are Beliefs About Essences†

14.1 Introduction

Putnam (1975) and Burge (1979) have made a convincing case that neither meanings nor beliefs are in the head. Most philosophers, it seems, have accepted their argument. Putnam explained that a subject's grasp of an expression's meaning is often insufficient to fix its reference, and that she needs help from her natural and social environment. Burge explained that having a belief, even in the de dicto sense, is really a relational property which may change merely when the implicit relatum, the linguistic community, changes.

To accept this, however, one does not necessarily need to accept all the anti-individualistic consequences Burge has drawn from these insights. On the contrary, these consequences have met much more reluctance. Many share the view, we do as well, that there must be something in the head, not only a brain, but also a mind, indeed a mind with internal or intrinsic representational or semantic properties. This view was also supported by arguments mainly concerning, on the one hand, psychological explanation and the causation of individual behavior, and on the other, self-knowledge. Of course, these arguments have been disputed, but the dispute has not shattered our prejudice.[1] Here, we would simply like to presuppose the correctness of this view without any further comments.

Thus, all those sharing the prejudice set out to characterize what's in the head, i.e., so-called narrow contents. Now, narrow contents are rather expressed by, or

---

† This paper was jointly written by Ulrike Haas-Spohn and me. It was originally published in: A. Newen, U. Nortmann, R. Stuhlmann-Laeisz (eds.), *Building on Frege. New Essays on Sense, Content, and Concept*, Stanford: CSLI Publications, 2001, pp. 287–316. It is reprinted here with kind permission of CSLI Publications and my wife.

[1] We had and have good company: Loar (1986), Fodor (1987, ch. 2) (though ch. 4 apparently got the upper hand in the end – cf. Fodor 1994), Perry (1988), Block (1991), Lewis (1994a), Chomsky (1995), Chalmers (2002), and others. Moreover, most of cognitive science certainly sees itself as an individualistic enterprise.

associated with, whole sentences. But sentences are composed of parts, basically a singular and a general term, and hence narrow contents seem to be composed in the same way. We reserve here the term "concept" for those entities which a subject expresses by, or internally associates with, singular and general terms (and maybe other expressions, too). Having a concept is hence *defined* to be an internal, non-relational property. In the absence of a generally agreed usage of the term "concept", this stipulation is certainly legitimate and it is often made. The terms "narrow content" and "concept" thus stand for the same thing; the only difference, which we do not strictly observe, however, lies in the associated kinds of expressions.

For internalists like us the existence of concepts and narrow contents is thus beyond doubt. The question is rather a constructive one: how precisely does one conceive of them? This is, as the title indicates, the topic of this paper. However, the offers so far are rather more problematic than impressive. We shall refer to two major options in the following.

First, the dominant view concerning the mind-body problem has been, and perhaps still is, functionalism. Functionalism is the view that internal mental states are functional states, i.e., to be individuated by the place they occupy within a large functional net spanned between perceptual input and behavioral output. Insofar as mental states have narrow content, their content is also characterized in a functional way. This gave rise to the program of so-called conceptual or functional role semantics which may thus be conceived as an attempt to establish internalism.[2]

Second, one may build upon the epistemological reinterpretation of Kaplan's character theory, which was not fully endorsed by Kaplan (1977), but acquired prominence through Fodor (1987, ch. 2), though it is first recognizable in Stalnaker (1978)[3] and Perry (1977). According to the character theory, semantics has to recursively specify a character for each expression, assigning to it its extension relative to a context (of utterance) and an index (or point of evaluation). And, according to the epistemological reinterpretation, the diagonal of the character of an expression represents the cognitive significance of, or the concept associated with, this expression.[4]

Here we shall pursue only the second approach via the epistemologically reinterpreted character theory. Our main reason is that functional role semantics failed to give a clear and precise theory of how concepts and narrow contents build up in a recursive way. By contrast, the character theory has a clearly specified formal structure which is easily connected with linguistic semantics; in particular, characters combine recursively in much the same way as intensions do in intensional semantics. Prima facie,

---

[2] Cf., e.g., Field (1977) and Block (1986). Functional role semantics is not necessarily individualistic, though. It is ambiguous between short-arm and long-arm functional roles; cf. Harman (1987).

[3] Stalnaker is certainly not an internalist, as his (1989) and (1990) clearly show. However, his (1978) may well be interpreted as making more internalist sense of the character theory as Kaplan did.

[4] Perhaps one should also mention the very dense account of Lewis (1986b, sect. 1.4) which is related to all three approaches mentioned, but not identical with any of them; to consider it seriously would, however, require a separate discussion.

these formal virtues are the overwhelming reason for our choice (although we are well aware that formal structure alone does not determine its interpretation).

However, there are difficulties with the character theory as well. If one considers their interpretational questions, two serious problems emerge, as Schiffer (1990) and Block (1991) have forcefully made clear. The first problem is that the character theory seems to be either inadequate or superfluous. Schiffer argues that the character theory cannot avoid having recourse to functional roles or states. But then it seems to be only a detour, since one could have explained narrow contents rather by directly appealing to functional roles. We call this Schiffer's problem. The second problem, set up by Block, is that the character theory can apparently take only one of two inadequate forms. Either it must specify narrow contents by reference to linguistic expressions themselves, i.e., fall prey to syntacticism. Or, it must specify narrow contents in a profoundly holistic way, i.e., fall prey to an unacceptable degree of holism. We call this Block's dilemma.

Thus it seems that the character theory, whatever its formal virtues, cannot get off of the ground unless it offers some good response to these challenges. This is, more specifically, the task we want to address here. We tackle it in four sections: Section 14.2 explains the epistemologically interpreted character theory and its problems in more detail, Section 14.3 presents the solution we want to propose, Section 14.4 explains it in a bit more detail, and Section 14.5 explains that this solution indeed avoids Block's dilemma as well as Schiffer's problem.

14.2 The Problems Specified

We cannot go on after this rough and general introduction without referring to some specific statement of the epistemologically reinterpreted character theory. Let us therefore briefly look at the treatment it receives in Haas-Spohn (1995), where it is dealt with in book length. We shall see that her account is also susceptible to the two problems just mentioned, but this will pave the way for improvement.

What is a character? A character is a function assigning to each possible context of utterance (context for short) an intension, which is, in turn, a function from points of evaluation (indices for short) to extensions. Equivalently, the character of an expression is a function assigning to each context and index the extension the expression has at this context and index. The characters of complex expressions build up recursively in the way familiar from intensional semantics.

We take a possible context c to be just a centered world, i.e. a triple $\langle s_c, t_c, w_c\rangle$ such that the subject s_c exists at time t_c in the world w_c and may (but need not) utter the relevant expression. A possible index i consists of all items which may be shifted by operators of the given language. Here, it will suffice to put only a possible world w_i into the index i.[5]

[5] Cf. also Lewis (1980b).

Sentences, in particular, are true or false at contexts and indices, according to their character. This entails a notion of truth at a context simpliciter. A sentence is true at the context *c* if and only if it is true at *c* and the index which consists of the context world w_c itself. The function assigning to each context the truth value the sentence has at the context is called the diagonal of the sentence.[6] Similarly, we may define the diagonals of other expressions. Note that this definition works only on the condition that for each item of indices there is a corresponding item of contexts, and note that our definitions meet this condition.

Now we can say what the epistemological reinterpretation of the character theory is supposed to be. Basically, it just consists in considering possible contexts at the same time as possible doxastic alternatives of some subject. Thus, what a subject believes is that she is in one of the contexts of a certain set of possible contexts. And if a subject believes a sentence to be true, she believes that she is in a context in which the sentence is true; that is, the sentence's diagonal is a superset of the set of the subject's doxastic alternatives.[7] All this agrees well with the characterization of contexts as centered worlds since centered worlds are known to be needed for the representations of beliefs de se and de nunc.[8]

Now, to be a bit more specific, consider a certain natural language *L* like English and some referring expression α of *L*; one best imagines α to be a name like "Aristotle" or a one-place predicate like "water", "table", or "red".[9] Then Haas-Spohn (1995, pp. 99, 136, 150) explained the (*objective*) *character* of α in *L* in the following way:

> $\|\alpha\|_L\,(c, i)$ = the object or the set of objects at the index *i* which is the same or of the same kind, i.e., has the same essential properties as the object or the objects from which the usage of α in the language *L* originates in the context *c*.[10]

The crucial term is here "the usage of α in *L*". In the context *c*, it stands for the whole communicative pattern in *c* associated in *L* with the expression α. However, what is essential to this pattern are not all of its ramifications it actually has in the context, but only the methods of identifying or recognizing the reference of α which are available to the community speaking *L*. These methods may be those of Putnam's experts for gold as opposed to the laymen, or those of Evans' producers of a name who are acquainted with its bearer as opposed to the consumers of the name,[11] or indeed those of almost everybody in the case of chairs and tables, in

---

[6] Diagonals are called primary intensions by Chalmer (2002). His secondary intensions are what we call simply intensions.

[7] Here, and elsewhere, we do not distinguish between a set and its characteristic function.

[8] Cf. Lewis (1979b) or Haas-Spohn (1995, sects. 2.2–2.3). But see here Chapter 15 for arguments that doxastic alternatives need a variable assignment as a further component.

[9] We shall not address definite and indefinite descriptions and all kinds of indexicals and demonstratives since they involve a number of further problems which we better avoid.

[10] Obviously, the "i.e." is only justified if the essential properties are necessary and sufficient for individuating the object or the kind. This may be false. But at least it seems true that they are necessary and that nothing else (except the pure thisness) achieves the individuation. So we may ignore the point in the following.

[11] Cf. Putnam (1975, pp. 235ff.) and Evans (1982, ch. 11).

which nobody has privileged knowledge. Thus, such usages are in principle well described in the relevant literature.

Two points are important about such usages as conceived by Haas-Spohn (1995). First, the expression α itself is not essential to its usage; the very same usage may be associated with another expression as well. This entails, in particular, that different languages may have the same usage of different expressions; this is crucial for their translatability.[12]

Secondly, the extension, the object or objects from which the usage originates, is also not essential to the usage; in different contexts or context worlds different objects may fit the same usage. In our world, H_2O fits the usage of "water". But for all we know, or knew 250 years ago, it may as well have been XYZ from which our usage of "water" originates. Likewise, in the actual context world our usage of "Aristotle" originates from the actual Aristotle. But there may be another context world in which somebody else had the same career as our Aristotle and has triggered the same usage of "Aristotle". In this way, then, the extension of α may vary with the context; and thus Kaplan's strategy of explaining the informativity of identity sentences between overt indexicals[13] may be carried over to hidden indexicals like "water" or "Hesperos". Hence, Haas-Spohn (1995) intends that a usage is something which may properly be called a communal concept which is internal to the relevant language community and does not change by merely changing the community's environment.[14]

The above explanation of the character of α in L is still incomplete since we have not yet specified its domain. Concerning indices, we may assume that all indices or possible worlds belong to its domain. Concerning contexts, however, the explanation presupposes that the very usage of α in the language L exists in the context; otherwise, the character of α in L is undefined simply because there is no origin of the usage if there is no usage.[15] Thus, if we understand a language to be the collection of all the usages of its terms, the recursive explanation of the characters of its expressions works only for those contexts in which the language exists.

So, what is, finally, the *diagonal* of the expression α in the language L? It is the function which is defined for all contexts in which the usage of α in L exists and which assigns to each context the extension α has there according to its usage. This indicates the heavy burden the notion of a usage has to carry, and, in view of this, the explanations given may well seem insufficient. We shall return to this point.

---

[12] In fact it is often the other way around. Translation *merges* the usages of different languages and thus *makes* them identical; cf. Haas-Spohn (1997, sect. 3). This is an insight which seems to put the indeterminacy of translation and related issues into a very different light.

[13] Cf. Kaplan (1977) who explains in sect. XVII how "this = that" may be informative and refrains in sect. XXII to generalize the method to "Hesperos = Phosphoros" because he considers names to be absolute. According to the above explanation, however, names are *hidden* indexicals.

[14] Anti-individualists will find this notion of a usage to be question-begging, whereas we attempt here to provide individualistic foundations to such communal concepts.

[15] The counterfactual question what the origin would have been if the usage had existed does not generally make good sense.

For the moment, however, we have to attend to another crucial point. Since usages are communal concepts which, as explained, summarize not what each individual knows, but what everybody together knows about the relevant extensions, they are unsuited for describing concepts and narrow contents, which are intended here to be internal to a given subject; the subject need not fully know about usages or communal concepts. This was indeed the basic point of Burge (1979): that a subject may have an incomplete or false linguistic understanding and still be amenable to de dicto belief ascriptions. So, how do we get down to the level of individual subjects?

A natural idea, indeed the one Haas-Spohn (1995) pursued, is the following. We repeat it here because it makes Block's dilemma very perspicuous. If a subject's knowledge of her own language may be incomplete, and indeed severely incomplete without clear lower boundary, then, it seems, we have to completely abstract away from such knowledge and to add it again for each subject according to her individual measure.[16] But what survives such abstraction? It seems the only thing we can hold fixed is the knowledge of the grammar, i.e. of the (purely morphologically conceived) words and their ways of composition. Thus we end up with what Haas-Spohn (1995, sect. 3.9) defines as *formal characters* which belong to a grammar G, the syntactic skeleton of a natural language:

$\|\alpha\|^G(c, i)$ = the object or the set of objects at the index i which is the same or of the same kind, i.e. has the same essential properties as the object or the objects from which the usage of α in the context language l_c originates – which is the language of s_c at t_c in w_c and has the grammar G.

In continuation of the parallel, the domain of a formal character consists first of all indices and second of all contexts in which the subject of the context speaks a language with the expression α or, indeed, with the whole grammar G. From this, *formal diagonals* are again easily derived.

Formal diagonals describe the minimal semantic knowledge accompanying the syntactic knowledge of the grammar. In order to know the formal diagonal of the expression α, one merely needs to know the triviality that α refers to whatever it is used for in one's language.

Thus, formal diagonals have at least some features desired by the internalist. Insofar as knowledge of grammar is internal, knowledge of formal diagonals is internal as well. Moreover, there is no problem of intersubjectivity since all subjects mastering the grammar G thereby master the same formal diagonals. However, if we identify concepts with formal diagonals, we clearly fall prey to syntacticism, one horn of Block's dilemma, since the words themselves, and only the words, are essential to concepts so understood. This is an understanding which is intuitively both too narrow and too wide at the same time. It is too narrow because it entails that speakers of different grammars must ipso facto have different concepts. And it is much too

[16] This strategy and the quantification over possible languages it involves goes back to Stalnaker (1978). Thus, the formal characters to be defined immediately are our way of capturing the idea behind Stalnaker's propositional concepts.

wide because any two persons associating whatever they want with the same word ipso facto have the same concept. By moving to formal characters, we have therefore lost the two virtues usages or communal concepts seemed to preserve.

This is no surprise because we have so far realized only the first part of our strategy, the step of maximal abstraction. However, a subject has beliefs about usages in her language just as beliefs about any other empirical matter, and only these beliefs add substance to the formal diagonals. Hence, we have to take the second step and enrich the picture by the subject's individual beliefs. Our first attempt to do so will turn out to be too coarse; but without it one cannot understand the later refinements.

For the representation of beliefs, we propose following the standard line formalized in doxastic logic. There, a subject's doxastic state is simply represented as a set of so-called doxastic alternatives, her *belief set*, and each proposition which is a superset of the belief set is then believed in that state. This representation has well-known problems: it neglects the fact that beliefs come in degrees; it cannot account for mathematical, but at best for empirical beliefs; it seems to presuppose logical omniscience since it assumes propositions to be believed regardless of how they are expressed; and so on. However, rival accounts are beset with other and no less grave problems. We therefore stick to this representation.[17]

In order to understand it properly, however, one needs to get clearer about what a doxastic alternative is supposed to be. We already said that it is simply a possible context $c = \langle s_c, t_c, w_c \rangle$. But what precisely does it mean that c is a doxastic alternative of a given subject s at a given time t in the world w? The usual explanation is that it means that all the beliefs of s at t in w do not exclude that s_c is s itself, t_c the present time t, and w_c the actual world w. This explanation is certainly correct, but not very helpful because it inherits all the ambiguities of the notion of belief of which it makes essential use. In particular, it leaves unclear whether or not a doxastic alternative can be understood in an individualistic way. The following explanation (cf. also Haas-Spohn 1995, p. 34) is more explicit:

> $c = \langle s_c, t_c, w_c \rangle$ is a doxastic alternative of s at t in w iff the following holds: suppose that s would fully investigate the world w_c – where this includes that it may take the perspective of every individual in w_c during its entire existence (so far it has a perspective at all), that it may learn all the languages there are in w_c, that it may subject each part of the world w_c to any investigations and experiments it can think, etc. – and suppose further that s would make its most considerate judgment about w_c after this investigation; then the assumption that it is now s_c at t_c in w_c is compatible with the doxastic state of s at t in w. More briefly, $c = \langle s_c, t_c, w_c \rangle$ is a doxastic alternative of s at t in w iff s' maximally experienced and considerate judgment about $\langle s_c, t_c, w_c \rangle$ is just an extension and not a revision of s' doxastic state at t in w.

[17] Spohn (1997a, sect. 5) distinguishes two large families of representations of belief, the computational and the semantic ones, each of which has their characteristic problems, and he argues that from the perspective of a rationality theory of belief semantic representations have primacy – which we join here.

The counterfactual suppositions in this explanation are impossible to satisfy in various respects. However, the explanation is not thereby rendered incomprehensible; it just drives common counterfactuals to the extreme, as philosophers are prone to do. The explanation conforms to the common idea of characterizing a belief state (or any kind of partial assignment) by the set of all its maximal extensions. What it adds to this idea is a fuller description of what is involved in such a maximal extension: neither a maximally consistent set of sentences of a given language, nor a totality of facts with facts being individuated from some external point of view, nor just a possible world in the sense of a maximally inclusive object, but rather our normal ways of belief formation using experience and judgment which are maximally expanded with respect to such a possible world.

It is philosophically highly important to have a clear picture of what is involved in such a maximal extension.[18] Let us point out three consequences for the moment.

First, the above explanation of doxastic alternatives does not entail that the subject has to be able to express her belief set in her own language. Some context may not qualify as a doxastic alternative simply because it does not feel or look like the actual world in some inexpressible way. Second, the explanation leaves entirely open how the subject's belief set (and its supersets) relates to our external belief ascriptions couched in our language. This is a question full of incompletely understood intricacies which we may, and are well advised to, leave aside. Third, this is most important for the rest of the paper, *to have a given belief set is an internal, non-relational property* according to our explanation. The fact that the context *c* is, or is not, a doxastic alternative of *s* at *t* in *w* clearly does not depend at all on anything in *w* outside *s* at *t*; it merely depends on *s*' internal disposition at *t* in *w*. In any case, we find it obvious that *s*' belief set as characterized above is entirely unaffected by exchanging all H_2O by XYZ, or exchanging *s*' linguistic community by a slightly different one which can change, in some other sense, *s*' water and arthritis beliefs into twater and tharthritis beliefs.

This characterization of a subject's beliefs will be used throughout this paper. The present purpose, however, is only a limited one; namely, to carry out the above idea of abstracting away usages as they actually are and to substantiate the formal diagonals thus obtained by the subject's beliefs about these usages. This raises the problem of distinguishing her beliefs about the usages of her linguistic community from all her other beliefs. Since the problem appeared unsolvable,[19] Haas-Spohn (1995, pp. 58f.) mentioned the possibility of restricting the domain of formal diagonals

[18] Christopher Peacocke pointed out to us that a lot of idealization is involved in what the subject's judgment would be under such counterfactual circumstances (perhaps it has only headache then and no judgment any more) and that the standards of idealization are not easily explained, maybe only in a circular way. Cf., e.g., the analogous critical discussion of the dispositional analysis of meaning in Kripke (1982, pp. 22–37). This is a legitimate worry which, however, we cannot further address here.

[19] See, however, footnote 30.

not to some selected beliefs, but to all beliefs, i.e. to the belief set of the subject. This has the desired effect that the subject's terms do not apply to any arbitrary objects so-called in the various contexts, but only to objects conforming to the subject's beliefs. In this way, the substance we have lost by introducing formal characters has returned to a subjectively relativized extent.[20]

Since formal diagonals, as well as belief sets, are internal, the restricted formal diagonals are internal as well. In this respect, they could qualify as concepts in the sense intended here. However, we are obviously stuck now with the other horn of Block's dilemma. If a subject's concepts are formal diagonals restricted to her belief set, any change in her beliefs changes her concepts; this is indeed an extremely holistic conception of concepts. Likewise, since any two subjects are almost guaranteed to have different concepts, communication and psychological generalizations seem like a mystery. Moreover, we have not gotten rid of the flaw that subjects having acquired different grammars are bound to have different concepts. All this seems unacceptable.

These considerations may suffice as a concrete exemplification of how Block's dilemma arises for a theory of internal concepts. So far, we have apparently avoided Schiffer's problem, since individualistic functional roles nowhere entered the picture, but we did so only by doing very badly on Block's dilemma. White (1982) has already anticipated a way out of that dilemma. His framework is quite similar to the one presented so far.[21] The domain of the partial characters he defines consists of what he calls contexts of acquisition, which are pairs of a possible world and some functional state the subject acquires in the world. In order to avoid the emptiness of syntacticism, White restricts the domain of the partial character of a given expression to equivalent contexts of acquisition having one and the same functional state as a component, and, by associating a separate functional state with each expression, White has prima facie avoided holism. However, these very sketchy remarks already show that it is the functional states which are doing the work here, and that the possible escape from Block's dilemma immediately leads into Schiffer's problem.

14.3 How to Define Concepts: A Proposal

Should we give up, hence, trying to explain concepts and narrow contents via the epistemologically reinterpreted character theory? No; we have not yet tried hard enough. So far, we have considered only two extreme options: the minimal option that the concept a subject associates with an expression contains only the trivial belief that the concept refers to whatever the word is used for, and the maximal option that the concept contains all beliefs of the subject, in particular all the beliefs

[20] Similarly, the propositional concepts of Stalnaker (1978) are not defined for all contexts whatsoever, but only within the context set consisting of the common presuppositions.

[21] For more detailed comparative remarks see Haas-Spohn (1995, pp. 141f.).

the subject has about the reference of the concept. This leaves open a huge range of middle courses which deserves inquiry.

Block (1991, p. 40) develops a nice picture by distinguishing between the lexicon entry and the encyclopedia entry of an expression. The encyclopedia entry of an expression corresponds to our maximal option; intuitively, however, it is rather the lexicon entry which corresponds to the associated concept. So, this is what we seek to characterize.[22] There seem to be two basic ideas, with ramifications, for driving a middle course towards an adequate notion of concepts.

One idea is to appeal to communal standards, e.g., to define a concept as containing just the social minimum of beliefs about its reference which is required for being recognized as a partner in communication. This is Putnam's idea of a stereotype. One may doubt whether such social standards of semantic competence exist in a salient way; but if they do, they certainly provide a useful notion for some purposes.

However, this idea seems to be the wrong one for our present purposes. If concepts are explained with reference to communal standards, then all competent speakers in the group automatically have the same concepts. This seems unwelcome for individualistic psychology, which should presumably allow for more conceptual variation across subjects. There is a further decisive objection which will be explained later on because it applies to other proposals as well.

The other basic idea, which appears sufficiently individualistic, is to appeal to the subject's recognitional capacities, i.e., to define a subject's concept of an object or a property as consisting of those features on the basis of which the subject recognizes the object or instantiations of the property. What this means, however, depends crucially on what one takes here recognition to mean.

There is room for interpretation since the strictest and simplest understanding of recognition does not work at all. The strictest understanding would be to say that a subject is able to recognize an object if and only if, under all possible circumstances, it is always and only the object itself which the subject takes to be the object. This is much too strict because we are hardly ever able to recognize objects in this sense; there are always circumstances under which we mistake the best known objects and even ourselves. Recall also how absurdly narrow Russell's acquaintance relation turned out to be. The same holds, mutatis mutandis, for the recognition of properties. Recognition therefore needs to be understood in a much weaker way. There are various options from which varying notions of a concept result.

Our recognitional capacities may first be seen in our normal means of recognition which work, according to our familiarity, fairly reliably in the circumstances we usually encounter. Something like this presumably comes closest to our intuitive notion of a recognitional capacity. However, it is utterly vague and entails that a

[22] The task reminds of the task of explicating the analytic/synthetic distinction which is confronted with the objections so forcefully initiated by Quine; indeed, Block's despair of drawing his distinction in a more precise way may be seen as a late effect of Quine's negative conclusion. However, this conclusion is still contested; moreover, the two tasks are not identical. So we should not be discouraged by these objections.

recognitional capacity may be something very varied. For instance, I[23] may identify my son on the basis of my belief that he is somewhere in the crowd before me and that nobody else in the crowd is likely to wear the same kind of satchel. The example shows that my normal means of recognition use a lot of specific knowledge that varies from situation to situation. By contrast, concepts are intuitively something more stable or invariant. It would certainly be strange to say that the fact that he has that kind of satchel belongs to my concept of my son.[24]

There seem to be two ways of gaining more specificity here. One way is to narrow down a subject's means of recognizing an object or an instantiation of a property to the way the object or the instantiation looks to the subject under various circumstances. This line leads to what are called the subject's perceptual concepts. The other way is to raise a subject's recognitional means from those she normally applies to the best and most considered means which she usually cannot, or does not care to, apply.

So far, the available options are very sketchy. But there is an argument which tells at even this sketchy stage which direction to pursue more thoroughly. The argument is this.

Clearly, we want our beliefs to be closely connected with our concepts since the contents of our beliefs should recursively build up from the concepts involved. For instance, in the primitive case of predication, we would like to characterize the content of a belief such as that a is F in the form of a truth condition, i.e., as something like the function from contexts to truth values given as follows:

> A subject's belief that a is F is *subjectively true* in a context c if everything and at least something that conforms to the subject's concept of a in c also conforms to the subject's concept of F in c.

The strange term "subjectively true" indicates that some explanation is still missing. For the moment, however, we may ignore this and take the adverb "subjectively" as redundant. We shall return to the point at the very end of the next section.

Moreover, one may stumble at the quantifier "everything and at least something". This is an attempt to do justice to the fact that there may be no or several objects in a single context c which conform to the subject's concept associated with the name "a". The attempt is certainly plausible. Finer analysis would show, however, that we here run into similar problems as does the counterpart theory of Lewis (1968) with non-identity-like counterpart relations. The corresponding proposal has

[23] There is a kind of pragmatic contradiction in writing a joint paper on subjective concepts, since the authors cannot use the stylistic device of taking themselves as an example. So we decided to use "I" in examples, and the reader is free to choose which of the authors he takes to be speaking.

[24] This is the first time that we slipped into talking of the concept of an object (or a property) instead of the concept associated with a given term – certainly a convenient, but also very dangerous slip, since it imports the de re/de dicto ambiguity and its proliferating consequences. Throughout, however, "the concept of a" is to mean the very same as the more clumsy phrase "the concept associated with 'a' ", though we are perfectly aware, of course, that the two phrases should be distinguished.

there proven to be insufficient, and more complicated solutions are required.[25] But we need not dwell on this point; our attempt will do for the rest of the paper.

The crucial point about the truth condition is that it seems to yield inadequate results when it is based on anything other than the subject's best and most considered means for recognizing *a*. For instance, if the subject's concept of *a* would consist in some communal stereotype of *a*, the subject could possibly believe that *a* does not satisfy its stereotype or that many things different from *a* satisfy this stereotype, and then the above truth condition assigns truth or falsity to the belief that *a* is *F* in contexts in which the subject would intuitively not count it as, respectively, true or false. The same holds in the case where the subject's concept of *a* consists of the criteria normally used to recognize *a*. Again, it seems possible that the subject knows or believes in a given situation that *a* does not currently have the features normally used for recognizing *a*, or that things different from *a* satisfy the criteria normally used for recognizing *a*, and then the above and the intuitive truth condition for the subject's belief that *a* is *F* diverge again. The only way to avoid this discrepancy seems to be basing the subject's concept of *a* on her best means for recognizing *a*, as we have proposed. The same holds, mutatis mutandis, for the property *F*.[26]

One may fear, though, that the best recognitional means available to a subject with respect to an object *a* or a property *F* come close to what Block called an encyclopedia entry. Should the subject not optimally use *all* her beliefs concerning *a* or *F* for recognizing *a* or an *F*? The answer is decidedly no. There are, for sure, many possible contexts in which the subject would recognize something as *a*, though it there lacks many properties the subject believes *a* to have. The subject has her ways, whatever they are, of distinguishing contexts which contain *a*, but with other than the believed properties, from contexts which do not contain *a* at all. This is a crucial assertion, without which the rest of the paper would not make any sense.

The following explanation captures this subjective distinction, or the subject's best recognitional means, or indeed the subject's concepts in a more explicit way:

> Let α be a name or a one-place predicate and $@ = \langle s, t, w\rangle$ the actual context (which may be any context). Then the concept $\beta_@(\alpha)$ which *s* associates with α at *t* in *w* is the function which assigns to each possible context $c = \langle s_c, t_c, w_c\rangle$ the set of objects in w_c which, according to *s*' judgment at *t* in *w*, might be the object, or instantiate the property, denoted by α in @.

Or to spell out the phrase "according to *s*' judgment" in analogy to the above explanation of doxastic alternatives: $x \in \beta_@(\alpha)(c)$, or *x* is a *doxastic counterpart* in *c* of what α denotes in @, iff the following holds: suppose that *s* would fully investigate

[25] See Hazen (1979) and Kupffer (2000, chs. 3 and 4).

[26] It should be observed that this proposal nicely parallels with how Haas-Spohn (1995) understands the usage of a name "*a*" or a predicate "*F*" in a given language *L*. We noted above that she follows the literature which tends to base such usages or communal concepts on the best judgmental standards available to the community of *L*. Hence it seems appropriate to do likewise in the individual case.

the world w_c – where this includes that it may take the perspective of every individual in w_c during its entire existence (so far it has a perspective at all), that it may learn all the languages there are in w_c, that it may subject each part of the world w_c to any investigations and experiments it can think, etc. – and suppose further that s would make its most considerate judgment about w_c including x after this investigation; then the assumption that x is the object, or instantiates the property, denoted by α in @ is compatible with s' judgment at t in w about what is denoted by α in @.

This may look imperspicuous to some and trivial to others. But its meaning and power will unfold in the following explanations and arguments.

14.4 Explanations

The final section will argue that this notion of a concept indeed helps the internalist against Schiffer's problem and Block's dilemma. This section is devoted to three kinds of explanations: some remarks about the features concepts have according to our definition; an explanation that the title of this paper is indeed appropriate; and a clarification of the relation of this definition to the proposals discussed in Section 14.2. So, let us first explain five more or less expected and instructive features of concepts which are entailed by our definition.

(1) Concepts are usually not egocentric. By this we mean that, usually, things can conform to one's concept associated with α in a context c without there being anything in c which could be oneself. Hence, insofar as modes of presentation and acquaintance relations have usually been thought to be egocentric, they are not concepts in the above sense.

(2) To which extent is the look, sound, or feel of things important for their conforming to one's concepts? It depends. Often it is conditionally important. Consider my concept of my son. Clearly, there could be many possible things in possible contexts which look perfectly like my son without possibly being my son according to my concept of him. Conversely, however, something could hardly be my son according to my concept without looking very much like him. Hardly! Of course, my son could look very different from his present look, not only actually, but also according to my concept of him. But if I encounter, in a possible context c, such a differently looking object, it could only be my son if there is something in the context c explaining why that object started or emerged to deviate from my son's look which is so well known to me. In this sense, the look of my son (the sound of his voice, etc.) is a conditional part of my concept of him. In a similar way, the look of species, substances, and also individual things is a conditional part of my concepts of them; for instance, no doxastic counterpart of the black ball-pen in my drawer could be red during its entire existence. But there are other cases as well. It seems, for instance, that the look of things is not essential for their conforming to the concept I associate with the word "table"; what is essential is only what is done with the things in the relevant context. If there are culturalized beings in the context which use a given object only for sitting down *at* it, then that object counts as a table

according to my concept, even if it never looks like a table. Conversely, if something looks like a table, but is only used as something else, say, for sitting *on* it, then my concept does not count it as a table, but, say, as a seat.

(3) To which extent does the place of objects enter into one's concepts of them? Again, it is very often conditionally important. According to my concept of him, my son could be (almost) anywhere in the universe. However, the context must then provide some plausible story of how he got there. Any object, however intrinsically similar to my son, could not be my son if it is far away from Earth, or Germany for that matter, during its entire existence. The same holds for many concepts of many other objects; after all, most objects we know are on the surface of Earth. The same may even hold for predicates. One may think, for instance, that a species which develops somewhere else in the universe, but, as it happens, interbreeds with our tigers, does still not consist of tigers. If so, one's concept of tigers includes their emergence on Earth.

Hence, very many of our concepts are, so to speak, geocentric. This entails the question what my concept of Earth may be. It seems to be quite poor. According to my concept, at least, the history of and on Earth so richly known to me is highly contingent to Earth; almost any planet of comparable size, age, and composition revolving around a sun of comparable size, age, and composition in the Milky Way could be Earth. And, of course, my concept of the Milky Way is even poorer, since it contains hardly more than the Milky Way being some spiral galaxy.

(4) Their causal origin is essential to many objects. This is also reflected in our concepts of them. For instance, nothing which is not procreated by us could be our son, and since I also believe so, my concepts of myself and my spouse enter into my concept of our son. The same holds with respect to ourselves and our parents. Of course, my concepts of our ancestors soon get very dim; still, all of them are part of my concept of my son. In fact, my son could not exist without history being pretty much as it is. Thus, a lot I believe about history enters into my concept of my son. This makes for a perhaps unexpected richness of that concept. In the same vein, my concept of things is quite poor when I know very little about their causal preconditions, as is the case, for instance, with Earth. In fact, what we just said about the conceptual role of location presumably reduces to the present point about causal origin. Our son could not be born outside Germany or Earth, unless we, or our parents, etc., travelled. The same holds, mutatis mutandis, for tigers and other kinds if their causal origin is essential for them.

(5) Do concepts involve social relations, are they mutually connected by communication? Yes, of course; there is a clear relation between the concept I associate with a certain expression and the concepts others associate with that expression, a relation which Putnam (1975) has described as division of linguistic labor. Consider my concept of an elm, to take one of Putnam's examples. Elms might exist without mankind; in such a context, the extension of my concept of an elm would alternatively contain elms, beeches, and, maybe, other deciduous trees, since I, by myself, cannot distinguish elms from beeches and, maybe, other trees. This may entail that my perceptual concept of an elm is the same as that of a beech, but it does not entail sameness of the two concepts in our sense. On the contrary, since I believe elms and

beeches to constitute different kinds, and since I am allowed to identify the various kinds of trees in that context, my concept of an elm has any one of these kinds as extension in this context, and my concept of a beech any other kind, though I do not know which.[27]

In other contexts there is an even clearer difference in the extensions of the two concepts, namely in contexts in which there is a linguistic community which generally resembles my actual community as I know it and which I observe during my full investigation of these contexts applying the term "elm" only to certain trees and not to others (to which I might have been inclined to apply it as well). Then there are two possible cases. Either these applications of the term "elm" contradict my concept of an elm so flatly – say, the community applies it to coniferous trees – that I conclude that this could not be my linguistic community after all and that its judgment cannot help mine in this matter. In this case, my judgment is as bad and the extension of my concept of an elm as wide as before. Or the linguistic community in the context behaves like mine in every relevant respect, and in particular with respect to the term "elm", so that I conclude that this community could be mine and that I may trust its judgment. In such a context, the extension of my concept of an elm is as narrow as the usage of the community and certainly different from the extension of my concept of a beech (though, of course, it would be compatible with my concept of an elm that this counterfactual community applies "elm" only to beeches).

In this way, the division of linguistic labor is reflected in subjective concepts. This entails in particular that referential and deferential aspects are often inextricably mixed in subjective concepts. The simple reason is that subjects often trust the judgment of their fellows more than their own. Of course, the degree to which semantic deference enters into subjective concepts may vary considerably. For instance, my concept of an Indian deity, say, or of multiple sclerosis, is so poor, that I would follow almost any opinion if it presents itself as a consistent opinion of our experts. In such cases, the deferential component of concepts is overwhelming. By contrast, I may be convinced that I know more or less as well as all others what tables are. In such cases, my own most considerate judgment is hardly helped by others, and the deferential component of my concept of tables largely vanishes. However, it seems that it never vanishes completely in concepts associated with linguistic concepts; it seems present even in the concept associated with the predicate "x looks red to me" in the phenomenal reading.[28]

Let us next explain the appropriateness of the title of our paper. Our aim was, we said, to drive a middle course between the minimal and maximal option, both of which we found to be inadequate. So, which beliefs are contained in the concept a subject associates with the expression α if they are more than that α has an

[27] This idea of alternative extensions in one and the same context is not mentioned in our above definition of concepts, because it entails additional complications. But it seems required in order to overcome the difficulties referred to in footnote 25.

[28] This point is made already by Austin (1962) (see his magenta example on p. 113). Cf. also Spohn (1997b and 1997/98, sect. 5) [here: ch. 13 and sect. 11.5].

extension and less than all beliefs about that extension? Our title gives a simple and informative answer which runs as follows.

G is an essential property of *a* if and only if it is metaphysically or ontologically necessary that *a* is *G*, i.e., if nothing which is not *G* could be (identical with) *a*. For instance, being human or having the parents our son has are essential properties of our son. This is the common definition; it is full of niceties, which we better skip over, however. We can extend it to a relation between properties. *G* is essential for *F* if and only if it is metaphysically necessary that every *F* is *G*. For instance, being unmarried is essential for being a bachelor (though it is not essential for bachelors, for no bachelor is essentially a bachelor), or consisting of hydrogen and oxygen is essential for being water.

Now, one may express our definition of concepts also in the following way. The concept a subject associates with "*a*" is the conjunction of all concepts *G*, or the strongest *G*, such that the subject believes that *G* is essential for *a*. Similarly, the concept a subject associates with "*F*" is the conjunction of all concepts *G*, or the strongest *G*, such that the subject believes that *G* is essential for *F*.

When one compares this with the original definition, it is rather obvious that this is an equivalent characterization. Indeed, it is trivial in view of the fact that being identical with *a* is the strongest essential property of *a*, and being *F* is the strongest property essential for being *F*. The characterization would become more interesting if we were to introduce restrictions on the metaphysical side, for instance, by excluding identity from genuine properties and relations; or on the epistemological side, for instance, by postulating that all concepts are ultimately qualitative in some suitable sense. We would in fact be prepared to make such restrictions, but it would take us too long to go into this issue.

Let us rather briefly check whether this characterization agrees with the five features of concepts just noted. What we said about the fact that beliefs about causal origin often are part of concepts fits perfectly, of course. We also stated that the look of objects or kinds often enters into our concepts of them. But, as a rule, looks are certainly inessential. However, we qualified our statement. Often, the look of an object or of a kind displays its essence provided that it is allowed to unfold its normal look; and it is only this complex concept which is part of the concept of an object or a kind. Finally, what about the deferential component of concepts? What others believe about an object or a kind is certainly not essential to it. Sure, but to the extent we trust others, we believe what they believe, and if we take the experts' beliefs about essences as trustworthy and they believe essences to be such and such, we also believe these essences to be such and such. So, the present characterization agrees well with the earlier observations.

Viewed in this way, is our proposal for defining concepts not a familiar one? We are not aware of this. To our knowledge it is mentioned only in Block (1995, sect. 4) where he attributes the view to two lines in Fodor (1987), discusses it on one page, and dismisses it right away. The paper is about one example, namely the concept a teen associates with the word "grug" which denotes beer in his assumed dialect. The teen knows very little about grug; he knows, e.g., that it comes in six-packs. Block simply assumes that this belief is part of the teen's concept of grug, and he is certainly right to claim that it is not essential to grug to come in six-packs. But Block has a different notion of concepts here. His notion seems to be the one we have already mentioned, namely that concepts are something like normal means of recognition, and the teen's poor means of recognizing grug refer to its packing. However, we have already argued that this is not the best notion of a concept, and indeed we would flatly deny that the belief that grug comes in six-packs is part of the teen's concept of grug. So, as we say, there does not seem to be much discussion of the line of thought we are proposing here.

Let us finally explain how the present definition of concepts relates to the two kinds of diagonals in Section 14.2. In a way, this is for our own records, but it also illuminates our definition in some important respects.

Recall that the objective diagonal of an expression α in a given language L was the function which is defined for all contexts in which the usage of α in L exists and which assigns to each such context the extension α has in that context. In particular, it assigns to the actual context the actual extension of α in the familiar sense. The formal diagonal of α as part of a grammar G did the very same. The only difference was that the formal diagonal has a larger domain consisting of all contexts in which the grammar G exists which might be realized in different languages. The specific language having the grammar G and spoken in the context was implicitly fixed by the subject of the context.

By contrast, concepts as defined above are functions defined for any contexts whatsoever (though most of them will be so alien that we find hardly anything in them conforming to our concepts). The all-important question is therefore: do they agree with diagonals on their common domain? And the crucial answer is: conditionally yes!

Imagine that the subject having the concept associated with α may investigate a context in which her actual language L exists with its very usage of α. Then we may expect that the subject judges that the context's linguistic community might indeed be her own, at least as far as the usage of α is concerned, that the community as a whole is more competent than she himself with respect to α, and that she should therefore follow the community's final judgment. In this way, semantic deference enforces an agreement between the extensions of the subjective concept and the public usage. This expectation may be wrong, however. The subject may also find the usage of α in L, as compared with her concept of α, so strange that she (falsely) concludes that this is not her actual linguistic community, rather than concluding that she is the victim of a severe misconception. Only then may the subject's judgment about α or α's deviate from that of the community. Semantic deference is thus an important ground for the subject's agreement with the community.

Note, however, that the extension of the subject's concept and the experts' concept may agree even in a context in which the subject does not defer to the experts or in which none of them exists at all. The context may be kind, so to speak; the subject may believe that her concept refers to a single natural kind, though she knows very little about that kind and the experts may know very much. But suppose the context provides only one natural kind which conforms to the little the subject knows about it. Then, only this kind is in the extension of the subject's concept, just as in the extension of the much better informed communal concept. And again the two agree.

Let us illustrate this with the two standard examples "water" and "arthritis". The actual extension of the concept Oscar presently associates with "water" consists of all H_2O and nothing else, even if Oscar knows nothing about chemistry. The primary reason for this is that Oscar believes water to be a natural kind amply instantiated in his environment and that there is no natural kind in the actual world which he would confuse with water in his maximally informed and considerate judgment. For the same reason, the actual extension of the concept which Oscar's ancestor associates with "water" 250 years ago also consists only of H_2O. Semantic deference becomes relevant in a context in which there are two kinds of liquid which Oscar by himself might take for water. If he finds there a linguistic community which might be his own and which acknowledges only one liquid to be water, then his subjective concept has only this liquid as extension in this context. If he finds there a community which is as indiscriminate as he is, then both liquids constitute alternative extensions of his concept in this context. And the same is true, if he finds that there are two trustworthy communities, as in Putnam's twin earth story, to which he might defer, and which refer, however, to different liquids. All this shows that there is, on the one hand, a lot of agreement in the extensions of various subjective and communal concepts at different times, and that, on the other hand, the differences among all these concepts show in suitable counterfactual contexts.

What about the actual extension of the concept Fritz associates with "arthritis"? This case is more delicate. If Fritz' belief that arthritis is an ailment which may occur in the thigh is conditional on the agreement of his community, then he will also defer to his actual community which denies this, and the extension of his subjective concept will coincide with that of the communal concept. If Fritz' belief about the essence of arthritis is unconditional, then he will not acknowledge the actual community to be his community, and his judgment will be unassisted. In this case, the subjective and the communal concept may diverge. But it may also be the case that, after fully investigating the actual world, his judgment is that arthritis occurs only in the joints because the investigation shows that there is a natural kind of appropriate ailments in the joints, but none of which extends to thighs.

If we return to comparing our definitions of diagonals and concepts, a further important difference emerges. Concerning objective diagonals in *L*, we said that the extension of α consists of the object(s) with the same essential properties as the object(s) from which the usage of α in *L originates*. This is a clear heritage of the causal theory of reference on which Haas-Spohn (1995) relies; and therefore the usage of α in *L* had to exist in the context *c* in order for α's extension being defined in *c*. By contrast, the extension of α in *c* according to a subjective concept consists of the object(s) in *c conforming* to the concept. There, the causal aspect has disappeared and with it the restriction of the concept's domain. But how then can the two functions, the objective diagonal and the subjective concept, agree within their common domain?

The question does not really concern subjective concepts. It rather points to a tension in our notion of a usage. On the one hand, a usage has, we said, an extension only where it exists and has causes. On the other hand, we said that a usage is something like a communal concept internal to the community, and then objects in any world should be able to conform to the usage. The tension hides a confusion of metaphysical and epistemological matters. Metaphysically, it is inessential to most objects or kinds of objects that they are actually conceived of, i.e. that they cause an intelligent species to form specific concepts. We also believe this. So, if a (communal or subjective) concept assembles beliefs about the essence of its reference, such a causing does not belong to it. Within an epistemological perspective, however, the belief in such a causing is an a priori companion of the concept; any community (or subject) which acquires a concept associated with some term thereby acquires the belief that the concept and the term refer to the object(s) in confrontation with which the concept was acquired. This is so at least to the extent in which a causal theory of reference applies. Hence, insofar as the concept or the usage exists in a context, its extension is described by our objective diagonals in the same way as by our definition in this section.[29]

Similar remarks apply to the comparison of formal diagonals with the concepts of a subject s. Again, the two functions agree for those contexts in which the subject s_c of the context speaks a language with the grammar G (otherwise the context would not be in the domain of the formal diagonal) and in which the community speaking that language in that context could be s' community as far as s believes (so that s can defer judgment to the community). And the causal ingredients in the formal diagonal give an a priori condition on the extension within this common domain, and thus do not constrain the extension as specified by the concept. What about the fact that the formal diagonal of α essentially involves the expression α itself, whereas the concept associated with α does not? Again, this does not create a difference within the common domain, since it is a further a priori condition on the concept associated with α that it is associated with α.[30]

---

[29] This does not seem to agree with Putnam's Twin Earth stories. Suppose the English and the Twin English community exist in the same world and associate the same internal communal concept with "water", as Putnam suggests. Everybody agrees that the concept has different extensions in the two communities, namely, respectively, H_2O and XYZ. But we seem to have to say that both extensions consist of all H_2O *and* all XYZ since both, H_2O and XYZ, conform to this concept. This is not so, however. According to the concept, its extension in this world consists *either* of all H_2O *or* of all XYZ. We don't know of which; if our extension is H_2O, theirs is XYZ, and vice versa. The decision is made by the context which, by being a centered world, says which community is in the center.

[30] A further thought which we owe to Manfred Kupffer: In Section 14.2, when restricting formal characters and their diagonals to the subject's belief set, we have, it may have appeared, given up too soon on distinguishing the subject's beliefs about her linguistic community and its meanings from her other beliefs. We may conceive of a doxastic alternative c in a richer way, consisting not only of an individual s_c, a time t_c, and a world w_c, but, given s_c has the language l_c at t_c in w_c, also of the objective character function $\|.\|_{l_c}$. A subject's belief set then consists only of doxastic alternatives thus enriched (because she believes to have a language), and only of those enriched by a character function which might be, for all she believes, the character function of her own language. In this way, we may explicitly distinguish the subject's beliefs about the meanings of her language, and we could explain the subject's concepts by restricting the formal character and its diagonal not to the subject's belief set, but only to the larger set of doxastic alternatives enriched by a suitable character function.

There is no conflict, though. Our previous considerations rather imply that concepts thus explained (= the diagonals of the larger set of enriched doxastic alternatives) agree with concepts in our sense on their common domain; the difference is only that concepts in our sense have, desirably, a wider domain. However, the agreement supports our case; it is nice to see that this different line of thought arrives essentially at the same result.

All this enables us, at last, to explain what is subjective about our above truth condition of a subject's belief recursively built up from the concepts involved. In contexts in which the subject can defer her judgment to the surrounding community, there is nothing subjective about the truth condition. To that extent, the truth condition is intersubjective and indeed objective (since the relevant contexts may be fully investigated, with no space for error left), i.e. to that extent the subject's belief that a is F is subjectively true if and only if the sentence "a is F" of her language is true. The difference shows in other contexts without an appropriate community. There, the poverty of a subject's concepts and a large divergence from the concepts of her community may come to the fore. Hence, there is a difference in subjective and objective truth conditions, as it should be, but not a critical one.

14.5 Individualism Rescued?

To what extent does the proposal explained in the previous sections promote the individualist's project? Four points are worth discussing.

(1) Our proposal provides something of a definition at all; this is more than what one usually finds in the literature. It does so mainly because it firmly rests on the epistemologically reinterpreted character theory, which has by far the best formal grip on these matters. This theory also provides concepts and narrow contents with a recursive structure essentially[31] following the recursive structure of the expressions with which they are associated. No negligible advantages.

(2) Again, the crucial point is, of course, that concepts are individualistic according to our definition in the same way as belief sets are individualistic according to our definition; to have a concept is an internal, non-relational property. Which function from contexts to extensions a subject associates with an expression depends solely on its internal cognitive state, does not presuppose the existence of anything outside the subject, and does in no way change when the environment of the subject changes without affecting her internal state. For instance, Oscar, Twin Oscar, and the (appropriate) Swamp Man would display precisely the same dispositions; they would respond in our huge counterfactual test in precisely the same way, and hence they have precisely the same concepts. Of course, agreement will usually be at most partial; the Frenchmen may associate with "Londres" the same concept as I associate with "London", while our concepts diverge elsewhere.

Defining concepts and contents in a narrow way is one thing; describing them is another. We have to build a theory of how concepts combine to contents, how contents become attitudinized, how perception acts upon the attitudes, how the attitudes result in action, and so on. By doing this we say how this huge array of counterfactuality integrates into factuality; conversely, this makes this array accessible

[31] See the qualification giving rise to footnote 25.

from the facts we observe on the street and in the lab. Of course, theory is vastly underdetermined by the data, here as everywhere. We have not said a word about how this theory goes and which ways of describing all these internal entities go along with it. But this would clearly be a different task, one which does not impair the internality of its starting point.

On the contrary, spelling out this theory would fully display the strategy of individualism, which consists of defining the momentary states (i.e. state types) of subjects in such a way that they are connected with past and future only through causal laws. By contrast, externalists take such connections to be part of the identity conditions of these states, by defining them either as being caused in a specific manner, as does the causal-information theoretic account of Dretske (1981), or as dispositions or attitudes analytically tied to their manifestations or intentional objects – a false understanding of (most) dispositions, as Spohn (1997c) argues. Even the functionalist is externalistically biased insofar as he defines a mental state by its functional role, by its place in a causal net extending from past to future. In our account, however, the narrow mental states of a subject are not defined by their causal ancestry, but rather as dispositions which are only causally related to their actual manifestations. It is only the envisaged rich theory which conjectures the functional role of these states; that role is not definitionally fixed to begin with. These remarks show at the same time that our proposal has not led us into Schiffer's problem; our proposal is so far independent of functionalist conceptions.

(3) The next question, then, is how we fare with respect to Block's dilemma. Here, it is clear that we have perfectly avoided the syntacticist horn of that dilemma. Which expression a subject associates with a concept is fully contingent and does in no way add to the identity of the concept. This entails in particular that members of different linguistic communities may nevertheless have the same concepts. Of course, the deferential component of a subject's concept makes reference to her own linguistic community, and this distinguishes concepts in unconnected languages. However, translation has the effect of merging the experts of the communities and thus of merging their usages or communal concepts, and thereby differences of subjective concepts due to deference vanish as well.

(4) Whether we are equally successful with respect to the holistic horn of Block's dilemma is less clear. This is the final point, to be discussed at more length. We shall not attempt to clear up the term "holism"; there seems little agreement on its precise meaning. However, it is very clear that concepts as we have explained them are thoroughly interconnected. It would be extremely important to study the architectonics of concepts in detail – though this is nothing we can achieve here. We see no reason, though, to expect the conceptual connections to be unidirectional, i.e. that there is a set of basic concepts from which all the other concepts are defined step by step, as Carnap (1928), for instance, has tried to establish in an exemplary way. Rather, all kinds of circular dependencies among concepts are to be expected. Concepts will certainly turn out to be holistic.

The essential reason for this holism is that, in the first place, ontology is holistic. There is rich ontological dependence among objects and properties; we mentioned the example that many objects and maybe even properties ontologically depend on Earth, i.e., could not exist or be instantiated, if Earth would not exist. Hence, if essences are thoroughly intertwined, beliefs about them, i.e. concepts, will be intertwined as well.

However, if we follow Block's and Fodor's definition of holism, concepts as we have explained them are not holistic. Block (1991) says "that narrow content is holistic if there is no principled difference between one's 'dictionary' entry for a word, and one's 'encyclopedia entry' " (p. 40). But the whole point of this paper was to propose such a principled difference! The lexicon entry for a word contains only one's beliefs about the essence of its reference, whereas the encyclopedia entry contains all other beliefs about the reference as well.

The case is similar with Fodor (1987). What he says about holism does not exactly fit our present discussion. He there defines that "meaning holism is the idea that the identity – specifically the intentional content – of a propositional attitude is determined by the *totality* of its epistemic liaisons" (p. 56). This does not exactly fit, first because Fodor addresses only the narrow content of propositional attitudes and not that of subsentential expressions, and second because the term "epistemic liaisons" refers to confirmatory or justificatory relations between propositions – something we have not touched at all. If, however, we straighten out the definition by taking the epistemic liaisons of a word to consist in the beliefs in which it occurs, we are back at Block's definition.

Let us look, hence, at what Fodor (1987) dubs the Ur-argument for meaning holism which runs as follows: "Step 1: Argue that at least some of the epistemic liaisons of a belief determine its intentional content. Step 2: Run a 'slippery slope' argument to show that there is no principled way of deciding *which* of the epistemic liaisons of a belief determine its intentional content. So either none does or they all do. Step 3: Conclude that they all do" (p. 60).

Fodor goes on to discuss three versions of the Ur-argument and tries to argue that in all three of them step 1 has erroneously been taken for granted. Given the above straightening out we have no quarrel with step 1, however. Rather, step 2 is faulty. There may be vagueness or indeterminateness in the beliefs about essences or perhaps even in the essences themselves. But there is no slippery slope.

However, it is not important whether or not concepts should be called holistic according to our definition; holism as such is not bad. The question is rather whether or not the unacceptable consequences for which holism is blamed in this area are avoided by our definition. Let us look at four such consequences.

A first bad consequence of holism appeared to be that belief change ipso facto meant conceptual change. This, however, is not so at all with our proposal. Take my concept of my son, again. I acquire new beliefs about him all day long and forget many old ones. But, according to our explanation, my concept of him has in no way changed in the last few years; all the beliefs I have acquired or forgotten concerned contingent matters and did not add to, or subtract from, my beliefs about his essence. The same holds, say, for my concept of tables. Almost every day I learn something about tables, for instance, at which places tablehood is instantiated. But my concept of tables is fixed since long ago.

A second bad consequence seemed to be that holism renders impossible intrapersonal and interpersonal psychological generalizations. This is an objection we never understood. Each individual constellation may be unique, but this does not prevent it from being subsumable under general laws. It was always clear that, strictly speaking, there is only one application for Newton's theory of gravitation, namely the whole universe. But this did not deprive it of its lawful character. Block (1991, p. 41) makes similar remarks to the effect that there is not really an objection here.

A third bad consequence of holism was said to be that it makes communication miraculous because the concepts of different subjects are almost guaranteed to differ, preventing them from understanding each other. There are several remarks to be made about this point.

To begin with, we are not sure whether subjects need to have the same concepts in order to understand each other. It rather seems to be sufficient to know which matter the others talk about, i.e. to which objects and properties they refer. As long as this is secured, it does not do much harm when we have a different grasp of the objects and properties referred to; communication may also serve to assimilate the differing grasps. In this perspective, sameness of concepts is required only insofar concepts are constitutive for ontology. This may indeed be a relevant aspect in abstract realms, but we do not think it has much relevance in everyday matters.

Still, it would be good to know the extent to which we have the same concepts according to our proposal. The answer is a mixed one. Take my son again. I know his grandparents, others don't. So, our earlier remarks imply that there are diverging concepts of him. Take Bill Clinton, by contrast. Most of us know him just from TV. Certainly, we have looked at TV at different times, and hence, we believe different things about him. But there is no reason to assume that our concepts of him differ in any way; we believe quite the same about Clinton's essence. Take tables. Again, there is no reason why our concepts of tables should differ despite our differing beliefs about tables. If we compare the functions from contexts to extensions which we associate with the word "table", our guess would be that the variance keeps well within the range of vagueness of that word. Take elms, finally. Presumably, many of us are still roughly in the poor state Putnam describes. But some of us may have been ashamed of this, and thus have informed themselves. Their concept of an elm, then, differs from that of the rest. Hence, there is neither a guarantee nor an impossibility of agreement in concepts.

However, one should observe that there is considerable conditional agreement. We argued in the previous section that subjective concepts and objective diagonals agree on their common domain (if the relevant condition is satisfied). Since this holds for all subjects, we find the same conditional harmony among their concepts.

These remarks do not add up to a satisfactory discussion of the question how communication is possible on the basis of concepts as beliefs about essences. But we may tentatively conclude that there is no clear evidence at all that a serious objection will be forthcoming here.

The fourth and final bad consequence of holism seems to be what Fodor (1987, p. 102) calls the disjunction problem, which is the problem of how error is possible – which it must clearly be – according to one's theory of meaning, content, or concepts.

This problem arises in particular for the causal-information theoretic account of Dretske (1981), and in this way it is treated by Fodor (1987,1990, chs. 3 and 4). However, the problem of error also plagues holistic accounts. Suppose Fodor's Ur-argument, quoted above, is sound. Then all the epistemic liaisons of a content which I believe, i.e. hold to be true, would be constitutive of that content. Now suppose I change these epistemic liaisons. Could this result in a different balance of reason for that content and even in a different judgment, e.g., that this content is really false? No, because it would be a new content which I would judge false; the old content would cease to exist. That is, the old content can exist only as held true. Similarly, if a concept is an encyclopedia entry in Block's sense. I believe all parts of that encyclopedia entry to be true. Now, for some reason, I want to change my mind and to discard some parts. Because they have proved wrong? No, we cannot put it this way. If I change my encyclopedia, I change my concepts, and my beliefs change content. So, again, I can put together my concepts only to form contents with a fixed truth assignment; all contents would be conceptual truths or falsehoods. These would be fatal consequences indeed.[32] Of course, I often err even by my own lights, and any adequate theory must be able to account for this.

It should be clear, however, from our comments on the first possible objection that our proposal has none of these absurd consequences and enables us to change our mind without changing our concepts. In particular, our explanation of concepts and our subjective truth condition for beliefs clearly allow us to have beliefs which are false by our own lights; our most considerate judgment may well falsify our actual judgment. There is no error problem for our account.

So, to sum up: have we escaped the holistic horn of Block's dilemma? Our discussion does perhaps not firmly establish a positive answer, but it shows, we think, that the prospects for our proposal are bright – all the more so as it was clear that the syntacticist horn of the dilemma was definitely avoided and that there was no danger of stumbling into Schiffer's problem.

---

[32] In fact, this issue was first raised in relation to the account of the meaning of theoretical terms in Kuhn (1962), pp. 111ff. and 198ff., and, for instance, Feyerabend (1965).

Chapter 15
Changing Concepts*

At the beginning of his paper (2005), Nenad Miscevic said that "empirical concepts have not received the epistemological treatment they deserve". When first reading this complaint I was surprised. Are the huge philosophical efforts to come to terms with concepts not primarily directed to empirical concepts? Miscevic insists, however, that concepts evolve, that we learn concepts and change concepts, and that this is most obvious in the case of empirical concepts like our concept of whales or our concept of water. I realized then that Miscevic has raised a most important question: *How can a concept change?* Why is this question important?

It is almost standard that a concept is, or may be represented as, an intension, i.e., as a function mapping possibilities to appropriate extensions. Disagreement starts when it comes to say what the possibilities are, which specific function a concept is, etc. It may also be that a concept rather is a two-dimensional entity, i.e., a function mapping two possibilities, possibly of different kinds, to extensions. This standard is widely agreed, and it is important to note that the standard is enforced by the fact that we want concepts to somehow build up propositions and that we want to somehow understand propositions as truth conditions.

However, the simple consequence is: different functions, different concepts. We can say that yesterday we had this concept and now we have that, but the standard conception does not allow us to say that yesterday's concept changed into today's. This consequence looks unacceptable.

Miscevic is right to point out that the problem is particularly important for an inquiry into the relation between concepts and apriority. If concepts may change, then, presumably, conceptual truths, i.e., truths in virtue of these concepts, may change as well. But this badly fits to the guiding idea that concepts are the source

*This little paper is first published here. It emerged from a commentary on a talk of Nenad Miscevic at the DFG conference *Concepts and the A priori* at the University of Konstanz on June 17–19, 2004, that appeared as Miscevic (2005). It hardly required any change for translating it from a commentary into a kind of appendix to Chapter 14.

of apriority. If conceptual truths change under the influence of experience, they are rather a posteriori and not a priori, as Miscevic (2005) has elaborated. The issue indeed threatens the presuppositions of the dominant approach to apriority.

So, how may we conceive of changing concepts? What the standard conception describes is presumably only *states* or *stages* of a concept. The question then is what holds the various stages of a concept together so as to form one concept persisting through its possible changes?[1] Let me discuss five different answers:

The *first* answer is that it is simply the *word* used at various times to express the various stages of a concept. There is a lot of truth in this answer. Still, it is definitely unsatisfactory, for three reasons. First, we should leave room for non-linguistic concepts not expressed by words. Secondly, it is not so clear what a word is. One may think that a word is identified by its morphophonological shape. But this shape may change, too. So, we would need an account of words as persisting through their morphophonological changes. We may expect, though, that linguistics provide such an account. Thirdly, however, there is the objection, decisive in my view, that no morphophonological individuation of words will do. One and the same word shape may stand for different concepts, by being used ambiguously, by being used in different languages, or by changing to an entirely different meaning.[2] This is a familiar point. If words are to individuate concepts, they have to be semantically interpreted words, and since their semantic interpretation roughly consists in the concepts, we have first to individuate concepts in order to individuate words in this semantic sense. Hence, the first answer is unhelpful.

A *second* answer is that the continuity of a concept lies in the continuity of its *possessor*. The answer is still incomplete; we would still have to say what makes for the continuity of a concept within the possessor. Anyhow, the proposal won't do. Whether concept-possessors are persons or entire linguistic communities, a concept does not live and die with its possessor. Moreover, if the possessor is somehow essential to a concept, it becomes difficult to explain how a concept can be

[1]In the German context the question raises quite different associations. There was and still is a very influential movement focussing on so-called Begriffsgeschichte that provided the methodological foundations for the journal *Archiv für Begriffsgeschichte* founded by Erich Rothacker in 1955 and for the well-known encyclopedia *Historisches Wörterbuch der Philosophie* initiated by Joachim Ritter, a renowned student of Rothacker, in the 1960s. "Begriffsgeschichte" may modestly mean the history of the usage of a philosophical word or term. The projects were, however, more ambitious; the idea rather was to present the history of the concepts themselves (within which concepts were certainly never conceived as functions from something into extensions). If Schröder (2000) is correct, the repeated criticism that it does not make sense to speak of a history of concepts beyond that of terms was never convincingly rebutted by that movement. That is, 50 years of German post-war philosophy did not really get beyond the first and the second deficient answer I am about to discuss. Not that I would help here concerning philosophical concepts. I am happy to talk about ordinary empirical concepts; philosophical concepts clearly are the most difficult ones for the semanticist.

[2]See the most illuminating discussion of Kaplan (1990/91) whom the question how words are individuated leads to quite similar considerations than the ones presented here.

shared by different possessors. There is a possibly useful notion of cultural identity according to which a cultural community is constituted by its shared concepts. Accordingly, we still form a cultural community with the ancient Greeks and Romans (to some extent at least). In this sense, it is indeed trivial that a concept is continuous with the community possessing it. However, the present proposal obviously becomes circular by this move.

A *third* idea proceeds from the observation that concepts have *aims*; concepts aim, we might say, at their subject matter. So, perhaps, the various stages of a concept are united by their common aim. To speak less metaphorically: What a concept is about, aims at, or attempts to grasp, is its actual extension or reference or rather, speaking two-dimensionally, its actual secondary or C-intension. And a concept may change while its reference remains fixed. Our notion of gold, e.g., has changed several times; still, it is always our notion of gold that always refers to the same stuff. When Putnam (1975) calls upon us: "Let's be realistic!", he refers exactly to this point, as Miscevic (2005) has again emphasized.[3]

I think this idea is on the right track; but it still won't do. Clearly, we, individuals as well as communities, may have two different concepts for the same subject matter, as long as we don't notice it; this is the familiar story about Hesperos and Phosphoros before the Babylonian discovery of their identity. Conversely, a concept may remain the same while changing its reference, as long as we don't notice it. This is the jade story. Originally, the reference of the Chinese jade concept was Chinese jade. Yet, after the massive import of substantially different, but phenomenologically indistinguishable Burmean jade into China the reference changed to Chinese *or* Burmean jade without, I contend, any change in the concept.

For both of these exceptions the qualification "as long as we don't notice it" was essential. So it seems it is not the actual aim or reference of a concept that counts, since we might be in error about the actual aim in some way or other. What counts, this is the *fourth* proposal I want to make, is rather the *believed reference* or the *intended aim* of the concept.[4]

This fits well to the conception of concepts in Haas-Spohn and Spohn (2001) [here: ch. 14]. We have argued there that my concept of an object or a property is what I believe this object or property to *be*. Since what an object or property *is* is determined by its essence, the collection of its essential or metaphysically necessary properties, this means that my concept of an object or property consists in my belief about its essential properties. And we have argued that this conception does not only fit to our concepts of natural kinds, on which Miscevic has focussed, but also

---

[3] To be sure, Putnam (1975) is reluctant to speak of concepts. But he pleads that the intension at least of natural kind terms is rigid, i.e., the projection of their actual extension to other possible worlds, and that this intension remains constant throughout possible changes of our grasp of that extension.

[4] Kaplan (1990/91) arrives at the corresponding conclusion concerning the identity of words through their history of usage.

to our concepts of objects and other properties. The only modification I have to add in view of the present considerations is that what we then called concepts are rather concept stages in the present sense.

Given this conception of concepts or rather concept stages, there is no fundamental mystery in conceptual change. Beliefs about essences may change just as any other beliefs, and thus, in principle, all the well-elaborated accounts of belief change apply to conceptual change as well. The details and the particular role of beliefs about essences within our overall net of inductively connected beliefs would certainly need most careful considerations. Still, the framework is well prepared by Spohn (1988) [here: ch. 1] and subsequent papers. The crucial point, of course, is that I have much more beliefs about an object or property than about its essence, and the former may change without the latter. Thus, not every belief change is a conceptual change. To avoid this badly holistic consequence was indeed our main goal in Haas-Spohn and Spohn (2001) [here: ch. 14].

What is it, then, that unites all these concept stages to one concept? It is my belief that they are all about the same subject matter. If I first believe that, necessarily, whales are fishes, and then believe that, necessarily, whales are mammals and thus change my concept of whales, I still take these beliefs to be about the same subject matter, namely about *those* animals, and hence as expressing various stages of the same concept. This is how the fourth proposal to individuate concepts and the specific conception of concept stages Haas-Spohn and Spohn (2001) [here: ch. 14] fit together.

How, then, should we relate analyticity and apriority to concepts thus conceived? Miscevic (2005) proposes to apply analyticity to concept stages, and thus the above assertions about whales become analytic relative to the relevant concept stages; i.e., "whales are mammals" is analytic relative to our present whale concept. Miscevic is right in calling this assertion a posteriori and thus arrives at paradoxical conclusions. I doubt that this is a wise terminological choice; I think we better relate analyticity to concepts and not to concept stages. At the end of this note, I shall touch upon the issue how much analyticity then remains.

Miscevic also distinguishes weak or superficial apriority related to concept stages and strong or deep apriority related to concepts, and he goes on to argue that all alleged conceptual apriority turns out to be only weak, thus suggesting that there is no strong apriority related to empirical concepts. I disagree; there are, I believe, also strongly a priori sentences. At least two kinds come to my mind:

The one kind is given by sentences of the form: "Whales are called 'whales'", or rather "whales are called 'whales' in my language". In a way, this is simply disquotation, but there is more to it. It brings out the a priori connection between a linguistically expressible concept and the word expressing it. This is the truth behind the first answer discussed above. We cannot identify a concept via an antecedently identified word; still, the one is a priori accompanied by the other. More importantly, this a priori sentence brings out that semantic deference is built into a concept right from the start. Whales are called "whales" not only by me, but by my teachers and by my linguistic community as well, and thus the power of determining what whales are is automatically deferred to my community.

The other kind of strongly a priori sentences has the form: "Most of what we take to be whales *are* whales". This has a Davidsonian ring. But it is not as general as asserting that most of our beliefs about whales are true; the a priori assumption is the more restricted one that most of our reference-fixing beliefs concerning whales are true. Moreover, the assumption does not ground in a theory of interpretation according to which a person can only be rationalized by the principle of charity as having mostly true beliefs. The point is rather that this assumption is the only base on which to change and develop our concept at hand; only when its believed reference is largely maintained, we may claim to have changed our old concept rather than to have acquired a new concept.[5]

Note, by the way, that the apriority of "most of what we take to be whales are whales" entails the apriority of "there are whales". Quine (1969a, p. 86) revolted against the analyticity of "there are dogs"; he then took the indistinguishability of "information that goes into understanding a sentence and information that goes beyond" as a further reason for abolishing analyticity. Apart from Quine's continuous refusal to distinguish analyticity and apriority, I think he is wrong. "There are dogs" is strongly a priori in Miscevic' sense. This is not quite to say that it is unrevisably a priori. But it is to say that we must believe that there are dogs as long as we have the concept of a dog; if we lose the belief (due to very strange circumstances), we lose the concept as well.

So, to resume, I contend, opposing Miscevic (2005), that there are some strongly a priori beliefs associated with a changing empirical concept. Indeed, I want to suggest that these beliefs are presupposed by the concept; otherwise, we could not meaningfully speak of that concept as a possibly changing one. Hence, these a priori beliefs embody my *fifth* and last idea for what it is that persists in, and thus individuates, changing concepts.

A final brief remark: We have seen that Miscevic relates analyticity to concept stages and thus arrives at paradoxical consequences. My observations suggest the question whether there are also strongly analytic sentences related to a concept. Only trivial ones, it seems, like "whales are whales", etc. In particular, if we follow Kripke in defining an analytic sentence as being a priori necessary, the above strongly a priori sentences associated with a concept turn out to be synthetic, since they are only contingently true. They are basically analogous to the sentence "I presently exist" which is the paradigm of an a priori, but contingently true sentence. Thus, we may perhaps vindicate old suspicions of Quine, Putnam, and others about the poverty of the notion of analyticity despite the richness of the notion of apriority.

---

[5] This a priori sentence applies to each moment of time. This allows for the peculiar case where this sentence is true at all times, though the many gradual changes may accumulate so that we end up with applying a concept to objects most or all of which did initially not fall under the concept. I am not quite sure whether we should really say in such a case that we have the same concept throughout. This is, however, a general and well-known ontological puzzle.

Chapter 16
The Intentional Versus the Propositional Structure of Contents*

16.1 The Thesis

The mind is a representing organ. It is somehow able to receive, store, retrieve, and express content. The hallmark of contentfulness is given by propositional attitudes such as desiring, expecting, fearing, and hoping. I commit here the common sin of taking belief to be the paradigmatic attitude and pretending it to be representative of all the other ones; I shall leave this pretension unchecked.

The issue then is: how to characterize belief contents, the objects to which the believer stands in the belief relation? One or, perhaps, the standard account is to conceive of contents as sets of doxastic possibilities (where "doxastic" so far only signals that the possibilities are used to characterize belief). A content consists of those possibilities that may be true according to it and excludes all the others; a content is a truth condition.

So, what are doxastic possibilities? Traditionally, they were assumed to be possible worlds w; this is what I call the *propositional* conception of contents. Then it was discovered that this won't do, doxastic possibilities should rather be conceived as centered worlds, i.e., triples $\langle w, s, t\rangle$ consisting of a possible world w, an object s existing in w, and a time t at which s exists in w. This allows dealing with the attitudes *de se* and *de nunc* that were argued to be irreducible to strictly propositional attitudes.

My thesis will be that this still won't do. It will be that doxastic possibilities should be conceived as quadruples $\langle w, s, t, \mathbf{d}\rangle$, where $\langle w, s, t\rangle$ is a centered world and $\mathbf{d} = \langle d_1, d_2, \ldots\rangle$ is a (finite or infinite) sequence of objects existing in w. I call this the intentional conception of contents.

---

* This is, in fact, my third attempt to explain and defend the thesis. In the first German version (Spohn 1997a) the thesis was embedded in a number of general observations about epistemology. This attempt was reduced to its core in the second English version (Spohn 1998). The present version is updated in various respects and will put the emphasis on what I call the third argument (in Section 16.5) that I had in mind from the beginning, but recognized only after the second version as being in the center of the dispute.

My terminology is perhaps not the happiest one; but I could not think of any better. It is unhappy, because the philosophical usage of "intentional", which is not the colloquial one, anyway, is at least ambiguous, an ambiguity originating from its modern founder Brentano (1874). In the wider sense, intentionality is just the distinctive feature of the mind. In that sense it is the directedness of the mind to something (external to it), its capacity to represent, to have or process contents, however they are conceived. In this sense, the propositional conception of contents is just one of several attempts to capture intentionality. There is, however, a narrower sense according to which intentionality more specifically denotes the directedness of the mind to (external) objects, something explicitly written into the intentional conception of contents in a way still to be uncovered, but not into the propositional conception that relates to (sets of) worlds and at best indirectly to objects insofar they may exist in worlds. In any case, it is this narrow sense that stands behind my terminological choice.[1]

My thesis may sound familiar, though perhaps unusually expressed. It is indeed essentially inspired by similar work, in particular by the path-breaking essays of Perry (1980), Kamp (1981), and Heim (1982). However, similar claims are usually embedded into an almost inextricable mixture of semantics and epistemology. By contrast, I intend the thesis to be a purely epistemological claim, and as such I have not seen it entertained in the literature.

Gaining a proper understanding of this contrast and thus of the thesis will require quite a lot of stage setting, disentangling, and explaining. This will take almost half of the paper, i.e., Section 16.2 for stage setting of a more general kind and Section 16.3 for deepening the specific dialectical background of the thesis. At the same time, this will elucidate the deep significance of my thesis.

I shall proceed with three arguments in favor of the thesis. I do not expect them to be conclusive; there hardly are conclusive arguments in philosophy. I hope, however, that they shift the balance of systematic reasons, strategic considerations, and aesthetic evaluations. More specifically, I advance two graphic arguments by way of example in Section 16.4. Not surprisingly, the arguments are not cogent, as is shown by an objection of Zimmermann (1999). This will shift the discussion to a

[1] The following quotation nicely displays Brentano's thesis of intentionality as the defining characteristic of the mental as well as the ambiguity (note that "inexistence", as it is used in the quote, does not mean "non-existence", but "existence in"): "Every mental phenomenon is characterized by what the Scholastics of the Middle Ages called the intentional (or mental) inexistence of an object, and what we might call, though not wholly unambiguously, reference to a content, direction toward an object, … or immanent objectivity. Every mental phenomenon includes something as object within itself, although they do not all do so in the same way. In presentation something is presented, in judgement something is affirmed or denied, in love loved, in hate hated, in desire desired and so on. This intentional in-existence is characteristic exclusively of mental phenomena. No physical phenomenon exhibits anything like this. We can, therefore, define mental phenomena by saying that they are those phenomena which contain an object intentionally within themselves" (Brentano 1874/1973, pp. 88f. in the English translation).

strategic level in Section 16.5. In various fields, not only the present one, there are those appealing to the so-called method of fine-grained descriptions in order to deal with problematic examples and those who find it more fruitful and illuminating to avoid this method. I shall argue for the latter position which, in the case at hand, means accepting the intentional conception of contents. This will be my third and main argument. The afterthoughts in Section 16.6 will emphasize the relevance of my conclusion.

16.2 Stage Setting

In order to assess the significance of the thesis we have to disentangle belief from language in several respects. This will be hard work. It is peculiar: the linguistic turn has taught philosophy so much; at the same time, though, it has obstructed the view to pure epistemology, so much so that the latter may appear to be a phantom. It is not, I believe. So let us clear our view in six preliminary steps.

(1) There are, basically, two ways of characterizing objects of belief; they are, roughly, either identified with propositions (in an as yet unspecified sense) or with sentences. That is, characterizations of objects of belief may either focus on semantic aspects, on the fact that beliefs are true or false; or they may focus on the fact that beliefs have to have some encoding, some representational structure making them apt for computation. Quine is certainly the champion of the latter view, though on the ground of his denial of meanings, while the more recent defenders of the view focus on beliefs as representations in the mind/brain encoded in some way and possibly in the language of thought, however language-like that is. Let me put this view to one side, without much argument.[2] It is simply that I am so much more impressed by the theoretical achievements of the broadly propositional view and do not see how the sentential view could ever compete with them.[3] Thereby, we disentangle belief from morphosyntactic features of language.

(2) Propositions are usually explained to be sentence meanings. Therein, of course, lies the semantic entanglement of epistemology that is much harder to grapple with. It entails that there are about as many characterizations of belief contents within the broadly propositional view as there are theories of meaning. There are

[2] Likewise, I shall ignore hybrid constructions like Carnap's suggestion in his (1947, sect. 14) to abstract objects of belief from intensional isomorphism, a suggestion profoundly elaborated by Bealer (1982).

[3] To give just one hint: Bayesianism, surely a great epistemological theory, is entirely built on the propositional view. I do not know of any workable probability theory defined for sentences and, that's essential, dispensing with intensionality, i.e., with the substitutivity of logical equivalents.

Fregean senses and thoughts,[4] Russellian singular and general propositions (cf. Russell 1910/11, 1918/19), and Carnapian intensions (cf. Carnap 1947); there are Hintikka's (1962) model sets, Kaplan's (1977) characters, Stalnaker's (1978) propositional concepts and their diagonals, properties as conceived in Lewis (1979b), the situations of Barwise and Perry (1981, 1983, ch. 9–10) and various constructions thereof, and so forth.

This manifold is slightly confusing. There is, however, a common basic idea behind it, at least since Carnap (1947): namely the idea to characterize propositions and thus contents via the exclusion of possibilities. When I believe, for instance, that the sun will rise tomorrow I exclude all possible cases in which it does not rise. This is not to say that I admit all cases in which the sun does rise; my further beliefs exclude many of them as well. However, if we consider all of my beliefs and all the cases excluded by them we arrive at a positive rest embracing the cases I admit as possible. These cases are called my doxastic alternatives, a term coined by Hintikka (1962, p. 49), and the set consisting of all my doxastic alternatives is called my belief set, which is a subset of the set of all doxastic possibilities.

The characterization of contents explicitly forms the technical basis of the standard system of doxastic logic; according to it one believes each superset of one's belief set, one disbelieves each set of possibilities disjoint with the belief set, and one is unopinionated about the rest. Consistency and deductive closure of beliefs is automatically built in into this account. Theories of graded beliefs, e.g., subjective probability theory, are based on the same idea. However, this idea is just a leitmotif. Due to its neutrality it seems to open a direct way to pure epistemology without semantic detour, but at the same time it hides the semantic entanglement we are in as badly as before.

(3) One way how language still creeps in is the characterization of (doxastic) possibilities or possible cases. When Carnap (1947) first implemented the basic idea he took possible cases to be state descriptions, linguistic entities. However, this made possibilities and thus belief too language-dependent. Speakers of different languages should be able to exclude the same cases, and there may be more possible cases to believe or to exclude than one is able to represent as a state description in a given vocabulary. Such problems suggest the conclusion that possible cases rather are complete possible worlds and not linguistically constituted in any way. Here I agree with the criticism in Lewis (1986b, ch. 3) of the various kinds of ersatzism. However, this leaves open so far whether we should understand possible worlds in a Wittgensteinian manner as in some sense maximal states of affairs, as repeatedly defended by Armstrong, e.g. in his (1997), or in a Lewisian manner as in some sense maximal individuals.[5]

---

[4] Frege (1918) is a problematic case, though. Since the only identity criterion for thoughts, i.e., for sentence meanings, he actually gives is an epistemological one, one cannot say that semantics is prior to epistemology for Frege. Cf. Kemmerling (1990, pp. 161ff.)

[5] Thus, I do not agree with Lewis (1986b, pp. 145–148), insofar as he tendentiously subsumes the Wittgensteinian manner under linguistic ersatzism via what he calls Lagadonian languages.

The issue is pressing, and step (6) below makes sense only with respect to Lewisian possible worlds, which I am hence inclined to assume.[6] However, this metaphysical issue is beyond the scope of this paper, and I would like to stay neutral. Indeed, my thesis will not be affected by the issue, as far as I see, and my arguments in its favor in Sections 16.4 and 16.5 work for both conceptions of possible worlds. I shall comment on the point whenever required.

(4) Even if we should have settled what possibilities are we still do not know well what it means for a subject to be characterized by a belief set, i.e., to exclude possibilities outside this set. Since we do not want to change the topic by revising concepts, but intend to grasp the ordinary notion of belief, it seems wise to look at how we talk about belief. This, however, gets us into another linguistic entanglement, the delicate distinction between belief and belief ascription.

For instance, it is our common practice to ascribe de re beliefs. My thesis may indeed have raised the suspicion that its motivation lies in that phenomenon. Quite to the contrary, though. The thesis has nothing to do with de re beliefs or belief in singular propositions. This could have been clear from my reference to doxastic alternatives and belief sets, since already Quine (1956) told us with his Ortcutt story that de re contents believed are almost inevitably contradictory and thus defy direct treatment in terms of the exclusion of possibilities. They can be related to doxastic alternatives and belief sets only indirectly by such maneuvers as have been proposed by Kaplan (1969) and Lewis (1979b, sect. XIII).

(5) The point runs deeper. Kripke (1972), Putnam (1975), and Burge (1979) have shown us that our de dicto belief ascriptions are de re in a way, too, either because there are many rigid designators, proper names and natural kind terms at least, for which de dicto coincides with de re, or because de dicto ascriptions implicitly contain a de re reference to communal linguistic practices (not necessarily known to the ascribee). Hence, Quine's point and its consequences generalize to de dicto belief ascriptions; this, I take it, is the upshot of Burge (1979) as well as Kripke (1979).

Burge (1979) has expressed the issue in a different way. He arrived at the anti-individualistic conclusion that believing that p is (usually) not an internal state of mind; it is a psychological state in the wide, not in the narrow sense, to use Putnam's (1975) terms. There may be mental states conforming to methodological solipsism, but the propositional attitudes so central to our psychology do not belong to them. Or in still other words: there are no narrow, only wide contents. By contrast, we must note that by hoping to represent beliefs and their contents in terms of doxastic alternatives we have already put our stakes on individualism. Contents thus characterized must be narrow contents. The reason is basically the Quinean one: the wide contents believed may easily be, and often are, contradictory, and hence they are not suited for representation in terms of doxastic alternatives.

[6] See also my speculations at the end of the introduction of this collection.

How are we to deal with this conflict? We might confine attention to restricted scenarios in which the difference between narrow and wide contents does not arise. This is probably the normal unreflective attitude towards these problems, though none we can maintain as philosophers. We might argue about the arthritis example of Burge (1979), the water example of Putnam (1975), and their variations. But I am not inclined to do so; I find them entirely plausible. We might start a philosophical argument about individualism. Then I had to write a different paper; so let me simply confess my individualism. Many think, of course, that this issue dooms the whole approach of understanding contents as sets of doxastic possibilities. But this would deprive my paper of its presupposition.

We should do none of this. We should rather draw the internalistic conclusion that a further disentanglement is required. There is no way of directly understanding doxastic alternatives and belief sets in terms of ordinary talk about belief. The relation between the basic internalistic characterization of belief in terms of doxastic alternatives and the common practice of de dicto and de re belief ascriptions can and must rather be construed in some indirect way. How exactly is, however, not our present task, all the more so as de dicto and de re ascriptions are not neatly separated, but thoroughly intermixed in a way hard to cope with for the semanticists of belief sentences.[7]

(6) However, this conclusion still leaves us with the task of offering some positive characterization of doxastic alternatives and belief sets, i.e., with the question: what does it mean for a subject to exclude a doxastic possibility? Again we must avoid linguistic answers. If possibilities were state descriptions, they could be excluded by denying them; but they aren't. Similarly, the method so dear to Quine of asking subjects for assent or dissent at best elicits de dicto beliefs. But we want to know about the exclusion of whole possible worlds and not of partial linguistic representations of them.

Indeed, I find the literature surprisingly silent on this question. Even Lewis (1986b, sect. 1.4, in particular pp. 35ff.) avoids a direct answer and prefers a functional characterization: the belief set and thus the beliefs of a subject (or her more finely gradated attitudes) are those that best systematize her behavior. Yes, certainly. The same spirit is found in the proposal of Beckermann (1996) to consider belief, as it were, as a magnitude taking propositions as values (just as length is a magnitude taking positive real numbers as values) and to devise a measurement theory for this magnitude by behavioral laws. Is there no more direct answer?

There is, and it is suggested by all these twin stories initiated by Putnam (1975) that invariably depend on substantially different possibilities that are nevertheless indistinguishable for the subject. As is more extensively argued in Haas-Spohn and Spohn (2001, sects. 2 and 3) [here: sect. 14.2–3] the distinguishability referred to is not a superficial one using only fast or sloppy, e.g., purely sensory methods, but the maximally thorough-going one using all our receptive, experimental, and

[7] See, e.g., Schlenker (1999) and Maier (2006). For a good survey see also MacKay and Nelson (2005).

judgmental powers to an ideal degree. That is, what is suggested is the following criterion (that one might call operational if it were not so excessively hypothetical).

Take a certain belief state of a subject. Suppose we somehow deep-freeze this state so that nothing is lost or added. Now confront the subject with a doxastic possibility, i.e., an entire possible world. In this world she is allowed to investigate everything everywhere. She may inspect all molecules under all kinds of microscopes, she may learn every language, take every perspective, etc. If there is anything in this world that the subject would not have expected according to her frozen belief state, then this world is excluded according to it and not a doxastic alternative. In other words: If the belief state the subject would get into through such a complete inquiry is merely an expansion of her frozen state and not a revision, then this world is a doxastic alternative, i.e., a member of her belief set.

For instance, on the basis of such a full investigation Putnam's Oscar at 1750 could, of course, distinguish H_2O from XYZ and Earth from Twin Earth, but they are not distinguished in his beliefs; if one is part of his doxastic alternatives, the other, is, too. Likewise for Burge's Fritz vis à vis worlds where "arthritis" means arthritis and worlds where "arthritis" means "tharthritis". This is the intended result.

This explanation makes sense only if possible worlds are understood as Lewisian ones. In her investigation the subject must grapple with the worlds, and hence they must be concrete objects to be grappled with. States of affairs or Wittgensteinian worlds cannot be inquired in this way, they can just be assumed or acknowledged. However, I do not want to press this point. The main argument of this paper should be independent of it.

Of course, this characterization of belief sets is not only unduly hypothetical, but also unduly idealized. Even if we ignore the entirely fictional character of this criterion, the test subject would often be unable to clearly say yes or no. She would often be unsure or indeterminate about many things. She will have only degrees of belief. The way and the order she would be presented with the alternatives would influence her response. And so on. However, as far as I see, such points have no force in the present context. We are not after experimental methodology. That would be a different task: to inquire the extent to which actually feasible discrimination tests could approximate this vastly counterfactual criterion.

Two things should be emphasized concerning this explanation of belief sets. It is, first, individualistic, as we said is required. Our criterion elicits the totality of the subject's cognitive life as his intrinsic disposition; it presupposes or holds fixed nothing external to the subject, and the elicited belief set can change only by changing the subject and not by merely changing his environment. Secondly, the criterion is sufficiently detached from language. Languages enter the picture only as parts of possible worlds; of course, no world could be a doxastic alternative if it did not contain a language and language users familiar to the subject.

Indeed, belief sets will be virtually indescribable. This is no surprise. Of course, our discriminatory capacities by far outrun our linguistic expressiveness. They even outrun the descriptive power of the most ingenious psychologists – whence our useful and well established practice of only somehow approximating belief sets by our de dicto and de re ascriptions. For instance, we have an extremely good capacity to recognize objects in every-day life and it would be even better in our counterfactual test. However, it is rarely perfect, we rarely know the essence of an object in order to infallibly identify it. Our recognitional capacity is so to speak non-rigid, and it is hard to say by which (relational) features it is guided. Still, because it works so well it is a small mistake to describe it with a rigid name for the subject recognized. All this is not to say, though, that belief sets are indescribable in principle. There is no reason why they should be. They are only so incredibly hard to describe.[8]

So far I have only explained how the propositional conception of contents is to be properly understood, by disentangling it from language in five ways: by distinguishing it from syntactic conceptions of the objects of belief, by reducing various semantic conceptions of contents to a neutral common core in terms of sets of possibilities, by assuming a non-linguistic characterization of these possibilities, by decoupling the basic conception of belief from our practice of belief ascriptions, and finally by giving also a non-linguistic characterization of what it means to exclude a possibility. Thereby the stage is set for the issue to be discussed in the paper, since the intentional conception to be argued for is nothing but a refinement of the propositional conception.

16.3 The Dialectical Background of the Thesis

Well, the common stage is set; our preparations need still more fine-tuning. My thesis has a familiar ring and it clearly originated from thinking about the familiar literature. However, it is by no means easy to discern similar from identical theses. Therefore we should look a bit more closely at the difference between the propositional and the intentional conception of contents and its dialectical battle-ground. In a way, this will result in a final subtle step of disentangling epistemology from semantics.

One similar thesis is the thesis about the indispensability and irreducibility of de se and de nunc attitudes, a deep point first advanced by Castañeda (1966) and powerfully reinforced by Perry (1979) and Lewis (1979b). I mentioned already in Section 16.1 that possible worlds w won't do as doxastic possibilities; they have to be at least centered worlds $\langle w, s, t\rangle$.[9] Of course, the above criterion for doxastic alternatives has to be modified accordingly: the subject's full investigation need not only check whether the world w as such conforms to her beliefs, but also whether

[8] Cf. also the extensive discussion under the title "disjunction problem", e.g., Fodor (1990, chs. 3–4), which deals with the same issue.

[9] Lewis (1979b) is able to further reduce centered worlds to properties, but only because he, contestedly, assumes first that each individual inhabits only one possible world and secondly that persons or subjects are (mereologically) composed of momentary person stages. I shall ignore this reduction in the sequel.

the possible object *s* might be she herself and whether the possible time *t* might be her present time in *w* according to her beliefs.

The point seems generally accepted, and I accept it, too. But there is reason for modesty. There was an argument between Lewis (1979b) and Stalnaker (1981) in which Stalnaker defended the narrow propositional conception according to which possible worlds are good enough as doxastic possibilities even in view of the examples apparently favoring Lewis' position, the most extreme one being the one of Jahwe and Zeus (cf. Lewis 1979b, sect. V). Lewis' position is certainly more intuitive and elegant, as is also argued by Haas-Spohn (1995, sect. 2.2). However, she makes clear that it is no more than that; ultimately, Stalnaker has an equivalent way of representing matters.

It may seem that the intentional refinement of the propositional conception is of the same kind as the *de se/de nunc* refinement and supported by the same kind of argument. I think this would be a misperception. Egocentricity or self-consciousness and object-directedness are two different phenomena, even though Kant (1781/87, pp. B274–279), in his *refutation of idealism*, has suggested a deep connection that is on the philosophical agenda since. And they require a different treatment. In any case, here I shall take the *de se/de nunc* refinement simply as tacitly understood. My arguments for the intentional refinement will differ from Perry's and Lewis' arguments for de se beliefs. Indeed, the case may be reversed. The argument for the intentional refinement may be adapted for de se beliefs, and thus the case against Stalnaker (1981) may be strengthened, though perhaps not decided.

The main source for similar theses, though, is two-dimensional semantics. The relation of my thesis to this most promising development in semantics is highly instructive, but not obvious. Let me explain it in a bit more detail. However, since I want to steer to the point relevant for this paper as directly as possible, I have to neglect a lot of interpretational variation and uncertainty in two-dimensional semantics.

The first full elaboration of two-dimensional semantics was Kaplan (1977).[10] His goal was to deal with the semantics of indexicals, demonstratives, and possibly other context-dependent expressions. For this purpose, the interpretation function to be recursively explained for a language must assign *characters* to expressions; the character $\|\alpha\|$ of α assigns an extension to α relative to a context of utterance and a circumstance of evaluation in Kaplan's terms, or relative to a context *c* and an index *i*, as I shall say following Lewis (1980b). Thus, $\|\alpha\|(c)(i)$ is the extension of α at *c* and *i*, and $\|\alpha\|(c)$ is the intension of α in the context *c* – or the secondary intension of Chalmers (1996, sect. 2.4) or the *C*-intension of Jackson (1998, ch. 2). In this way Carnap's framework of intensions and extensions is preserved by Kaplan.

Now an important issue is the structure of contexts *c* and indices *i*. Contexts *c* must collect those contextual features on which the intension of α as used in *c* may depend. Indices *i* must be so structured as to get the semantic recursion running. Which parameters of contexts and indices need to be assumed is a ramified

[10] It was, though, a larger group at the Philosophy Department of UCLA that predominantly developed the theoretical field since the late 1960s.

and continuing discussion. It is strongly suggested, though, that we need the following index parameters in i: a world w_i for treating modality, a time t_i in order to deal with tenses and temporal quantifiers, and a variable assignment or a sequence $\mathbf{d}_i$ of objects in order to treat objectual quantifiers. The latter point is already Tarski's deep insight that it is not truth, but satisfaction that can be recursively defined for first-order languages. Moreover, it is strongly suggested that we need the following context parameters in c: a context world w_c since we are able to contextually refer to practically every feature of the world, a speaker s_c and an utterance time t_c for localizing potential utterances in w_c, i.e., for interpreting "I" and "now", and again a sequence $\mathbf{d}_c$ of objects for interpreting demonstratives. The latter point is contested, though. Montague (1974, chs. 3 and 4) explicitly chooses the latter option, although he still struggles with disentangling the roles of contexts and indices within his points of reference. Kaplan (1977, sect. XV) prefers to take what he calls demonstrations as parts of contexts instead of demonstrated objects. In (1989, pp. 582–584) Kaplan has changed his opinion and adopts a view that I would capture by enriching a context c by a sequence $\mathbf{d}_c$ of objects as interpreted below. This is also the conclusion of Haas-Spohn (1995, sect. 4.7). The topic is an intricate one, and there is no point in starting a discussion here. I listed here what I called strong suggestions only in order to relate them to my topic, as I am about to do.

The lists of context and index parameters may seem a bit arbitrary; each follows its own apparent needs.[11] Theoretical pressure is produced by the notion of utterance truth. A sentence is true in a context and at an index. An utterance is a sentence in a context, and it is true if and only if the sentence is true in that context and at the very same context taken as an index. The latter step is called diagonalization. Thus, utterance truth conditions are generated by diagonalizing the characters of the sentences uttered. Kaplan (1977, p. 547) explicitly introduces this notion in order to explain validity or logical truth for his logic of demonstratives: a sentence is logically (or a priori) true iff its utterance is true in all contexts. This is, he says on pp. 538f., how much he can capture of apriority by his logic of demonstratives.

This produces theoretical pressure because it constrains our lists of context and index parameters. Whenever we assume an index parameter, we also need a corresponding context parameter; otherwise, diagonalization is not defined. The converse need not hold. There may be more context than index parameters. For instance, we clearly require the contextual subject s_c, the speaker; but the evidence that we also require a subject s_i in the index has remained unclear.[12]

There are two ways how we can deal with this theoretical pressure. Either, we can take it as well founded and thus as an argument for postulating an appropriate context parameter whenever we have found reason for assuming a given index parameter. Or we can treat diagonalization as a hypothesis in need of confirmation and getting confirmed when we list context and index parameters according to their

[11] We then find also surprising proposals such as that of Lewis (1980b, sects. 3 and 5) to include standards of precision among the context and possibly also among the index parameters.

[12] See, however, Schlenker (1999, ch. 3) for a strong case in favor of this requirement.

independent needs and find them admitting diagonalization. Of course, these two ways are not so clearly separated. Either way, the theoretical pressure advances theoretical coherence.

Much stronger theoretical pressure is produced by what is called the epistemological reinterpretation of Kaplan's character theory initiated by Stalnaker (1978). Many think that the formal similarity between Kaplan's and Stalnaker's work is superficial and in fact utterly misleading. I don't think so. Haas-Spohn (1995, sects. 2.1, 3.9, and 4.4) gives a convincing account of how to understand Stalnaker's propositional concepts as a continuation of Kaplan's characters and how thus to explain the existing differences. I am obviously touching here a long and deep discussion that we cannot pursue here.[13] Putting all niceties aside, let me bluntly state what I take to be the gist of the epistemological reinterpretation: It is that possible contexts at the same time serve as doxastic possibilities; both have the same structure. Altogether we have a very powerful constraint:

> *The Congruence Principle*: Each index parameter is a context parameter, and the context parameters are exactly those characterizing doxastic possibilities.

For instance, when we assume contexts to be characterized as triples $\langle w_c, s_c, t_c\rangle$ and account for de se and de nunc attitudes by taking centered worlds $\langle w, s, t\rangle$ as doxastic possibilities, we accurately conform to this principle.

If we accept the principle, the distinction of Chalmers (2006) between a contextual and an epistemic understanding of two-dimensional semantics would collapse. The utterance truth condition of a sentence is at the same time the narrow content associated by a subject with that sentence. Thus, in this reinterpretation diagonalization acquires great epistemological importance; and utterance truth conditions are rather called diagonal intensions or primary intensions (Chalmers 1996, sect. 2.4) or A-intensions (Jackson 1998, ch. 2). All in all we are tempted by a beautiful offer: horizontals are for metaphysics, diagonals are for epistemology, and two-dimensional semantics unites both in one framework. Perhaps too beautiful to be true.

Anyway, it should be clear by now why I am telling all this. The theoretical framework I have sketched provides great argumentative resources in relation to the thesis I want to defend. For instance, if Tarski is right about indices and the Congruence Principle is true, doxastic possibilities must contain sequences of objects. Or a bit closer to the point: given the Congruence Principle, arguments about how to deal with demonstratives automatically turn into arguments about the thesis.

This is why I said that the thesis is so hard to discern from similar ones. The arguments we usually find in the literature are semantic ones. If the Congruence Principle is presupposed in these arguments, they may be taken to support my thesis; if not, they argue for something slightly different. What is actually going on is often not so clear, however.

[13] The most careful discussion of this issue, i.e., of a contextual versus an epistemic understanding of the first dimension of two-dimensional semantics, is found in Chalmers (2006).

To be a bit more specific: Kamp (1981) has initiated so-called discourse representation theory that has acquired great linguistic significance in the meantime (cf. Kamp and Reyle 1993). Kamp (1981, p. 282) says that he intends his account to "bear on the nature of mental representation and the structure of thought". So, all the model building in the discourse representation structures (DRS's) is to represent what is internally going on in the mind of the speaker/hearer, i.e., epistemic meanings or the diagonals in the two-dimensional picture. A crucial role in the DRS's is played by the so-called discourse referents or parameters, and those appearing in the so-called principal discourse representation may be identified, I think, with the places of the sequence of objects being part of doxastic possibilities according to my thesis. However, this identification would have to be argued, and the indispensability of discourse referents as such is rather a semantic issue. The same remarks apply to the very similar so-called file change semantics developed by Heim (1982). She clearly intends files to be states of information, i.e., as something of an epistemic nature. Again, though, her goal is to promote semantics, and she puts forward exclusively semantic arguments. All this does not automatically determine its relation to pure epistemology.[14]

Therefore I would like to state expressly that I want to uncouple my thesis from the Congruence Principle and all the two-dimensional theorizing. My intention is to entirely stay on the epistemological side and to argue for the thesis in a purely epistemological way. Of course, the two-dimensional picture is always in the background. To a good extent it is this background in which the thesis unfolds its significance, and the perspective that the thesis may provide confirmation for the Congruence Principle and thus connect up with the above-mentioned semantic developments is exciting. However, all this is to be background, not part of my thesis and my argument. I leave it to the reader to judge whether I shall have succeeded in my intention.

16.4 Two Arguments for the Thesis and an Objection

Let me resume our focal thesis: According to the intentional conception of contents a subject's belief system has addresses or file cards or discourse parameters for objects. When the subject encounters, perceptually or linguistically mediated, an object she takes interest in she creates a new address or file card. All subsequent information she takes to be about the same object will then be stored at this address. Of course, since she may misidentify objects she may store information at the wrong address, and since she may not recognize an object she may have two addresses for the same object. We must always reckon with this ontological-epistemological backlash. Doxastic possibilities also allow for relational and for

[14] The notion of a discourse referent seems to go back to Karttunen (1969). He also uses the picture of a file. However, it was only Kamp (1981) and Heim (1982) who crucially advanced the long semantic struggle with pronouns and definite and indefinite noun phrases.

general information not stored at specific addresses. Thus, the formal model in terms of doxastic possibilities is broader than the vivid picture of a file suggests. In fact, Heim (1982, p. 287) defines a file precisely as a set of doxastic possibilities in the intentional sense.

This description also indicates how my quasi-operational criterion for the exclusion of doxastic possibilities is to be modified for the intentional conception: The quadruple $\langle w, s, t, \mathbf{d}\rangle$ is a doxastic alternative of a subject at a certain time if and only if she would admit after the most scrupulous investigation of w and all objects in w from all perspectives available in w that s conforms to her self-image, t to her image of the present time, the objects $d_1, d_2, \ldots$ in $\mathbf{d}$ to the images stored at her addresses 1, 2, ..., and w to her picture of world, that is, if the doxastic state she would arrive at after that investigation of $\langle w, s, t, \mathbf{d}\rangle$ would be an expansion and not a revision of her present state.[15]

How, then, may we argue for the thesis? To begin with, it is noticeable that there seem to be no arguments for related theses confined to static scenarios, to the beliefs of a single person at a single time. This may have the following reason: Suppose we understand belief contents in the intentional way, i.e., as satisfaction conditions of open formulae (if we could linguistically represent the contents). This allows an easy derivation of belief contents in the propositional sense, i.e. truth conditions. Logicians usually associate open formulae with universal closures; but this is obviously inappropriate in our case. To believe a satisfaction condition rather means to believe that there exist objects corresponding to the information stored at the various addresses, and this amounts to the existential closure of the satisfaction condition. Now, it seems plausible and arguable that a static theory of belief would be concerned only with truth conditions and cannot by itself discriminate different satisfaction conditions having the same truth condition. However, I am not aware that that argument has actually been carried through.

In any case, all the existing arguments in the vicinity of the intentional conception refer to dynamic scenarios in some way or other. This is even true of the arguments for the irreducibility of beliefs de se and de nunc. However, I do not see how to transfer these arguments to our case. As already indicated, discourse representation theory and file change semantics are rather occupied with finding adequate semantic representations of texts and discourses as they evolve. So, again I do not see how to turn their arguments about anaphoric reference and related phenomena into an argument about pure epistemology. Only Perry (1980) directly addresses belief and its change or preservation and discusses various dynamic examples showing the need for what he calls a file. Let me adapt his kind of examples to my somewhat different framework; I shall explain our differences afterwards.

Typically, changes in beliefs are driven by perception, and typically we use indexical descriptions for perceived objects. The girl about ten meters left of me just hurt her knee by falling from her skateboard – this is what I just saw and what

[15] The numbering of the addresses is inessential. What matters is the assignment of the possible objects in a doxastic possibility to the somehow well distinguished addresses of the belief state.

I came to believe. There are two ways to describe the increment of my beliefs: according to the propositional conception my old belief set is conjoined with the truth condition of the sentence "The girl about ten meters to the left of me just hurt her knee". (Let us ignore that the content of my perception obviously exceeds the content of this sentence.) According to the intentional conception my prior doxastic state is enriched by adding a new address and there storing the information "is a girl, is about ten meters left of me, just hurt her knee". So far both descriptions seem equally acceptable.

The story continues, however. I realize that I know the girl; she is my neighbor's daughter. I shall soon have forgotten the indexical description; perhaps there were several girls around, and it is just too tedious to memorize where all of them were placed. So, the other day all I remember is that my neighbor's daughter hurt her knee. However, since I still rely on a description of the girl the situation did not really change. The only difference to the first case is that according to the intentional conception the new information will be stored at an old address, namely the address that already contains the information "daughter of my neighbor". So, again, there is no reason to prefer one description to the other.

However, you will not be surprised to read that the full story goes like this: My neighbor actually has two daughters who are identical twins. Despite numerous encounters I am still unable to tell them apart. In this case it is plausible to maintain that I have exactly the same information about both girls. Let us summarize this information by the rather complex concept F. So, according to the propositional conception my prior doxastic state before the incident (as far as these girls are concerned) may be characterized by the following proposition:

(P1) $\{w \mid w \vDash \bigvee x \bigvee y\ (x \neq y \wedge Fx \wedge Fy)\}$.

According to the intentional conception this state is best captured by an open formula, i.e. by the following content:

(I1) $\{\langle w, x, y\rangle | \langle w, x, y\rangle \vDash x \neq y \wedge Fx \wedge Fy\}$

(where the sequence **d** of objects is reduced to the pair $\langle x, y\rangle$ that only matters). Now I said I remember from the incident I observed that one of the girls hurt her knee, i.e., for short, that she has property G. According to the propositional conception my posterior doxastic state some time after the incident is represented thus:

(P2) $\{w | w \vDash \bigvee x \bigvee y (x \neq y \wedge Fx \wedge Fy \wedge Gx)\}$.

According to the intentional conception the new state is represented by one of the following sets:

(I2a) $\{\langle w, x, y\rangle | \langle w, x, y\rangle \vDash x \neq y \wedge Fx \wedge Fy \wedge Gx\}$, or

(I2b) $\{\langle w, x, y\rangle | \langle w, x, y\rangle \vDash x \neq y \wedge Fx \wedge Fy \wedge Gy\}$, or

(I2c) $\{\langle w, x, y\rangle | \langle w, x, y\rangle \vDash x \neq y \wedge Fx \wedge Fy \wedge (Gx \dot{\vee} Gy)\}$.

(I2a) and (I2b) apply if, for whatever reason, I come to store the information at a specific address. (I2c) applies if I have no idea which of the two girls was injured. As I have told the story so far, the latter case will seem to be the more plausible, indeed the only possible one. However, I hope to make clear below that the former cases may not be disregarded.

This scenario constitutes the setting of my first argument. How should we describe the increment in belief? According to the intentional conception the increment (I1/2) may be simply conjoined. That is, in the three variants the increments are:

$$\text{(I1/2a)}\quad \{\langle w,x,y\rangle \mid \langle w,x,y\rangle \vDash Gx\}, \text{or}$$
$$\text{(I1/2b)}\quad \{\langle w,x,y\rangle \mid \langle w,x,y\rangle \vDash Gy\}, \text{or}$$
$$\text{(I1/2c)}\quad \{\langle w,x,y\rangle \mid \langle w,x,y\rangle \vDash Gx \veebar Gy\}.$$

And in each variant we just have (I1) $\cap$ (I1/2a,b,c) = (I2a,b,c).

The case is not so simple, however, with the propositional conception. The introductory versions of the story still allowed the conjunctive addition of:

$$\text{(P1/2}'\text{)}\quad \{w \mid w \vDash G(\iota xFx)\}$$

The last problematic version, though, was so constructed that this idea is blocked, because I do not know any identifying description of the girls and thus the description ιxFx does not refer according to my beliefs. What to do? Logically speaking, the whole posterior doxastic state

$$\text{(P1/2}''\text{)}\quad \{w \mid w \vDash \bigvee x \bigvee y (x \neq y \wedge Fx \wedge Fy \wedge Gx)\}$$

may be taken as the increment in belief; but intuitively the increment is not that big. The other extreme is to take the material implication

$$(P1/2''')\quad \{w \mid w \vDash \bigvee x \bigvee y (x \neq y \wedge Fx \wedge Fy) \rightarrow \bigvee x \bigvee y (x \neq y \wedge Fx \wedge Fy \wedge Gx)\}$$

as increment; but we thereby ascribe a surprisingly complex logical form to a rather simple information. One might also try something in between these extremes, for which, however, no simple logical form is in sight, either. So, here is the first argument: In the example the informational increment intuitively appears to be a simple conjunctive addition but the increment cannot be captured as such within the propositional conception, in contrast to the intentional conception which is able to do so.[16]

The argument is certainly not conclusive. Christoph Lumer pointed out to me (personal communication) that one could easily introduce a definite description for

[16] Similarly, Heim (1982, p. 305) assumes that the file change brought about by continuing a text with an atomic formula just consists in conjoining the satisfaction condition of that formula to the prior file. This inspired me to the above argument.

a pair of individuals ("*the* twins of my neighbor") and that the propositional conception could adequately represent the increment by using that definite description ("one of *the* twins of my neighbor hurt her knee"). A different point: Ede Zimmermann mentioned to me (personal communication) that propositions, being sets of worlds, have no logical form and that the argument therefore makes no sense, strictly speaking. Nevertheless, the argument certainly points to a difficulty.

The second argument refers to the same scenario. It starts from an observation already made, namely that the intentional conception allows for three different increments (I1/2) in information from (I1) to (I2). These increments result in three different belief states. However, their existential closures are logically equivalent; it does not make a logical difference whether Gx or Gy or $Gx \vee Gy$ is added as a conjunct within the scope of the existential quantifiers $\bigvee x \bigvee y$. So, according to the propositional conception there is a unique posterior belief state. What is intuitively more adequate: a unique increase or the unfolding into three possibilities? I would like to shift our intuitions to the latter.

For this purpose, let me introduce a second piece of information about one of the girls consisting in the predicate H. I do not think of another observation. This would not bring substantial news because the intentional conception would again allow three ways to account for the new piece of information and the propositional conception would do so as well, since the first piece of information about the injured knee already destroyed the symmetry of the bound variables. I am rather thinking of a case in which I suddenly remember, say, that one of the twins has a liver spot under her left eye and that this mark in principle allowed me to distinguish them, even though I mostly confused them, not thinking of the distinguishing mark.

So, suppose H is the concept of having a liver spot under her left eye and that, within the intentional conception, the free variable x represents the address for the girl with the liver spot. This conception allowed three ways for accounting for the perception about the injured knee. Because of my recollection we now have to add the conjunct Hx in each case. So, there are again three possibilities to account for the resulting doxastic state:

(I3a) $\{\langle w,x,y\rangle | \langle w,x,y\rangle \vDash x \neq y \wedge Fx \wedge Fy \wedge Gx \wedge Hx\}$, or

(I3b) $\{\langle w,x,y\rangle | \langle w,x,y\rangle \vDash x \neq y \wedge Fx \wedge Fy \wedge Gy \wedge Hx\}$, or

(I3c) $\{\langle w,x,y\rangle | \langle w,x,y\rangle \vDash x \neq y \wedge Fx \wedge Fy \wedge (Gx \vee Gy) \wedge Hx\}$.

As stated, though, the increment is the same in all three cases:

(I2/3) $\{\langle w,x,y\rangle | \langle w,x,y\rangle \vDash Hx\}$.

And again, we have (I2a,b,c) $\cap$ (I2/3) = (I3a,b,c), respectively.

Note the important fact that this continuation of the story also supports my claim above that there are three ways to account for the first increase in information. At first blush it seemed that I could only add the information $Gx \vee Gy$ because I did not have any clue which of the twins injured her knee. However, a mark like the liver spot might *cause* me to store the information at a specific address even if I am not aware of the mark and could not tell afterwards why I did so.

The propositional conception leads to a different treatment of my recollection. According to this conception there are three possible final doxastic states:

(P3a) $\{w|w \vDash \bigvee x \bigvee y(x \neq y \wedge Fx \wedge Fy \wedge Gx \wedge Hx)\}$, or

(P3b) $\{w|w \vDash \bigvee x \bigvee y(x \neq y \wedge Fx \wedge Fy \wedge Gx \wedge Hy)\}$, or

(P3c) $\{w|w \vDash \bigvee x \bigvee y(x \neq y \wedge Fx \wedge Fy \wedge Gx \wedge (Hx \dot{\vee} Hy))\}$,

where (P3c) is logically equivalent to

(P3d) $\{w|w \vDash \bigvee x \bigvee y(x \neq y \wedge Fx \wedge Fy \wedge (Gx \dot{\vee} Gy) \wedge Hx)\}$,

Which variant applies depends on whether the recollection concerns the girl with the injured knee, as in (P3a); the other girl, as in (P3b); or none of them specifically, as in (P3c). The increments thus are:

$$(P2/3a,b,c) = (P3a,b,c) - (P2), \text{respectively},$$

which are three different increments. Hence, what seemed to be one specific recollection is here split into three possible recollections. One might suggest that the order of changes in my doxastic states should be reversed, i.e., that the recollection has to come first and that the observation joins; in this case the recollection would bring about a unique change and the observation would result in three possible changes. This would then correspond to what you get according to the intentional conception. Indeed, past observations are sometimes reinterpreted in the light of emerging recollections. However, this is not always the case, and in my version of the story it did not play a role.

So, this is the second argument: According to the propositional conception the observation leads to a unique change of my doxastic state and the recollection may then take three different forms. Intuitively, however, it is just the other way round, and so it is represented by the intentional conception. Hence, the propositional conception gives an incorrect account of the succession of observation and recollection and can render it correct only by artificially reversing the real succession of events.

I mentioned already that these arguments are basically of the same kind as those of Perry (1980). A minor difference is that he worries about preservation or the continuity of belief, whereas I argue with the adequate representation of belief expansion. The major difference is that he is not operating with the internalistic conception of narrow contents characterized by the exclusion of possibilities. He rather considers whether the continuity of belief consists in the preservation of the situation believed (in his technical sense) or in the retention of text by which the belief can be expressed at the various times, and he finds convincing examples against both proposals as well as against various combinations of them, thus motivating the introduction of the new theoretical concept of a file, as he says on p. 328. However, situations in Perry's sense are wide contents, and the criterion of retention of text again involves us in issues of linguistic meaning, whereas I wanted to dissociate myself from both these notions. This is why it is difficult to compare our frameworks and to decide whether or not our arguments come exactly to the same.

However, the gist of the arguments is always the same. It is about the availability of names or definite descriptions or equivalent devices. That is why I chose the twin story. Similarly, Perry emphasizes again and again that he cannot find them in his examples that, by the way, also use such ingredients as confusion and failing memory. This is also the point of the arguments in the linguistic cases that doubt, for instance, the adequacy of the proposal of Evans (1980) to treat certain occurrences of anaphoric pronouns as E-type, i.e., as definite descriptions; cf. the critical discussion of Heim (1982, sects. I.1.4 and I.2.3).

Zimmermann (1999, pp. 359ff.), objects. He wants to defend the availability of definite descriptions and thinks that I have misrepresented the example. When I see the girl falling from her skateboard, I do not only see that she has G, i.e., that she has hurt her knee; rather I see her instantiating an enormously complex property G^* (most of which may escape my descriptive capacities). So, my perception actually moves me into the doxastic state (as far as the example is concerned):

$$\text{(P2*)} \quad \{w \mid w \vDash \bigvee x \bigvee y (x \neq y \wedge Fx \wedge Fy \wedge G^*x)\}$$

Of course, (P2*) entails (P2), since G^* entails G.

Moreover, Zimmermann says, and I agree, that in that situation there must have been some perceivable property R of the girl because of which I recognize her to be one of my neighbor's daughters of whom I believe the whole of F. Clearly, R must be part of G^*. So, in that situation I have also the background belief:

$$\text{(PB)} \quad \{w \mid w \vDash \bigwedge x (G^*x \rightarrow Rx) \wedge \bigwedge x (Rx \leftrightarrow Fx)\}$$

However, on this background the increment I have learnt in moving from (P1) to (P2*) has a simple representation:

$$\text{(P1/2*)} \quad \{w \mid w \vDash x \bigvee G^*x\}$$

It is easily checked that indeed (P2*) = (P1) ∩ (PB) ∩ (P1/2*). In this way the difficulties with (P1/2a,b,c) disappear; there is no need to worry about them.

The difficulty with the recollection in the continuation of my story dissolves in the same way. We may well assume the recollection to have a unique content:

$$\text{(P2/3)} \quad \{w \mid w \vDash \bigvee x (Fx \wedge Hx) \wedge \bigvee x (Fx \wedge \neg Hx)\}.$$

As desired, it is rather the observation taking three variants; the perceived totality G^* may contain H or $\neg H$ or neither. That is, G^* may be such that I believe after the observation:

(P2*a) $\{w \mid w \vDash \bigwedge x(G^*x \rightarrow Hx)\}$

(P2*b) $\{w \mid w \vDash \bigwedge x(G^*x \rightarrow \neg Hx)\}$

(P2*c) $\{w \mid w \vDash \bigvee x(G^*x \wedge Hx) \wedge \bigvee x(G^*x \wedge \neg Hx)\}$.

So, we have (P2*) $\cap$ (P2*a,b,c) $\cap$ (P2/3) $\subseteq$ (P3*a,b,c), respectively (only "$\subseteq$" because the G in (P3a,b,c) is weaker than G^*). The trifurcate effect of the recollection is thus explained by the possible shapes of the observation, in no worse a way than according to the intentional conception.

Zimmermann concludes that the propositional and the intentional conception are explanatory equally successful and that hence the first is to be preferred because of its greater simplicity.

16.5 The Method of Sufficiently Fine-Grained Descriptions

The objection is well taken, I think; there is no direct way to refute it. Mutatis mutandis, it may apply also to Perry's examples, though it need not carry over to the linguistic arguments that are different. What the objection does, then, is to shift our argument to a more strategic level, which I take to be the proper level of our dispute well-prepared by the previous section. In fact, it is a nice exemplification of what may be called the method of sufficiently fine-grained description, a method widely applied in philosophy, whereas my arguments intimated to avoid this method, though in a way not yet made explicit. Let me unfold this strategic issue in this section.

The point of the objection was to find a sufficiently rich property G^* that entailed both the property G of hurting one's knee on which I had originally focused and the property R sufficient to recognize the twins. It seems easy to find that fine-grained G^* in my every-day example, and it seems plausible that when confronted with ever more contrived examples one will succeed in the same way with even more fine-grained descriptions. By contrast, my arguments were meant to stay away from that strategy by focusing on that girl's hurting her knees as the only perceptual information remaining, by initially abstracting from the discriminating liver-spot, etc. On this coarse level of description my arguments certainly hold good.

This is a discursive pattern we often find in philosophy: Under the force of certain arguments one feels compelled to resort to more fine-grained descriptions of the cases at hand. Sometimes, one may even observe an absurd race between examples and escapes. The escapes seem to be the winning strategy, and I am happy to grant that they work (though there remain doubtful cases). However, one is so compelled only by being caught in a certain theoretical framework. And everyone would be happier, I assume, when being provided with theoretical resources freeing us from these argumentative forces and absurd races. Thus, there is a challenge to find these alternative theoretical means. I am convinced that in the end it is always possible to meet this challenge and that the alternative theories always turn out to be more satisfactory.

I have found at least four quite varied examples of this discursive pattern, and even though it may look like changing the topic, I think it is really instructive to study these examples, in their own right, but also with respect to our present case.

The historically first example I am aware of is *decision theory*, i.e., the groundbreaking account of Savage (1954) and in particular its Section 5.5 on small worlds the moral of which is, I think, still not fully appreciated. Savage faced a straightforward problem: In a decision situation one should take into account every item considered relevant to one's decision. Of course, one should! However, if one takes this demand seriously, one soon sees there is no end to the relevancies. The consequences and even, more narrowly, the favorable and unfavorable consequences of one's decision indefinitely extend into the future, the circumstances on which these consequences depend as well get broader and broader, and the decision at hand turns out not to be separable from all one's future decisions. Thus, Savage (1954, p. 83) ends up considering that "a person has only one decision to make in his whole life". He finds the consideration "stimulating", but also "highly unrealistic" and "unwieldy"; one might also find stronger words. His problem then "is to say as clearly as possible what constitutes a satisfactory isolated decision situation".

He solved it with his theory of small worlds. What he did there was to show how to reduce a decision model referring to fine-grained states of world, acts, and consequences to a provably equivalent decision model working with more coarse-grained states of the world, acts, and consequences; "provably equivalent" here means "to provably lead to the same decision". So, in effect, there were two problems, that of isolating independent decision situations and that of reducing grand to small decision situations, and despite his rhetoric he rather solved the second. Moreover, his solution was not perfectly general; as he was well aware, it worked only under certain restrictive assumptions. However, if one changes to the decision models of Fishburn (1964), the reduction works generally and without constraints, for me the ultimate reason to prefer Fishburn's over Savage's modeling.[17]

What is remarkable about this is that the postulate of equivalent reducibility of grand-world to small-world decision models is a substantial and consequential postulate. As just mentioned, one consequence concerns the precise format of decision models. Another consequence, and one that is very insufficiently appreciated as far as I see, concerns the decision rule that is required to be invariant under reductions. Savage, of course, applies the decision role of maximizing expected utility, and the natural reduction method is just the one that keeps this decision rule invariant; in fact, the decision rule is nothing but the maximal reduction in which only the possible actions and nothing else is considered (cf. Spohn 1976/78, sect. 3.6 and 1982, pp. 246–249). However, for other decision rules there are no good

[17] All this is fully explained in Spohn (1976/78, sects. 2.3, 3.5, and 3.6).

reduction methods that respect their invariance. To mention a familiar example: In a sufficiently fine-grained description each decision of mine (staying at home, turning on the radio, etc.) might have the worst possible consequence, i.e., result in getting killed. Given this fine-grained description, the maximin decision rule absurdly dictates indifference between all my options, and equivalent reduction would have to preserve this indifference. I wonder which decision rules are compatible with the postulate of equivalent reducibility and whether another justification of maximizing expected utility might be forthcoming in this way.

My second example is *learning by conditionalization*. For centuries, the only formal account of learning was Bayes' theorem or simple conditionalization with respect to the proposition learned. Jeffrey (1965, ch. 11), however, opened our mind by proposing his rule of generalized conditionalization, according to which what is learned is not a proposition, but some new probability distribution over some propositional partition induced by experience. The idea was to allow for the case of uncertain evidence and thus to avoid the old foundationalist presupposition that evidence is always certain. Jeffrey's rule then made a specific proposal for how to change one's subjective probabilities in the light of such uncertain evidence.

Levi (1967) started an argument with Jeffrey. He thought that Jeffrey's generalization would be superfluous and unjustified: unjustified because only certain evidence can justify the doxastic changes induced by it, and superfluous because one can always find an evidential proposition which is learned for sure and which induces the distribution representing uncertain evidence according to Jeffrey. Here it is again, our discursive pattern: Levi appeals to sufficiently fine-grained descriptions, whereas Jeffrey wants to avoid them.

We need not follow the argument about justification. Still, Levi may be right, we may always find a sufficiently detailed evidential proposition and thus represent learning by simple conditionalization. Of course, these evidential propositions soon outrun our linguistic descriptions. Often, the best we can say about our evidence is that the scene before us appeared to us in such and such a way (which is not an especially helpful proposition to conditionalize on). But even if we grant Levi's argument, the point is simply that simple conditionalization is not invariant under variations of descriptive granularity. Doxastic changes that can be described by simple conditionalization within a fine-grained propositional framework cannot be so described within coarsenings of that framework. By contrast, Jeffrey's generalized conditionalization is provably invariant in this way; a generalized conditionalization turns into another by coarsening the propositional framework. This is why I find Jeffrey's rule theoretically superior to Levi's insistence on traditional conditionalization. Indeed, we have here a particularly clear exemplification of our discursive pattern.

A rich field of application of the method of sufficiently fine-grained description is *causation*, my third illustration. There are in fact two variants, the method of fine-graining causal chains and the method of fine-graining events (i.e., causes and effects). As already observed by Salmon (1980)[18] and many others, these are the main

[18] He speaks of "the method of more detailed specification of events" and "the method of interpolated causal links".

methods of dealing with recalcitrant examples. Let me focus here on just one problem case, the problem of (symmetric) causal overdetermination. This is indeed a problem for almost all theories of (deterministic) causation. Regularity theories tend to be too liberal; they find overdetermination where there really is none. But this may be counted against regularity theories. Conversely, counterfactual analyses tend to be too restrictive, to allow no overdetermination whatsoever. Hence it has become popular to explain away overdetermination: if we describe the allegedly overdetermined effect in a sufficiently detailed way, we see that it would not have realized in exactly this way, if one of the allegedly overdetermining causes had not occurred. Thus these causes turn out to be joint contributory causes; that is the normal way of causation. Again, one can consider more and more contrived examples. Perhaps the strategy of sufficiently fine-graining the effects always succeeds.[19]

However, the counterfactual analysts pursue this strategy not because it would be so natural, but because they are captives of their theoretical framework that seems to leave them no other choice in dealing with overdetermination. Change the framework, and the dialectics of the case is completely changed. At least this is what I have proposed since Spohn (1983a). I prefer to analyze causation in terms of ranking functions instead of counterfactual conditionals; for a recent attempt to defend my analysis see Chapter 3. Section 3.5 there explains how this analysis can allow for overdetermination in a straightforward and appropriate way. It is not committed to artificially shifting or expanding the description of the problem cases.

The issue is certainly more complex than just displayed. However, there is no point in attempting to develop the complexities here; see Chapter 3. Still, the sketch I have given seems basically fair. We again have the choice between one theory being forced to invoke fine-grained descriptions and another theory not being so forced. And again I have no doubt that the latter is more fruitful even if the former remains defensible.

My last example is closest to our concerns; it is the debate between Lewis and Stalnaker about the representation of *de se beliefs*. As mentioned in Section 16.3, Lewis (1979b) accounted for de se beliefs by taking centered worlds as doxastic alternatives, and Stalnaker (1981) argued that worlds would do. Again, the debate was about fine-graining. Lewis suggested coarse-graining by assuming poor Lingens to have completely forgotten who he is, and Stalnaker enforced fine-graining by pointing out that Lingens' perceptual perspective would still be detailed enough to ensure self-identification. Lewis then introduced his ultimate, desperate example of the two gods propositionally omniscient, but not knowing who they are. It is hard to figure out the details of the case. Somehow, divine knowledge must be very different from human knowledge; and so the force of that example remains unclear. Still, Stalnaker countered with claiming purely haecceitistic differences between worlds – again a desperate move.

[19] For all this cf. Lewis (1986d, pp. 207–212). Interestingly, Lewis is not always in favor of applying the method of fine-grained descriptions. In (2000, pp.183f.) he explicitly refuses to fine-grain causal chains in order to reduce so-called cases of preemption by trumping to cases of preemption by cutting.

It seems that the argument cannot be conclusively decided; softer arguments are all there are. Again, though, one might wish to entirely avoid that gambit of ever more fine-grained propositional structure. Lewis is able to do so with his richer structure of centered worlds, whereas Stalnaker must pay for his poorer structure of doxastic possibilities by assuming sufficiently fine-grained propositions.

Of course, the four examples could acquire their full force only when we discussed them much more carefully. However, I am confident that such scrutiny would confirm the conclusions reached. They all point into the same direction. And when we take the same direction concerning our topic, then it is clear what my third argument for the intentional and against the propositional conception of content is. It is this:

According to the propositional conception a typical piece of experience or information is that an object described in a certain way falls under a certain concept. This works provided the doxastic subject has a definite description of the relevant object. However, relative to smaller or more coarse-grained propositional or conceptual frameworks such descriptions may easily cease to exist. This is simply the effect of the coarser framework and does not depend on complicated stories about (almost) indistinguishable twins. The point of the stories about forgetting or neglecting information in my first two arguments was simply to illustrate the variation of descriptive granularity. Now, if definite descriptions get lost, the increase in information cannot be accounted for by the propositional conception in its typical way. This account is simply not invariant under the granularity of doxastic possibilities.

By contrast, the intentional conception avoids this difficulty. According to it a typical piece of evidence or information is that some concept is attached to a certain address or file card; i.e., that the object represented by that address falls under that concept. This does not depend on whether or not this address can be qualitatively distinguished from other addresses within a given conceptual framework.

As already stated, addresses or file cards are *not rigid*. Rarely, the information stored at them will be objectively identifying. Doxastic possibilities will usually contain different objects at the same position in their sequence of objects. However, addresses are *stable* or invariant across conceptual changes, refinements as well as coarsenings. This is their raison d'être. If narrow contents were just general propositions built from narrow concepts, be they linguistically expressible or not, this stability could not be achieved. Russell (1910/11) also acknowledged singular propositions as belief contents referring to objects of acquaintance. He supposed, however, this reference to be rigid (although he did not use this term); he had better assumed that it is only stable in the sense explained.

Let me summarize this section in a still more general and abstract way. Including our focal case we have five examples in which the independence of descriptive granularity seems theoretically superior to and more fruitful than the appeal to sufficiently fine-grained descriptions. This leads me to speculate about a general principle of philosophical psychology:

> *The Invariance Principle*: The propositional attitudes, their contents, and their static and dynamic laws must be so conceived as to be invariant under coarse- and fine-graining of the underlying conceptual and propositional framework.

Let me emphasize that this principle is neutral with respect to the nature of possible worlds that are (part of the) doxastic possibilities making up propositions. The point is not that the possible worlds themselves may be fine- or coarse-grained, as Savage's metaphor of the grand and small worlds may suggest. This would indeed make sense only with respect to Wittgensteinian possible worlds, but not with respect to Lewisian possible worlds. One may cut objects in pieces, but one cannot coarse-grain them. The invariance principle rather alludes to coarse- and fine-graining of the propositional algebra constructed over the set of doxastic possibilities. This is well compatible with the latter being maximally specific.

I am wondering about general justifications of the invariance principle. Here, we must be content with having provided ample inductive support for it. In my case, the principle entails the intentional conception of contents, if my argument is correct. If we accept the former, we should accept the latter.

16.6 Some Afterthoughts

Accepting the intentional conception has profound consequences. In Section 16.3 I had sketched the immediate dialectical background of my thesis. Afterwards, I had abstracted from it and confined myself to pure epistemology. Successfully, I hope: I have talked about belief change, more precisely about belief expansion, and about the invariance principle, but not about meaning, linguistic concepts or the like. However, if we reinstall the background and if my argument goes through as a purely epistemological one, this has clear consequences for two-dimensional semantics in its epistemological reinterpretation, and it serves as confirmation of the congruence principle. It may thus also be taken as supporting the related semantic theories by Kamp (1981) and Heim (1982).

As such it also allows alternative and perhaps more plausible accounts of the logical form of various problem cases, for instance of de re belief ascriptions or of the puzzle of intentional identity in so-called Hob-Nob sentences created by Geach (1967).[20] Quine has repeatedly reminded us, e.g. in (1960, sect. 32) that there are not only propositional, but also objectual attitudes like seeking, hunting, thinking of somebody. He tended to translate them into propositional attitudes, but it seems that within the intentional conception they can be taken for what they are.

Indeed, the implication is a more fundamental one, I think. The intentional conception appears to undermine the so-called context principle, i.e., the principle of the primacy of sentence meaning over word meaning. Frege already put it thus: "It is only in the context of a sentence that a word has a meaning" (1884, sect. 62). This principle came to play a prominent role in the philosophy of language. It did so in Quine's theory of meaning and translation, where meaning is primarily a

[20] I agree with the account given by Kamp (1984/85, sect. VII; 1990, sect. 5), where the central notion is that of sharing a discourse referent.

matter of items capable of direct confrontation with experience, i.e., of observation sentences and more holistic theoretical constructions (cf. Quine 1960, chs. 1–2). The principle occurs in Davidson's theory of interpretation that makes essential use of the principle of charity and thus constructs the meanings of expressions (of a subject or a community) with an eye on the truth of the beliefs expressed by utterances of complete sentences (cf. Davidson 1984, chs. 2 and 9–11). The context principle also inspired various forms of skepticism, e.g., Quine's thesis of the indeterminacy of translation one version of which is basically the thesis of the inscrutability of reference (cf. Quine 1960, ch. 2), and several proposals in its wake (e.g., Putnam 1980). Ultimately, the principle and its applications rest on the assumption that doxastic attitudes are propositional attitudes the content of which can be only judged as true or false. By contrast, the intentional conception allows us to also consider the reference or, in Kamp's terms, external anchoring of the addresses or file cards figuring in doxastic alternatives. Therefore, philosophical accounts that rest on the context principle seem in urgent need of reconsideration.

This applies in particular to the account of meaning developed by Grice (1957); he as well relies upon the principle of the primacy of sentence meaning. Burge (1979, p. 109) already remarked that his anti-individualistic conception of the attitudes undermines the reductive Gricean program (cf. also Spohn 2003b), and Schiffer, once a dedicated defender of Gricean intention-based semantics, devotes his whole (1987) to demolish this approach. This paper did not attempt to argue with such anti-individualistic tendencies, though it confessed its individualistic spirit. In any case, the need to reconsider the Gricean program and to restate its defeasible parts in the light of the intentional conception of contents seems obvious.

These remarks could give only a few hints, not more. How exactly the intentional conception bears out these suggestions is another issue. But if only half of these remarks hold good, this would strongly underscore the relevance of the thesis defended here, beyond the importance it has in itself.

Bibliography

Almog, Joseph (1981), "Dthis and Dthat: Indexicality Goes Beyond That", *Philosophical Studies* 39, 347–381.
Almog, Joseph (1984), "Would You Believe That?", *Synthese* 58, 1–37.
Armstrong, David M. (1968), *A Materialist Theory of the Mind*, London: Routledge.
Armstrong, David M. (1983), *What is a Law of Nature?*, Cambridge: Cambridge University Press.
Armstrong, David M. (1997), *A World of States of Affairs*, Cambridge: Cambridge University Press.
Arntzenius, Frank, and Ned Hall (2003), "On What We Know About Chance", *British Journal for the Philosophy of Science* 54, 171–179.
Aronson, Jerrold L. (1971), "On the Grammar of 'Cause' ", *Synthese* 22, 414–430.
Austin, John L. (1962), *Sense and Sensibilia*, Oxford: Clarendon Press.
Barwise, Jon, and John Etchemendy (1987), *The Liar: An Essay on Truth and Circularity*, Oxford: Oxford University Press.
Barwise, Jon, and John Perry (1981), "Situations and Attitudes", *Journal of Philosophy* 78, 668–691.
Barwise, Jon, and John Perry (1983), *Situations and Attitudes*, Cambridge, MA: MIT Press.
Bauer, Heinz (1968), *Wahrscheinlichkeitstheorie und Grundzüge der Maßtheorie*, Berlin: de Gruyter.
Bealer, George (1982), *Quality and Concept*, Oxford: Oxford University Press.
Beauchamp, Tom, and Alexander Rosenberg (1981), *Hume and the Problem of Causation*, Oxford: Oxford University Press.
Beckermann, Ansgar (1996), "Is There a Problem About Intentionality?", *Erkenntnis* 45, 1–23.
Benkewitz, Wolfgang (1999), "Belief Justification and Perception", *Erkenntnis* 50, 193–208.
Benkewitz, Wolfgang (forthcoming), *Wahrnehmen, Glauben und Gegenstände. Eine historisch-systematische Untersuchung*, Heidelberg: Synchron.
Binkley, Robert W. (1968), "The Surprise Examination in Modal Logic", *Journal of Philosophy* 65, 127–136.
Black, Robert (1998), "Chance, Credence, and the Principal Principle", *British Journal for the Philosophy of Science* 49, 371–385.
Blackburn, Simon (1993), *Essays in Quasi-Realism*, Oxford: Oxford University Press.
Block, Ned (1986), "Advertisement for a Semantics for Psychology", in: P. A. French, T. E. Uehling, and H. K. Wettstein (eds.), *Midwest Studies in Philosophy* Vol. X, *Studies in the Philosophy of Mind*, Minneapolis, MN: University of Minnesota Press, pp. 615–678.
Block, Ned (1991), "What Narrow Content is Not", in: B. Loewer and G. Rey (eds.), *Meaning in Mind. Fodor and His Critics*, Oxford: Blackwell, pp. 33–64.
Block, Ned (1995), "Ruritania Revisited", in: E. Villanueva (ed.), *Philosophical Issues* Vol. 6, Atascadero: Ridgeview, pp. 171–187.

BonJour, Laurence (1985), *The Structure of Empirical Knowledge*, Cambridge, MA: Harvard University Press.
BonJour, Laurence (1997), "Haack on Justification and Experience", *Synthese* 112, 13–23.
Boynton, Robert M. (1979), *Human Color Vision*, New York: Holt, Rinehart & Winston.
Brentano, Franz (1874/1973), *Psychologie vom empirischen Standpunkt*, Leipzig: Duncker & Humblot 1874; English translation: *Psychology from an Empirical Standpoint*, London: Routledge & Kegan Paul.
Bunzl, Martin (1979), "Causal Overdetermination", *Journal of Philosophy* 76, 134–150.
Burge, Tyler (1979), "Individualism and the Mental", in: P. A. French, T. E. Uehling jr., and H. K. Wettstein (eds.), *Midwest Studies in Philosophy* Vol. IV, *Metaphysics*, Minneapolis, MN: University of Minnesota Press, pp. 73–121.
Carnap, Rudolf (1928), *Der logische Aufbau der Welt*, Hamburg: Meiner.
Carnap, Rudolf (1936/37), "Testability and Meaning", *Philosophy of Science* 3, 419–471 and 4, 1–40.
Carnap, Rudolf (1947), *Meaning and Necessity*, 2nd ed., Chicago, IL: Chicago University Press (1956).
Carnap, Rudolf (1956), "The Methodological Character of Theoretical Concepts", in: H. Feigl and M. Scriven (eds.), *Minnesota Studies in the Philosophy of Science* Vol. I, Minneapolis, MN: University of Minnesota Press, pp. 38–76.
Carnap, Rudolf (1966), *Philosophical Foundations of Physics*, New York: Basic Books.
Carnap, Rudolf (1971/80), "A Basic System of Inductive Logic", Part I in: R. Carnap and R. C. Jeffrey (1971), pp. 33–165; Part II in: R. C. Jeffrey (1980), pp. 7–155.
Carnap, Rudolf, and Richard C. Jeffrey (eds.) (1971), *Studies in Inductive Logic and Probability* Vol. I, Berkeley, CA: University of California Press.
Carroll, John W. (1994), *Laws of Nature*, Cambridge: Cambridge University Press.
Cartwright, Nancy (1979), "Causal Laws and Effective Strategies", *Noûs* 13, 419–437; also in: N. Cartwright (1983), pp. 21–43.
Cartwright, Nancy (1983), *How the Laws of Physics Lie*, Oxford: Clarendon Press.
Cartwright, Nancy (1988), "Regular Associations and Singular Causes", in: B. Skyrms and W. L. Harper (1988), pp. 79–97.
Cartwright, Nancy (1989), *Nature's Capacities and Their Measurement*, Oxford: Clarendon Press.
Cartwright, Nancy (2001), "Modularity: It Can – and Generally Does – Fail", in: M. C. Galavotti, P. Suppes, and D. Costantini (eds.), *Stochastic Causality*, Stanford: CSLI Publications, pp. 65–84.
Cartwright, Nancy (2002), "In Favor of Laws That Are Not *Ceteris Paribus* After All", *Erkenntnis* 57, 425–439.
Cartwright, Nancy (2003), "What is Wrong with Bayes Nets?" in: H. Kyburg and M. Thalos (eds.), *Probability Is the Very Guide of Life*, La Salle, IL: Open Court, pp. 253–276.
Castañeda, Hector-Neri (1966), "'He': A Study in the Logic of Self-Consciousness", *Ratio* 8, 130–157.
Chaitin, Gregory J. (1966), "On the Length of Programs for Computing Finite Binary Sequences", *Journal of the Association of Computing Machinery* 13, 547–569.
Chalmers, David J. (1996), *The Conscious Mind*, Oxford: Oxford University Press.
Chalmers, David J. (2002), "The Components of Content", revised version in: D. Chalmers (ed.), *Philosophy of Mind: Classical and Contemporary Readings*, Oxford: Oxford University Press, pp. 608–633.
Chalmers, David J. (2006), "The Foundations of Two-Dimensional Semantics", in: M. García-Carpintero and J. Macià (eds.), *Two-Dimensional Semantics*, Oxford: Clarendon Press, pp. 55–140.
Chisholm, Roderick M. (1957), *Perceiving. A Philosophical Study*, Ithaca: Cornell University Press.
Chomsky, Noam (1995), "Language and Nature", *Mind* 104, 1–61.
Church, Alonzo (1940), "On the Concept of a Random Sequence", *Bulletin of the American Mathematical Society* 46, 130–135.
Cohen, Jonathan L. (1977), *The Probable and the Provable*, Oxford: Clarendon Press.

Collins, John (2000), "Preemptive Prevention", *Journal of Philosophy* 97, 223–234.
Collins, John, Ned Hall, and Laurie A. Paul (eds.) (2004), *Causation and Counterfactuals*, Cambridge, MA: MIT Press.
Cresswell, Max J. (1980), "Jackson on Perception", *Theoria* 46, 123–147.
Davidson, Donald (1969), "The Individuation of Events", in: N. Rescher et al. (eds.), *Essays in Honor of Carl G. Hempel*, The Netherlands, Dordrecht: Reidel, pp. 216–234.
Davidson, Donald (1984), *Inquiries into Truth and Interpretation*, Oxford: Oxford University Press.
Davies, Martin, and Lloyd Humberstone (1981), "Two Notions of Necessity", *Philosophical Studies* 58, 1–30.
Davis, Wayne A. (1988), "Probabilistic Theories of Causation", in: J. H. Fetzer (1988), pp. 133–160.
Dawid, A. Philip (1979), "Conditional Independence in Statistical Theory", *Journal of the Royal Statistical Society* (Series B) 41, 1–31.
de Finetti, Bruno (1937), "La Prévision: Ses Lois Logiques, Ses Sources Subjectives", *Annales de l'Institut Henri Poincaré* 7; English translation: "Foresight: Its Logical Laws, Its Subjective Sources", in: H. E. Kyburg Jr. and H. E. Smokler (eds.), *Studies in Subjective Probability*, New York: Wiley (1964), pp. 93–158.
Diaconis, Percy, and David Freedman (1980), "De Finetti's Theorem for Markov Chains", *Annals of Probability* 8, 115–130.
Donnellan, Keith (1966), "Reference and Definite Descriptions", *Philosophical Review* 75, 281–304.
Dretske, Fred (1981), *Knowledge and the Flow of Information*, Oxford: Blackwell.
Earman, John, and John Roberts (1999), "*Ceteris Paribus*, There is No Problem of Provisos", *Synthese* 118, 439–478.
Earman, John, John Roberts, and Sheldon Smith (2002), "*Ceteris Paribus* Lost", *Erkenntnis* 57, 281–301.
Eells, Ellery, and Elliot Sober (1983), "Probabilistic Causality and the Question of Transitivity", *Philosophy of Science* 50, 35–57.
Ellis, Brian (1979), *Rational Belief Systems*, Oxford: Blackwell.
Evans, Gareth (1979), "Reference and Contingency", *The Monist* 62, 161–189.
Evans, Gareth (1980), "Pronouns", *Linguistic Inquiry* 11, 337–362.
Evans, Gareth (1982), *The Varieties of Reference*, Oxford: Clarendon Press.
Fair, David (1979), "Causation and the Flow of Energy", *Erkenntnis* 14, 219–250.
Fetzer, James H. (ed.) (1988), *Probability and Causality*, The Netherlands, Dordrecht: Reidel.
Fetzer, James H. (2002), "Propensities and Frequencies: Inference to the Best Explanation", *Synthese* 132, 27–61.
Feyerabend, Paul (1962), "Explanation, Reduction, and Empiricism", in: H. Feigl and G. Maxwell (eds.), *Minnesota Studies in the Philosophy of Science* Vol. III, Minneapolis, MN: University of Minnesota Press, pp. 28–97.
Feyerabend, Paul (1965), "Problems of Empiricism", in: R. G. Colodny (ed.), *Beyond the Edge of Certainty*, Englewood Cliffs, NJ: Prentice-Hall, pp. 145–260.
Field, Hartry (1972), "Tarski's Theory of Truth", *Journal of Philosophy* 69, 347–375.
Field, Hartry (1977), "Logic, Meaning, and Conceptual Role", *Journal of Philosophy* 74, 379–409.
Fishburn, Peter C. (1964), *Decision and Value Theory*, New York: Wiley.
Fodor, Jerry A. (1987), *Psychosemantics. The Problem of Meaning in the Philosophy of Mind*, Cambridge, MA: MIT Press.
Fodor, Jerry A. (1990), *A Theory of Content and Other Essays*, Cambridge, MA: MIT Press.
Fodor, Jerry A. (1994), *The Elm and the Expert*, Cambridge, MA: MIT Press.
Frege, Gottlob (1879), *Begriffsschrift, eine der arithmetischen nachgebildete Formelsprache des reinen Denkens*, Halle; Reprint: Darmstadt: Wissenschaftliche Buchgesellschaft (1964).
Frege, Gottlob (1884), *Die Grundlagen der Arithmetik*; German-English edition: *The Foundations of Arithmetic*, Oxford: Blackwell (1959).

Frege, Gottlob (1892), "Sinn und Bedeutung", *Zeitschrift für Philosophie und Philosophische Kritik*, N.F. 100, 25–50; English translation in: M. Beaney (ed.), *The Frege Reader*, Oxford: Blackwell (1997), pp. 151–171.

Frege, Gottlob (1918), "Der Gedanke. Eine logische Untersuchung", *Beiträge zur Philosophie des deutschen Idealismus* 1, pp. 58–77; English translation: "Thought", in: M. Beaney (ed.), *The Frege Reader*, Oxford: Blackwell (1997), pp. 325–345.

Friedman, Michael (1974), "Explanation and Scientific Understanding", *Journal of Philosophy* 71, 5–19.

Gabbay, Dov M., and Philippe Smets (eds.) (1998–2000), *Handbook of Defeasible Reasoning and Uncertainty Management Systems* Vol. 1–5, The Netherlands, Dordrecht: Kluwer.

Gaifman, Haim (1988), "A Theory of Higher Order Probabilities", in: B. Skyrms and W. L. Harper (1988), pp. 191–219.

Gärdenfors, Peter (1979), "Conditionals and Changes of Belief", *Acta Philosophica Fennica* 30, 381–404.

Gärdenfors, Peter (1981), "An Epistemic Approach to Conditionals", *American Philosophical Quarterly* 18, 203–211.

Gärdenfors, Peter (1984), "Epistemic Importance and Minimal Changes of Belief", *Australian Journal of Philosophy* 62, 136–157.

Gärdenfors, Peter (1988), *Knowledge in Flux. Modeling the Dynamics of Epistemic States*, Cambridge, MA: MIT Press.

Gärdenfors, Peter, and Hans Rott (1995), "Belief Revision", in: D. M. Gabbay et al. (eds.), *Handbook of Logic in Artificial Intelligence and Logic Programming* Vol. 4, Oxford: Oxford University Press, pp. 36–132.

Geach, Peter T. (1967), "Intentional Identity", *Journal of Philosophy* 74, 627–632.

Geiger, Dan, and Judea Pearl (1988), "Logical and Algorithmic Properties of Conditional Independence and Qualitative Independence", *Technical Report* No. R-97-IIL, Cognitive Systems Laboratory, University of California, Los Angeles, Oct. 1988; published in: *The Annals of Statistics* 21 (1993) 2001–2021.

Giere, Ronald (1980), "Causal Systems and Statistical Hypotheses", in: L. J. Cohen and M. Hesse (eds.), *Applications of Inductive Logic*, Oxford: Clarendon Press, pp. 251–270.

Gillies, Donald (2000), *Philosophical Theories of Probability*, London: Routledge.

Glymour, Clark (2002), "A Semantics and Methodology for *Ceteris Paribus* Hypotheses", *Erkenntnis* 57, 395–405.

Glymour, Clark, Richard Scheines, Peter Spirtes, and Kevin Kelly (1987), *Discovering Causal Structure*, New York: Academic.

Goldstein, Matthew (1983), "The Prevision of a Prevision", *Journal of the American Statistical Association* 78, 817–819.

Goldszmidt, Moisés, and Judea Pearl (1992), "Default Ranking: A Practical Framework for Evidential Reasoning, Belief Revision and Update", in: B. Nebel, C. Rich, and W. Swartout (eds.), *Proceedings of the Third International Conference on Knowledge Representation and Reasoning*, San Mateo, CA: Morgan Kaufmann, pp. 661–672.

Good, Irving J. (1961–63), "A Causal Calculus", *British Journal for the Philosophy of Science* 11, 305–318, 12, 43–51, and 13, 88.

Good, Irving J. (1980), "A Further Comment on Probabilistic Causality: Mending the Chain", *Pacific Philosophical Quarterly* 61, 452–454.

Good, Irving J. (1988), "Causal Tendency: A Review", in: B. Skyrms and W. L. Harper (1988), pp. 23–50.

Grice, H. Paul (1957), "Meaning", *Philosophical Review* 66, 377–388.

Haas-Spohn, Ulrike (1995), *Versteckte Indexikalität und subjektive Bedeutung*, Berlin: Akademie-Verlag; English translation:
http://www2.sfs.uni-tuebingen.de/Alumni/Dissertationen/ullidiss/index.html

Haas-Spohn, Ulrike (1997), "The Context Dependency of Natural Kind Terms", in: W. Künne, A. Newen, and M. Anduschus (eds.), *Direct Reference, Indexicality and Propositional Attitudes*, Stanford: CSLI Publications, pp. 333–349.

Haas-Spohn, Ulrike, and Wolfgang Spohn (2001), "Concepts Are Beliefs About Essences", in: A. Newen, U. Nortmann, and R. Stuhlmann-Laeisz (eds.), *Building on Frege. New Essays on Sense, Content, and Concept*, Stanford: CSLI Publications, pp. 287–316; also in this volume, ch. 14.

Hájek, Alan (2007), "The Reference Class Problem is Your Problem Too", *Synthese* 156, 563–585.

Hall, Ned (1994), "Correcting the Guide to Objective Chance", *Mind* 103, 505–517.

Hall, Ned (2000), "Causation and the Price of Transitivity", *Journal of Philosophy*, 97, 198–222.

Hall, Ned (2004), "Two Mistakes About Credence and Chance", in: L. Jackson and G. Priest (eds.), *Lewisian Themes: The Philosophy of David K. Lewis*, Oxford: Oxford University Press, pp. 94–112.

Hall, Ned, and Laurie A. Paul (2003), "Causation and the Preemption", in: P. Clark and K. Hawley (eds.), *Philosophy of Science Today*, Oxford: Clarendon Press, pp. 100–130.

Halpern, Joseph Y. (2001), "Conditional Plausibility Measures and Bayesian Networks", *Journal of AI Research* 14, 359–389.

Hansson, Sven Ove (1998), "Revision of Belief Sets and Belief Bases", in: D. M. Gabbay and P. Smets (1998–2000), Vol. 3, pp. 17–75.

Hardin, Clifford (1988), *Color for Philosophers*, Indianapolis, IN: Hackett Publishing.

Harman, Gilbert (1987), "(Nonsolipsistic) Conceptual Role Semantics", in: E. LePore (ed.), *New Directions in Semantics*, London: Academic, pp. 55–81.

Harper, William L. (1976), "Rational Belief Change, Popper Functions and Counterfactuals", in W. L. Harper and C. A. Hooker (eds.), *Foundations of Probability Theory, Statistical Inference, and Statistical Theories of Science* Vol. I, The Netherlands, Dordrecht: Reidel, pp. 73–115.

Hart, Herbert L. A., and Tony Honoré (1959), *Causation in the Law*, 2nd ed., Oxford: Clarendon Press (1985).

Hazen, Allen (1979), "Counterpart-Theoretic Semantics for Modal Logic", *Journal of Philosophy* 76, 319–338.

Heim, Irene (1982), *The Semantics of Definite and Indefinite Noun Phrases*, Ph.D. dissertation, University of Massachusetts, published as Paper No. 73 of the SFB 99, University of Konstanz (1982) and at: New York, Garland (1988).

Heim, Irene (1991), "Artikel und Definitheit", in: A. von Stechow and D. Wunderlich (eds.), Handbuch der Semantik. Berlin: de Gruyter, pp. 487–535.

Hempel, Carl G. (1965), *Aspects of Scientific Explanation and Other Essays in the Philosophy of Science*, New York: Free Press.

Hempel, Carl G. (1988), "Provisoes: A Problem Concerning the Inferential Function of Scientific Theories", *Erkenntnis* 28, 147–164.

Hesslow, Germund (1976), "Two Notes on the Probabilistic Approach to Causality", *Philosophy of Science* 43, 290–292.

Hilbert, David R. (1987), *Color and Color Perception. A Study in Anthropocentric Realism*. CSLI Lecture Notes No. 9, Stanford.

Hild, Matthias (1998a), "Auto-Epistemology and Updating", *Philosophical Studies* 92, 321–361.

Hild, Matthias (1998b), "The Coherence Argument against Conditionalization", *Synthese* 115, 229–258.

Hild, Matthias, and Wolfgang Spohn (2008), "The Measurement of Ranks and the Laws of Iterated Contraction", *Artificial Intelligence* 172, 1195–1218.

Hild, Matthias (forthcoming), *Introduction to Induction. On the First Principles of Reasoning*.

Hintikka, Jaakko (1962), *Knowledge and Belief*, Ithaca, NY: Cornell University Press.

Hintikka, Jaakko, and Illka Niiniluoto (1976), "An Axiomatic Foundation for the Logic of Inductive Generalization", in: M. Przelecki, K. Szaniawski, and R. Wójcicki (eds.), *Formal Methods in the Methodology of Empirical Sciences*, The Netherlands, Dordrecht: Reidel, pp. 57–81.

Hitchcock, Christopher (2001), "The Intransitivity of Causation Revealed in Equations and Graphs", *Journal of Philosophy* 98, 273–299.

Hoefer, Carl (1997), "On Lewis's Objective Chance: 'Humean Supervenience Debugged' ", *Mind* 106, 321–334.
Humburg, Jürgen (1971), "The Principle of Instantial Relevance", in: R. Carnap and R. C. Jeffrey (1971), pp. 225–233.
Hume, David (1739), *A Treatise Concerning Human Nature*. Cited from the 2nd edition by P. H. Nidditch of the edition by L. A. Selby-Bigge, Oxford: Oxford University Press (1978).
Hume, David (1777), *An Enquiry Concerning Human Understanding*. Cited from the 3rd edition by P. H. Nidditch of the edition of L. A. Selby-Bigge of David Hume's Enquiries, Oxford: Oxford University Press 1975.
Humphreys, Paul (1980), "Cutting the Causal Chain", *Pacific Philosophical Quarterly* 61, 305–314.
Hunter, Daniel (1991), "Maximum Entropy Updating and Conditionalization", in: W. Spohn, B. C. van Fraassen, and B. Skyrms (eds.), *Existence and Explanation. Essays Presented in Honor of Karel Lambert*, The Netherlands, Dordrecht: Kluwer, pp. 45–57.
Hunter, Daniel (1996), "On the Relation Between Categorical and Probabilistic Belief", *Noûs* 30, 75–98.
Jackson, Frank (1998), *From Metaphysics to Ethics*, Oxford: Clarendon Press.
Jackson, Frank, and Robert Pargetter (1987), "An Objectivist's Guide to Subjectivism About Color", *Revue Internationale de Philosophie* 41, 127–141.
Jeffrey, Richard C. (1965), *The Logic of Decision*, 2nd ed., Chicago, IL: University Press, (1983).
Jeffrey, Richard C. (1977), "A Note on the Kinematics of Preference", *Erkenntnis* 11, 135–141.
Jeffrey, Richard C. (ed.) (1980), *Studies in Inductive Logic and Probability* Vol. II, Berkeley, CA: University of California Press.
Jeffrey, Richard C. (2004), *Subjective Probability. The Real Thing*, Cambridge: Cambridge University Press.
Jensen, Fin V. (1996), *An Introduction to Bayesian Networks*, London: UCL Press.
Joyce, James M. (1999), *The Foundations of Causal Decision Theory*, Cambridge: Cambridge University Press.
Kamp, Hans (1981), "A Theory of Truth and Semantic Representation", in: J. Groenendijk, T. Janssen, and M. Stokhof (eds.), *Formal Methods in the Study of Language*, Amsterdam: Amsterdam Centre, pp. 277–322.
Kamp, Hans (1984/85), "Content, Thought, and Communication", *Proceedings of the Aristotelian Society, New Series* 85, 239–261.
Kamp, Hans (1990), "Prolegomena to a Structural Account of Belief and Other Attitudes", in: C. A. Anderson and J. Owens (eds.), *Propositional Attitudes. The Role of Content in Logic, Language, and Mind*, Stanford: CSLI-Lecture Notes No. 20, pp. 27–90.
Kamp, Hans, and Uwe Reyle (1993), *From Discourse to Logic. Introduction to Modeltheoretic Semantics of Natural Language, Formal Logic and Discourse Representation Theory*, The Netherlands, Dordrecht: Kluwer.
Kant, Immanuel (1781/87), *Kritik der reinen Vernunft*; English translation: *Critique of Pure Reason*, London: Macmillan (1929).
Kaplan, David (1969), "Quantifying In", in: D. Davidson and J. Hintikka (eds.), *Words and Objections*, The Netherlands, Dordrecht: Reidel, pp. 206–242.
Kaplan, David (1977), "Demonstratives. An Essay on the Semantics, Logic, Metaphysics, and Epistemology of Demonstratives and Other Indexicals", manuscript, published in: J. Almog et al. (eds.), *Themes from Kaplan*, Oxford: Oxford University Press (1989), pp. 481–563.
Kaplan, David (1978), "Dthat", in P. Cole (ed.), *Syntax and Semantics, Vol. 9: Pragmatics*. New York: Academic, pp. 221–243.
Kaplan, David (1989), "Afterthoughts", in: J. Almog et al. (eds.), *Themes from Kaplan*, Oxford: Oxford University Press, pp. 565–614.
Kaplan, David (1990/91), "Words", *Proceedings of the Aristotelian Society* 64, 93–119.

Karttunen, Lauri (1969), "Discourse Referents", manuscript, in: J. D. McCawley (ed.), *Syntax and Semantics Vol. 7: Notes from the Linguistic Underground*, New York: Academic (1976), pp. 363–385.
Kelly, Kevin (1999), "Iterated Belief Revision, Reliability, and Inductive Amnesia", *Erkenntnis* 50, 11–58.
Kelly, Kevin (forthcoming), "Ockham's Razor, Empirical Complexity, and Truth finding".
Kemmerling, Andreas (1990), "Genau dieselbe Überzeugung", in: Forum für Philosophie in Bad Homburg (ed.), *Intentionalität und Verstehen*, Frankfurt a. M.: Suhrkamp, pp. 153–196.
Kiiveri, Harry, Terry P. Speed, and John B. Carlin (1984), "Recursive Causal Models", *Journal of the Australian Mathematical Society* (Series A) 36, 30–52.
Kim, Jaegwon (1973), "Causation, Nomic Subsumption, and the Concept of an Event", *Journal of Philosophy* 70, 217–236.
Kim, Jaegwon (1984), "Concepts of Supervenience", *Philosophy and Phenomenological Research* 45, 153–176.
Kitcher, Philip (1981), "Explanatory Unification", *Philosophy of Science* 48, 507–531
Klaua, Dieter (1969), *Allgemeine Mengenlehre*, Berlin: Akademie-Verlag.
Köhler, Eckehart (2004), "Physical Intuition as Inductive Support", in: M.C. Galavotti and F. Stadler (eds.), *Induction and Deduction in the Sciences*, The Netherlands, Dordrecht: Kluwer, pp. 151–167.
Kripke, Saul A. (1972), "Naming and Necessity", in: D. Davidson and G. Harman (eds.), *Semantics of Natural Language*, The Netherlands, Dordrecht: Reidel, pp. 253–355, 763–769; ext. ed. Oxford: Blackwell 1980.
Kripke, Saul A. (1979), "A Puzzle About Belief", in: A. Margalit (ed.), *Meaning and Use*, The Netherlands, Dordrecht: Reidel, pp. 239–283.
Kripke, Saul A. (1982), *Wittgenstein on Rules and Private Language*, Oxford: Blackwell.
Kuhn, Thomas S. (1962), *The Structure of Scientific Revolutions*, Chicago, IL: Chicago University Press, 2nd ed. (1970).
Kupffer, Manfred (2000), *Counterparts and Qualities*, Ph.D. thesis, University of Konstanz.
Kyburg, Henry E. jr. (1980), "Acts and Conditional Probabilities", *Theory and Decision* 12, 149–171.
Lambert, Karel (1988), "Prolegomenon zu einer Theorie des wissenschaftlichen Verstehens", in: G. Schurz (ed.), *Erklären und Verstehen in der Wissenschaft*, Munich: Oldenbourg. pp. 299–319.
Lambert, Karel (1991), "On Whether an Answer to a Why-Question is an Explanation If and Only If It Yields Understanding", in: G. Brittan jr. (ed.), *Causality, Method, and Modality*, The Netherlands, Dordrecht: Kluwer, pp. 125–142.
Lange, Marc (2000), *Natural Laws in Scientific Practice*, Oxford: Oxford University Press.
Lange, Marc (2002), "Who's Afraid of Ceteris-Paribus Laws? Or: How I Learned to Stop Worrying and Love Them", *Erkenntnis* 57, 407–423.
Lauritzen, Steffen L. (1982), *Lectures on Contingency Tables*, 2nd ed., Aalborg: Aalborg University Press.
Lehrer, Keith (1990), *Theory of Knowledge*, 2nd ed., London: Routledge, Boulder, CO: Westview Press (2000).
Levi, Isaac (1967), "Probability Kinematics", *British Journal for the Philosophy of Science* 18, 197–209.
Levi, Isaac (1980), "Potential Surprise: Its Role in Inference and Decision-Making", in: L. J. Cohen and M. Hesse (eds.), *Applications of Inductive Logic*, Oxford: Clarendon Press, pp. 1–27.
Levi, Isaac (1983), "Truth, Fallibility, and the Growth of Knowledge", in R. S. Cohen and M. W. Wartofsky (eds.), *Language, Logic, and Method*, The Netherlands, Dordrecht: Reidel, pp. 153–174.
Lewis, David (1968), "Counterpart Theory and Quantified Modal Logic", *Journal of Philosophy* 65, 113–126.
Lewis, David (1973a), *Counterfactuals*, Oxford: Blackwell.
Lewis, David (1973b), "Causation", *Journal of Philosophy* 70, 556–567; with postscripts also in: D. Lewis (1986a), pp. 159–172.

Lewis, David (1976), "Probabilities of Conditionals and Conditional Probabilities", *Philosophical Review* 85, 297–315.
Lewis, David (1979a), "Counterfactual Dependence and Time's Arrow", *Noûs* 13, 455–476.
Lewis, David (1979b), "Attitudes *De Dicto* and *De Se*", *Philosophical Review* 88, 513–543; with postscripts also in: D. Lewis (1983), pp. 133–156.
Lewis, David (1980a), "A Subjectivist's Guide to Objective Chance", in: R. C. Jeffrey (1980), pp. 263–293; with postscripts also in Lewis (1986a), pp. 83–132.
Lewis, David (1980b), "Index, Context, and Content", in: S. Kanger and S. Öhman (eds.), *Philosophy and Grammar*, The Netherlands, Dordrecht: Reidel, pp. 79–100.
Lewis, David (1980c), "Veridical Hallucination and Prosthetic Vision", *Australasian Journal of Philosophy* 58, 239–249.
Lewis, David (1983), *Philosophical Papers* Vol. I, Oxford: Oxford University Press.
Lewis, David (1986a), *Philosophical Papers* Vol. II, Oxford: Oxford University Press.
Lewis, David (1986b), *On the Plurality of Worlds*, Oxford: Blackwell.
Lewis, David (1986c), "Causal Explanation", in: D. Lewis (1986a), pp. 214–240.
Lewis, David (1986d), "Postscripts to 'Causation'", in: D. Lewis (1986a), pp. 172–213.
Lewis, David (1994a), "Reduction of Mind", in: S. Guttenplan (ed.), *A Companion to the Philosophy of Mind*, Oxford: Blackwell, pp. 412–431.
Lewis, David (1994b), "Humean Supervenience Debugged", *Mind* 103, 473–490; also in: D. Lewis, *Papers in Metaphysics and Epistemology*, Cambridge: Cambridge University Press (1999), pp. 224–247.
Lewis, David (1997), "Finkish Dispositions", in: *Philosophical Quarterly* 47, 143–158.
Lewis, David (2000), "Causation as Influence", *Journal of Philosophy* 97, 182–197; extended version in: J. Collins, N. Hall, and L. A. Paul (2004), pp. 75–106.
Loar, Brian (1986), "Social Content and Psychological Content", in: R. Grimm and D. Merrill (eds.), *Contents of Thought*, Tucson, AZ: University of Arizona Press, pp. 99–110.
Loewer, Barry (1996), "Humean Supervenience", *Philosophical Topics* 24, 101–127.
Logue, James (1995), *Projective Probability*, Oxford: Oxford University Press.
MacKay, Thomas, and Michael Nelson (2005), "Propositional Attitude Reports", *Stanford Encyclopedia of Philosophy*, http://plato.stanford.edu/
Mackie, John L. (1965), "Causes and Conditions", *American Philosophical Quarterly* 2, 245–264.
Mackie, John L. (1974), *The Cement of Universe*, Oxford: Oxford University Press.
Maier, Emar (2006), *Belief in Context. Towards a Unified Semantics of de re and de se Attitude Reports*, Ph.D. thesis, University Nijmegen.
Martel, Iain (2003), "Indeterminism and the Causal Markov Condition", Working Paper Series *Philosophy and Probability* No. 4, Philosophy, Probability, and Modeling Research Group, University of Konstanz.
Martin, Charles B. (1994), "Dispositions and Conditionals", *Philosophical Quarterly* 44, 1–8.
Maund, Barry (1986), "The Phenomenal and Other Uses of 'Looks'", *Australasian Journal of Philosophy* 64, 170–180.
McGinn, Colin (1983), *The Subjective View. Secondary Qualities and Indexical Thoughts*, Oxford: Oxford University Press.
Meek, Christopher, and Clark Glymour (1994), "Conditioning and Intervening", *British Journal for the Philosophy of Science* 45, 1001–1021.
Mellor, David H. (1974), "In Defense of Dispositions", *Philosophical Review* 83, 157–181.
Mellor, David H. (1988), "On Raising the Chances of Effects", in: J. H. Fetzer (1988), pp. 229–239.
Merin, Arthur (1996), *Die Relevanz der Relevanz. Fallstudie zur Semantik der englischen Konjunktion* "But", Habilitationsschrift. University of Stuttgart; distributed as: Arbeitsberichte des SFB340, Nr. 142, Universities of Stuttgart and Tübingen; english translation forthcoming.
Mill, John S. (1843), "System of Logic", in: J. M. Robson (ed.), *Collected Works of John Stuart Mill*, Toronto: University of Toronto Press (1973).

Miller, David (1966), "A Paradox of Information", *British Journal for the Philosophy of Science* 17, 59–61.

Miller, David (1995), "Propensities and Indeterminism", in: A. O'Hear (ed.), *Karl Popper: Philosophy and Problems*, Cambridge: Cambridge University Press, pp. 121–147.

Miscevic, Nenad (2005), "Empirical Concepts and A Priori Truth", *Croatian Journal of Philosophy* 14, 289–315.

Montague, Richard (1974), *Formal Philosophy*, New Haven, CT: Yale University Press.

Mühlhölzer, Felix (1988), "On Objectivity", *Erkenntnis* 28, 185–230.

Mühlhölzer, Felix (1989), *Objektivität und Erkenntnisfortschritt. Eine Antwort auf Thomas S. Kuhn*, unpublished Habilitationsschrift, University of Munich.

Nayak, Abhya (1994), "Iterated Belief Change Based on Epistemic Entrenchment", *Erkenntnis* 41, 353–390.

Nida-Rümelin, Martine (1993), *Farben und phänomenales Wissen*, Wien: VWGÖ.

Nida-Rümelin, Martine (1996), "Pseudonormal Vision: An Actual Case of Qualia Inversion?", *Philosophical Studies* 82, 145–157.

Nida-Rümelin, Martine (1997), "The Character of Color Terms: A Phenomenalist View", in: W. Künne, A. Newen, and M. Anduschus (eds.), *Direct Reference, Indexicality and Propositional Attitudes*, Stanford: CSLI Publications, pp. 381–402.

Niiniluoto, Ilkka (1972), "Inductive Systematization: Definition and a Critical Survey", *Synthese* 25, 25–81.

Nozick, Robert (1969), "Newcomb's Problem and Two Principles of Choice", in: N. Rescher et al. (eds.), *Essays in Honor of Carl G. Hempel*, The Netherlands, Dordrecht: Reidel, pp. 114–146.

Nute, Donald (1980), *Topics in Conditional Logic*, The Netherlands, Dordrecht: Reidel.

Olsson, Erik (1999), " 'Cohering With' ", *Erkenntnis* 50, 273–291.

Otte, Richard (1981), "A Critique of Suppes' Theory of Probabilistic Cansality", *Synthese* 48, 167–189.

Otte, Richard (1985), "Probabilistic Causality and Simpson's Paradox", *Philosophy of Science* 52, 110–125.

Papineau, David (1985), "Probabilities and Causes", *Journal of Philosophy* 82, 57–74.

Papineau, David (1989), "Pure, Mixed, and Spurious Probabilities and Their Significance for a Reductionist Theory of Causation", in: P. Kitcher and W. C. Salmon (eds.), *Minnesota Studies in the Philosophy of Science* Vol. XIII, *Scientific Explanation*, Minneapolis, MN: University of Minnesota Press, pp. 307–348.

Peacocke, Christopher (1997), "Metaphysical Necessity: Understanding, Truth and Epistemology", *Mind* 106, 521–574.

Pearl, Judea (1988), *Probabilistic Reasoning in Intelligent Systems: Networks of Plausible Inference*, San Mateo, CA: Morgan Kaufmann.

Pearl, Judea (1998), "Graphical Models for Probabilistic and Causal Reasoning", in: D. M. Gabbay and P. Smets (1998–2000), Vol. I, pp. 367–389.

Pearl, Judea (2000), *Causality. Models, Reasoning, and Inference*, Cambridge: Cambridge University Press.

Peckhaus, Volker (1997), *Logik, Mathesis universalis und allgemeine Wissenschaft. Leibniz und die Wiederentdeckung der formalen Logik im 19. Jahrhundert*, Berlin: Akademie-Verlag.

Perry, John (1977), "Frege on Demonstratives", *Philosophical Review* 86, 474–497.

Perry, John (1979), "The Problem of the Essential Indexical", *Noûs* 13, 3–21.

Perry, John (1980), "A Problem About Continued Belief", *Pacific Philosophical Quarterly* 61, 317–332.

Perry, John (1988), "Cognitive Significance and New Theories of Reference", *Noûs* 22, 1–18.

Piantanida, Thomas P. (1974), A Replacement Model of X-linked Recessive Colour Vision Defects. *Annals of Human Genetics* 37, 393–404.

Pitcher, George (1971), *A Theory of Perception*, Princeton, N.J.: Princeton University Press.

Plantinga, Alvin (1993), *Warrant: The Current Debate*, Oxford: Oxford University Press.

Pollock, John L. (1976), *Subjunctive Reasoning*, The Netherlands, Dordrecht: Reidel.

Popper, Karl R. (1934/69), *Logik der Forschung*, 3rd ed., Tübingen: Mohr (1969).

Popper, Karl R. (1990), *A World of Propensities*, Bristol: Thoemmes.

Prior, Elizabeth W., Robert Pargetter, and Frank Jackson (1982), "Three Theses About Dispositions", *American Philosophical Quarterly* 19, 251–257.

Putnam, Hilary (1965), "How Not to Talk About Meaning", in: R. Cohen and M. Wartofsky (eds.), *Boston Studies in the Philosophy of Science* Vol. 11, New York: Humanities Press, pp. 117–131.

Putnam, Hilary (1975), "The Meaning of 'Meaning'", in: H. Putnam, *Philosophical Papers, Vol. II: Mind, Language and Reality*, Cambridge: Cambridge University Press, pp. 215–271.

Putnam, Hilary (1980), "Models and Reality", *Journal of Symbolic Logic* 45, 464–482; also in: H. Putnam (1983a), pp. 1–25.

Putnam, Hilary (1983a), *Realism and Reason. Philosophical Papers* Vol. 3, Cambridge: Cambridge University Press.

Putnam, Hilary (1983b), "Why There Isn't a Ready-Made World", in: H. Putnam (1983a), ch. 12.

Putnam, Hilary (1983c), "Why Reason Can't Be Naturalized", in: H. Putnam (1983a), ch. 13.

Quine, Willard V.O. (1956), "Quantifiers and Propositional Attitudes", *Journal of Philosophy* 53, 177–187.

Quine, Willard V.O. (1960), *Word and Object*, Cambridge, MA: MIT Press.

Quine, Willard V.O. (1969a), "Epistemology Naturalized", in: W.V.O. Quine, *Ontological Relativity and Other Essays*, New York: Columbia University Press, pp. 69–90.

Quine, Willard V.O. (1969b), "Natural Kinds", in: W.V.O. Quine, *Ontological Relativity and Other Essays*, New York: Columbia University Press, pp. 114–138.

Railton, Peter (1978), "A Deductive-Nomological Model of Probabilistic Explanation", *Philosophy of Science* 45, 206–226.

Ramsey, Frank P. (1929), "General Propositions and Causality", in: D. H. Mellor (ed.), *Foundations. Essays in Philosophy, Logic, Mathematics and Economics*, London: Routledge & Kegan Paul (1978), pp. 133–151.

Rescher, Nicholas (1973), *The Coherence Theory of Truth*, Oxford: Oxford University Press.

Rescher, Nicholas (1976), *Plausible Reasoning*, Van Gorcum: Assen.

Rescher, Nicholas (1985), "Truth as Ideal Coherence", *Review of Metaphysics* 38, 795–806.

Rosenthal, Jacob (2004), *Wahrscheinlichkeiten als Tendenzen. Eine Untersuchung objektiver Wahrscheinlichkeitsbegriffe*, Paderborn: Mentis.

Rott, Hans (1991), *Reduktion und Revision. Aspekte des nichtmonotonen Theorienwandels*, Frankfurt a. M.: Lang.

Rott, Hans (2001), *Change, Choice and Inference. A Study of Belief Revision and Nonmonotonic Reasoning*, Oxford: Oxford University Press.

Rott, Hans (2003), "Coherence and Conservatism in the Dynamics of Belief. Part II: Iterated Belief Change Without Dispositional Coherence", *Journal of Logic and Computation* 13, 111–145.

Russell, Bertrand (1910/11), "Knowledge by Acquaintance and Knowledge by Description", *Proceedings of the Aristotelian Society* 11, 108–128; also in: B. Russell, *Mysticism and Logic and Other Essays*, London: Allen & Unwin (1963), pp. 152–167.

Russell, Bertrand (1918/19), "The Philosophy of Logical Atomism", *The Monist* 28 (1918), 495–527, and 29 (1919) 32–63, 190–222, 345–380; also in: R. C. Marsh (ed.), *Logic and Knowledge*, London: Allen & Unwin, pp. 177–281.

Salmon, Wesley C. (1966), *The Foundations of Scientific Inference*, Pittsburgh, PA: Pittsburgh University Press.

Salmon, Wesley C. (1970), "Statistical Explanation", in: R. G. Colodny (ed.), *Nature and Function of Scientific Theories*, Pittsburgh, PA: Pittsburgh University Press, pp. 173–231.

Salmon, Wesley C. (1978), "Why Ask, 'Why?'? An Inquiry Concerning Scientific Explanation", *Proceedings and Addresses of the American Philosophical Association* 51, pp. 683–705.

Salmon, Wesley C. (1980), "Probabilistic Causality", *Pacific Philosophical Quarterly* 61, pp. 50–74.

Salmon, Wesley C. (1984), *Scientific Explanation and the Causal Structure of the World*, Princeton, NJ: Princeton University Press.

Salmon, Wesley C. (1988a), "Dynamic Rationality: Propensity, Probability, and Credence", in: J. H. Fetzer (1988), pp. 3–40.

Salmon, Wesley C. (1988b), "Intuitions – Good and Not-So-Good", in: B. Skyrms and W. L. Harper (1988), pp. 51–71.

Salmon, Wesley C. (1989), "Four Decades of Scientific Explanation", in P. Kitcher and W. C. Salmon (eds.), *Minnesota Studies in the Philosophy of Science* Vol. XIII, Minneapolis, MN: University of Minnesota Press, pp. 3–219.

Sartwell, Crispin (1992), "Why Knowledge is Merely True Belief", *Journal of Philosophy* 89, 167–180.

Savage, Leonard J. (1954), *The Foundations of Statistics*, 2nd ed., New York: Wiley (1972).

Schaffer, Jonathan (2000), "Trumping Preemption", *Journal of Philosophy* 97, 165–181.

Schaffer, Jonathan (2003), "Principled Chances", *British Journal for the Philosophy of Science* 54, 27–41.

Schiffer, Stephen (1987), *Remnants of Meaning*, Cambridge, MA: MIT Press.

Schiffer, Stephen (1990), "Fodor's Character", in: E. Villanueva (ed.), *Information, Semantics, and Epistemology*, Oxford: Blackwell, pp. 77–101.

Schlenker, Philippe (1999), *Propositional Attitudes and Indexicality. A Cross-Categorical Approach*, Ph.D. dissertation, MIT Press.

Schröder, Winfried (2000), "Was heißt 'Geschichte eines philosophischen Begriffs'?", *Archiv für Begriffsgesschichte*, Sonderheft 42, pp. 159–172.

Schurz, Gerhard (1995), "Theories and Their Applications: A Case of Nonmonotonic Reasoning", in: W. Herfel et al. (eds.), *Theories and Models in Scientific Processes*, Amsterdam: Rodopi, pp. 269–293.

Schurz, Gerhard (2002), "Ceteris Paribus Laws", *Erkenntnis* 57, 351–372.

Searle, John R. (1958), "Proper Names", *Mind* 67, 166–173.

Shackle, George L. S. (1961/69), *Decision, Order, And Time in Human Affairs*, 2nd ed., Cambridge: Cambridge University Press (1969).

Shafer, Glenn (1976), *A Mathematical Theory of Evidence*, Princeton, NJ: Princeton University Press.

Shafer, Glenn (1996), *The Art of Causal Conjecture*, Cambridge, MA: MIT Press.

Shenoy, Prakash (1991), "On Spohn's Rule for Revision of Beliefs", *International Journal of Approximate Reasoning* 5, 149–181.

Silverberg, Arnold (1996), "Psychological Laws and Non-Monotonic Logic", *Erkenntnis* 44, 199–224.

Skyrms, Brian (1980), *Causal Necessity*, New Haven, CT: Yale University Press.

Skyrms, Brian (1983), "Three Ways to Give a Probability Assignment a Memory", in J. Earman (ed.), *Testing Scientific Theories. Minnesota Studies in the Philosophy of Science* Vol. X, Minneapolis, MN: University of Minnesota Press, pp. 157–161.

Skyrms, Brian (1984), *Pragmatics and Empiricism*, New Haven, CT: Yale University Press.

Skyrms, Brian (1990), *The Dynamics of Rational Deliberation*, Cambridge, MA: Harvard University Press.

Skyrms, Brian, and William Harper L. (eds.) (1988), *Causation, Chance, and Credence*, The Netherlands, Dordrecht: Kluwer.

Spirtes, Peter, Glymour, Clark, and Richard Scheines (1993), *Causation, Prediction, and Search*, Berlin: Springer.

Spohn, Wolfgang (1975), "An Analysis of Hansson's Dyadic Deontic Logic", *Journal of Philosophical Logic* 4, 237–252.

Spohn, Wolfgang (1976/78), *Grundlagen der Entscheidungstheorie*, Dissertation submitted at the University of Munich 1976, published: Kronberg/Ts.: Scriptor, 1978: out of print, pdf-version: http://www.uni-konstanz.de/FuF/Philo/Philosophie/Spohn/spohn_files/GE.Buch.gesamt.pdf; (references refer to the published version).

Spohn, Wolfgang (1977), "Where Luce and Krantz Do Really Generalize Savage's Decision Model", *Erkenntnis* 11, 113–134.

Spohn, Wolfgang (1980), "Stochastic Independence, Causal Independence, and Shieldability", *Journal of Philosophical Logic* 9, 73–99.

Spohn, Wolfgang (1982), "How to Make Sense of Game Theory", in: W. Stegmüller, W. Balzer, and W. Spohn (eds.), *Philosophy of Economics*, Berlin: Springer, pp. 239–270; reprinted in: Y. Varoufakis, A. Housego (eds.), *Game Theory: Critical Concepts in the Social Sciences, Vol. IV, Discontents*, London: Routledge, pp. 213–241.

Spohn, Wolfgang (1983a), *Eine Theorie der Kausalität*, unpublished Habilitationsschrift, University of Munich, available as pdf-file:
http://www.uni-konstanz.de/FuF/Philo/Philosophie/Spohn/spohn_files/Habilitation.pdf

Spohn, Wolfgang (1983b), "Deterministic and Probabilistic Reasons and Causes", *Erkenntnis* 19, 371–396.

Spohn, Wolfgang (1983c), "Probabilistic Causality: From Hume via Suppes to Granger", in: M. C. Galavotti and G. Gambetta (eds.), *Causalità e Modelli Probabilistici*, Bologna: Editrice CLUEB, pp. 69–87.

Spohn, Wolfgang (1986), "The Representation of Popper Measures", *Topoi* 5, 69–74.

Spohn, Wolfgang (1987), "A Brief Remark on the Problem of Interpreting Probability Objectively", *Erkenntnis* 26, 329–334.

Spohn, Wolfgang (1988), "Ordinal Conditional Functions. A Dynamic Theory of Epistemic States", in: W. L. Harper and B. Skyrms (eds.), *Causation in Decision, Belief Change, and Statistics*, The Netherlands, Dordrecht: Kluwer, pp. 105–134; also in this volume, ch. 1.

Spohn, Wolfgang (1990a), "Direct and Indirect Causes", *Topoi* 9, 125–145; also in this volume, ch. 2.

Spohn, Wolfgang (1990b), "A General Non-Probabilistic Theory of Inductive Reasoning", in: R. D. Shachter, T. S. Levitt, L. N. Kanal, and J. F. Lemmer (eds.), *Uncertainty in Artificial Intelligence* 4, Amsterdam: Elsevier, pp. 149–158.

Spohn, Wolfgang (1991), "A Reason for Explanation: Explanations Provide Stable Reasons", in: W. Spohn, B. C. van Fraasen, and B. Skyrms (eds.), *Existence and Explanation. Essays Presented in Honor of Karel Lambert*, The Netherlands, Dordrecht: Kluwer, pp. 165–196; also in this volume, ch. 9.

Spohn, Wolfgang (1992/93), *Namen. Oder: das Einfachste ist das Schwierigste. Oder: eine Einführung in die Sprachphilosophie*, notes of a lecture held in winter term 1992/93, 103 pp., see:
http://www.uni-konstanz.de/FuF/Philo/Philosophie/Spohn/spohn_files/Sprachphilosophie.pdf

Spohn, Wolfgang (1993a), "Causal Laws are Objectifications of Inductive Schemes", in: J. Dubucs (ed.), *Philosophy of Probability*, The Netherlands, Dordrecht: Kluwer, pp. 223–252; also in this volume, ch. 5.

Spohn, Wolfgang (1993b), "Wie kann die Theorie der Rationalität normativ und empirisch zugleich sein?", in: L. H. Eckensberger and U. Gähde (eds.), *Ethische Norm und empirische Hypothese*, Frankfurt a. M.: Suhrkamp, pp. 151–196.

Spohn, Wolfgang (1994), "On the Properties of Conditional Independence", in: P. Humphreys (ed.), *Patrick Suppes: Scientific Philosopher, Vol. I, Probability and Probabilistic Causality*, The Netherlands, Dordrecht: Kluwer, pp. 173–194.

Spohn, Wolfgang (1997a), "Über die Gegenstände des Glaubens", in: G. Meggle (ed.), *Analyomen 2. Proceedings of the 2nd Conference "Perspectives in Analytical Philosophy", Vol. I: Logic, Epistemology, Philosophy of Science*, Berlin: de Gruyter, pp. 291–321.

Spohn, Wolfgang (1997b), "The Character of Color Predicates: A Materialist View", in: W. Künne, A. Newen, and M. Anduschus (eds.), *Direct Reference, Indexicality and Propositional Attitudes*, Stanford: CSLI Publications, pp. 351–379; also in this volume, ch. 13.

Spohn, Wolfgang (1997c), "Begründungen a priori – oder: ein frischer Blick auf Dispositionsprädikate", in W. Lenzen (ed.), *Das weite Spektrum der Analytischen Philosophie. Festschrift für Franz von Kutschera*, Berlin: de Gruyter, pp. 323–345; English translation in this volume, ch. 12.

Spohn, Wolfgang (1997/98), "How to Understand the Foundations of Empirical Belief in a Coherentist Way", *Proceedings of the Aristotelian Society, New Series* 98, 23–40; also in this volume, ch. 11.

Spohn, Wolfgang (1998), "The Intentional versus the Propositional Conception of the Objects of Belief", in: C. Martinez, U. Rivas, and L. Villegas-Forero (eds.), *Truth in Perspective. Recent Issues in Logic, Representation and Ontology*, Aldershot: Ashgate, pp. 271–291.

Spohn, Wolfgang (1999a), "Ranking Functions, AGM Style", in: B. Hansson et al. (eds.), *Internet Festschrift for Peter Gärdenfors*, Lund, http://www.lucs.lu.se/spinning/.

Spohn, Wolfgang (1999b), "Lewis' Principal Principle ist ein Spezialfall von van Fraassens Reflexion Principle", in: J. Nida-Rümelin (ed.), *Rationalität, Realismus, Revision*, Berlin: de Gruyter, pp.164–173.

Spohn, Wolfgang (1999c), "Two Coherence Principles", *Erkenntnis* 50, 155–175.

Spohn, Wolfgang (2000a), "Wo stehen wir heute mit dem Problem der Induktion?", in: R. Enskat (ed.), *Erfahrung und Urteilskraft*, Würzburg: Königshausen & Naumann, pp. 151–164.

Spohn, Wolfgang (2000b), "Deterministic Causation", in: W. Spohn, M. Ledwig, and M. Esfeld (eds.), *Current Issues in Causation*, Paderborn: Mentis, pp. 21–46.

Spohn, Wolfgang (2000c), "A Rationalization of Cooperation in the Iterated Prisoner's Dilemma", in: J. Nida-Rümelin and W. Spohn (eds.), *Practical Rationality, Rules, and Structure*, The Netherlands, Dordrecht: Kluwer, pp. 67–84.

Spohn, Wolfgang (2001a), "Bayesian Nets Are All There Is To Causal Dependence", in: M. C. Galavotti, P. Suppes, and D. Costantini (eds.), *Stochastic Dependence and Causality*, CSLI Publications, Stanford, pp. 157–172.

Spohn, Wolfgang (2001b), "Vier Begründungsbegriffe", in: T. Grundmann (ed.), *Erkenntnistheorie. Positionen zwischen Tradition und Gegenwart*, Paderborn: Mentis, pp. 33–52.

Spohn, Wolfgang (2002), "Laws, Ceteris Paribus Conditions, and the Dynamics of Belief", *Erkenntnis* 57, 373–394; also in this volume, ch. 6.

Spohn, Wolfgang (2003a), "Carnap Versus Quine, or Aprioristic Versus Naturalized Epistemology, or a Lesson from Dispositions", in: T. Bonk (ed.), *Language, Truth and Knowledge. Contributions to the Philosophy of Rudolf Carnap*, The Netherlands, Dordrecht: Kluwer, pp. 167–177.

Spohn, Wolfgang (2003b), "Burge macht uns weis: ein Zirkel bei Grice", in: U. Haas-Spohn (ed.), *Intentionalität zwischen Subjektivität und Weltbezug*, Paderborn: Mentis, pp. 137–143.

Spohn, Wolfgang (2003c), "Dependency Equilibria and the Causal Structure of Decision and Game Situations", *Homo Oeconomicus* XX, 195–255.

Spohn, Wolfgang (2005a), "Enumerative Induction and Lawlikeness", *Philosophy of Science* 72. 164–187; also in this volume, ch. 7.

Spohn, Wolfgang (2005b), "Induktion", in: W. Spohn, P. Schröder-Heister, and E. Olsson (eds.), *Logik in der Philosophie*, Heidelberg: Synchron-Wissenschaftsverlag (2005), pp. 137–159.

Spohn, Wolfgang (2005c), "Five Questions on Formal Philosophy", in: V. F. Hendricks and J. Symons (eds.), *Formal Philosophy. Aim, Scope, Direction*, Copenhagen: Automatic Press, pp. 169–192.

Spohn, Wolfgang (2005d), "Anmerkungen zum Begriff des Bewusstseins", in: G. Wolters, M. Carrier (eds.), *Homo Sapiens und Homo Faber. Festschrift für Jürgen Mittelstraß*, Berlin: de Gruyter, pp. 239–251.

Spohn, Wolfgang (2006), "Causation: An Alternative", *British Journal for the Philosophy of Science* 57, 93–119; also in this volume, ch. 4.

Spohn, Wolfgang (forthcoming a), "Chance and Necessity: From Humean Supervenience to Humean Projection", in: E. Eells and J. Fetzer (eds.), *The Place of Probability in Science*, Chicago, IL: Open Court, also in this volume, ch. 8.

Spohn, Wolfgang (forthcoming b), "A Survey of Ranking Theory", in: F. Huber and C. Schmidt-Petri (eds.), *Degrees of Belief. An Anthology*, Oxford: Oxford University Press.

Spohn, Wolfgang (in preparation), *Ranking Theory. A Tool for Epistemology*.

Stalnaker, Robert C. (1970), "Pragmatics", *Synthese* 22, 272–289; also in: D. Davidson and G. Harman (eds.), *Semantics of Natural Language*, The Netherlands, Dordrecht: Reidel (1972), pp. 380–397.

Stalnaker, Robert C. (1978), "Assertion", in: P. Cole (ed.), *Syntax and Semantics* Vol. 9: *Pragmatics*, New York: Academic, pp. 315–332.

Stalnaker, Robert C. (1981), "Indexical Belief", *Synthese* 49, 129–151.
Stalnaker, Robert C. (1984), *Inquiry*, Cambridge, MA: MIT Press.
Stalnaker, Robert C. (1987), "Semantics for Belief", *Philosophical Topics* 15, 177–190.
Stalnaker, Robert C. (1989), "On What's in the Head", in: J. E. Tomberlin (ed.), *Philosophical Perspectives* 3, *Philosophy of Mind and Action Theory*, pp. 287–316.
Stalnaker, Robert C. (1990), "Narrow Content", in: C. A. Anderson and J. Owens (eds.), *Propositional Attitudes. The Role of Content in Logic, Language, and Mind*, Stanford: CSLI Publications, pp. 131–145.
Stegmüller, Wolfgang (1960), *Hauptströmungen der Gegenwartsphilosophie, Band 1*, 2nd ed.; 7th edition, Stuttgart: Kröner (1989).
Stegmüller, Wolfgang (1970), *Probleme und Resultate der Wissenschaftstheorie und Analytischen Philosophie, Band II, Theorie und Erfahrung, 1. Teilband*, Berlin: Springer.
Strawson, Galen (1989), "Red and 'Red'" *Synthese* 78, 193–232.
Strevens, Michael (1995), "A Close Look at the 'New' Principle", *British Journal for the Philosophy of Science* 46, 545–561.
Studeny, Milan (1989), "Multiinformation and the Problem of Characterization of Conditional Independence Relations", *Problems of Control and Information Theory* 18, 3–16.
Sturgeon, Scott (1998), "Humean Chance: Five Questions for David Lewis", *Erkenntnis* 49, 321–335.
Suppe, Frederick (ed.) (1977), *The Structure of Scientific Theories*, Urbana, IL: University of Illinois Press.
Suppes, Patrick (1970), *A Probabilistic Theory of Causality*, Amsterdam: North-Holland.
Suppes, Patrick (1984), *Probabilistic Metaphysics*, Oxford: Blackwell.
Teller, Paul (1976), "Conditionalization, Observation, and Change of Preference", in: W. L. Harper and C. A. Hooker (eds.), *Foundations of Probability Theory, Statistical Inference, and Statistical Theories of Science* Vol. I, The Netherlands, Dordrecht: Reidel, pp. 205–259.
Thau, Michael (1994), "Undermining and Admissibility", *Mind* 103, 491–503.
Tooley, Michael (1987), *Causation*, Oxford: Oxford University Press.
Vallentyne, P. (1996), "Response-Dependence, Rigidification, and Objectivity", *Erkenntnis* 44, 101–112.
van Brakel, Jan (1990), "Units of Measurement and Natural Kinds: Some Kripkean Considerations", *Erkenntnis* 33, 297–317.
van Fraassen, Bas C. (1976), "Representation of Conditional Probabilities", *Journal of Philosophical Logic* 5, 417–430.
van Fraassen, Bas C. (1980a), *The Scientific Image*, Oxford: Oxford University Press.
van Fraassen, Bas C. (1980b), "A Temporal Framework for Conditionals and Chance", *The Philosophical Review* 89, 91–108.
van Fraassen, Bas C. (1984), "Belief and the Will", *Journal of Philosophy* 81, 235–256.
van Fraassen, Bas C. (1989), *Laws and Symmetry*, Oxford: Clarendon Press.
van Fraassen, Bas C. (1995), "Belief and the Problem of Ulysses and the Sirens", *Philosophical Studies* 77, 7–37.
Vendler, Zeno (1967), "Causal Relations", *Journal of Philosophy* 64, 704–713.
von Kutschera, Franz (1972), *Wissenschaftstheorie* Vol. I and II, Munich: Wilhelm Fink Verlag.
von Kutschera, Franz (1982), *Grundfragen der Erkenntnistheorie*, Berlin: de Gruyter.
von Kutschera, Franz (1994), "Zwischen Skepsis und Relativismus", in: G. Meggle and U. Wessels (eds.), *Analyomen 1. Perspectives in Analytical Philosophy*, Berlin: de Gruyter, pp. 207–224.
von Mises, Richard (1919), "Grundlagen der Wahrscheinlichkeitsrechnung", *Mathematische Zeitschrift* 5, 52–99.
von Stechow, Arnim, and Dieter Wunderlich (eds.) (1991), *Semantik. Ein internationales Handbuch der zeitgenössischen Forschung*, Berlin: de Gruyter.
Vranas, Peter (2004), "Have Your Cake and Eat It Too: the Old Principal Principle Reconciled With the New", *Philosophy and Phenomenological Research* 69, 368–382.
Ward, Barry (2002), "Humeanism Without Humean Supervenience: A Projectivist Account of Laws and Possibilities", *Philosophical Studies* 107, 191–218.

Ward, Barry (2005), "Projecting Chances: A Humean Vindication and Justification of the Principal Principle", *Philosophy of Science* 72, 241–261.

Weiskrantz, Lawrence (1980), "Varieties of Residual Experience", *Quarterly Journal of Experimental Psychology*, 32, 365–386.

White, Stephen L. (1982), "Partial Character and the Language of Thought", *Pacific Philosophical Quarterly* 63, 347–365.

Wittgenstein, Ludwig (1922), *Tractatus logico-philosophicus*, London: Routledge & Kegan Paul.

Wittgenstein, Ludwig (1953), *Philosophical Investigations*, translated by G. E. M. Anscombe, Oxford: Blackwell.

Zimmermann, Thomas E. (1991), "Kontextabhängigkeit", in: A. von Stechow and D. Wunderlich (1991), pp. 156–229.

Zimmermann, Thomas E. (1999), "Remarks on the Epistemic Rôle of Discourse Referents", in: L. S. Moss, J. Ginzburg, and M. de Rijke (eds.), *Logic, Language, and Computation* Vol. II, Stanford: CSLI Publications, pp. 346–368; also in: H. Kamp and B. Partee (eds.), *Context-Dependence in the Analysis of Linguistic Meaning*, Amsterdam: Elsevier, pp. 521–537.

Name Index

Subject Index

MIX
Papier aus verantwortungsvollen Quellen
Paper from responsible sources
FSC® C105338

If you have any concerns about our products,
you can contact us on
ProductSafety@springernature.com

In case Publisher is established outside the EU,
the EU authorized representative is:
Springer Nature Customer Service Center GmbH
Europaplatz 3, 69115 Heidelberg, Germany

Printed by Libri Plureos GmbH
in Hamburg, Germany